D1130761

GRAMICIDIN AND RELATED ION CHANNEL-FORMING PEPTIDES

The Novartis Foundation is an international scientific and educational charity (UK Registered Charity No. 313574). Known until September 1997 as the Ciba Foundation, it was established in 1947 by the CIBA company of Basle, which merged with Sandoz in 1996, to form Novartis. The Foundation operates independently in London under English trust law. It was formally opened on 22 June 1949.

The Foundation promotes the study and general knowledge of science and in particular encourages international co-operation in scientific research. To this end, it organizes internationally acclaimed meetings (typically eight symposia and allied open meetings, 15–20 discussion meetings, a public lecture and a public debate each year) and publishes eight books per year featuring the presented papers and discussions from the symposia. Although primarily an operational rather than a grant-making foundation, it awards bursaries to young scientists to attend the symposia and afterwards work for up to three months with one of the other participants.

The Foundation's headquarters at 41 Portland Place, London W1N 4BN, provide library facilities, open every weekday, to graduates in science and allied disciplines. The library is home to the Media Resource Service which offers journalists access to expertise on any scientific topic. Media relations are also strengthened by regular press conferences and book launches, and by articles prepared by the Foundation's Science Writer in Residence. The Foundation offers accommodation and meeting facilities to visiting scientists and their societies.

Information on all Foundation activities can be found at http://www.novartisfound.org.uk

Novartis Foundation Symposium 225

GRAMICIDIN AND RELATED ION CHANNEL-FORMING PEPTIDES

1999

JOHN WILEY & SONS, LTD

Chichester · New York · Weinheim · Brisbane · Toronto · Singapore

Published in 1999 by John Wiley & Sons Ltd,
Baffins Lane, Chichester,
West Sussex PO19 1UD, England

National 01243 779777
International (+44) 1243 779777
e-mail (for orders and customer service enquiries): cs-books@wiley.co.uk
Visit our Home Page on http://www.wiley.co.uk
or http://www.wiley.com

Other Wiley Editorial Offices

John Wiley & Sons, Inc., 605 Third Avenue,
New York, NY 10158-0012, USA

WILEY-VCH Verlag GmbH, Pappelallee 3,
D-69469 Weinheim, Germany

Jacaranda Wiley Ltd, 33 Park Road, Milton,
Queensland 4064, Australia

John Wiley & Sons (Asia) Pte Ltd, 2 Clementi Loop #02-01,
Jin Xing Distripark, Singapore 129809

John Wiley & Sons (Canada) Ltd, 22 Worcester Road,
Rexdale, Ontario M9W 1L1, Canada

Novartis Foundation Symposium 225
ix+273 pages, 57 figures, 13 tables

Library of Congress Cataloging-in-Publication Data

Gramicidin and related ion channel-forming peptides/[editors, Derek J. Chadwick and Gail Cardew].
p. cm. – (Novartis Foundation symposium ; 225)
Includes bibliographical references and index.
ISBN 0471 98846 4
1. Gramicidins Congresses. 2. Ionophores Congresses. 3. Ion channels Congresses. I. Chadwick, Derek. II. Cardew, Gail. III. Novartis Foundation. IV. Symposium on Gramicidin and Related Ion Channel-forming Peptides (1998 : London, England) V. Series.
QP552.G7G73 1999
572′.65–dc21 99-32308
CIP

British Library Cataloguing in Publication Data

A catalogue record for this book is available from the British Library

ISBN 0 471 98846 4

Typeset in 10½ on 12½ pt Garamond by Dobbie Typesetting Limited, Tavistock, Devon.
Printed and bound in Great Britain by Biddles Ltd, Guildford and King's Lynn.
This book is printed on acid-free paper responsibly manufactured from sustainable forestry, in which at least two trees are planted for each one used for paper production.

Contents

Symposium on Gramicidin and related ion channel-forming peptides, held at the Novartis Foundation, London, 17–19 November 1998

This symposium is based on a proposal made by Bonnie Wallace

Editors (and organizers): Derek J. Chadwick and Gail Cardew

Participants

A. Arseniev Shemaykin and Ovchinnikov Institute of Bioorganic Chemistry, Russian Academy of Sciences U1, Miklukho-Maklaya 16/10, Moscow 117871, Russia

B. Bechinger Max-Planck-Institut für Biochemie, Am Klopferspitz 18a, 82152 Martinsreid, Germany

D. D. Busath Zoology Department, Brigham Young University, Provo, UT 84602, USA

B. A. Cornell Cooperative Research Centre for Molecular Engineering and Technology, 126 Greville Street, Chatswood, NSW 2067, Australia

P.-J. Corringer Neurobiologie Moléculaire, Unité de recherche associée au Centre National de la Recherche Scientifique D1284, Institut Pasteur, 25 rue du Docteur Roux, 75724 Paris Cedex 15, France

T. A. Cross National High Magnetic Field Laboratory, Florida State University, Tallahassee, FL 32310, USA

J. Davis Department of Physics, University of Guelph, Guelph, Ontario, Canada N1G 2W1

C. Dempsey Department of Biochemistry, University of Bristol, School of Medical Sciences, University Walk, Bristol BS8 1TD, UK

R. Eisenberg Department of Molecular Biophysics, Rush Medical Center, Chicago, IL 60612, USA

F. Heitz Centre de Recherches de Biochimie Macromoléculaire (CNRS), Route de Mende, 34293 Montpellier Cedex 5, France

J. Hinton Department of Chemistry and Biochemistry, 115 Chemistry Building, University of Arkansas, Fayetteville, AR 72701–1201, USA

S. B. Hladky Department of Pharmacology, University of Cambridge, Tennis Court Road, Cambridge CB2 1QJ, UK

H.W. Huang Physics Department, Rice University, Houston, TX 77251-1892, USA

E. Jakobsson Department of Molecular and Integrative Physiology, Beckman Institute for Advances in Sciences and Technology, University of Illinois, Urbana, IL 61801, USA

P. C. Jordan Department of Chemistry, MS-015, Brandeis University, PO Box 9110, Waltham, MA 02254-9110, USA

J. A. Killian Department of Biochemistry of Membranes, University of Utrecht, Padualaan 8, 5384 CH Utrecht, The Netherlands

R. E. Koeppe II Department of Chemistry and Biochemistry, University of Arkansas, Fayetteville, AR 72701, USA

A. Ring Norwegian Defence Research Establishment, Department of Environmental Toxicology, Box 25, 2007 Kjeller, Norway

J. P. Rosenbusch Department of Microbiology, Biozentrum, University of Basel, CH-4056, Basel, Switzerland

B. Roux Membrane Transport Research Group (GRTM), Departments of Physics and Chemistry, Université de Montreal, Case Postale 6128, Succursale Centre-Ville, Montreal, Quebec, Canada H3C 3J7

M. S. P. Sansom Laboratory of Molecular Biophysics, The Rex Richards Building, Department of Biochemistry, University of Oxford, South Parks Road, Oxford OX1 3QU, UK

F. Separovic School of Chemistry, University of Melbourne, Parkville, Victoria 3052, Australia

O. Smart (*Bursar*) School of Biochemistry, University of Birmingham, Edgbaston, Birmingham B15 2TT, UK

W. Stein Laboratory of Cell Biology, NCI/NIH, Bldg 37, Rm 1B 28, 37 Convent Drive, MSC 4255, Bethseda, MD 20892-4255, USA

B. A. Wallace (*Chair*) Department of Crystallography, Birkbeck College, University of London, Malet Street, London WC1E 7HX, UK

G. A. Woolley Department of Chemistry, University of Toronto, 80 St George Street, Toronto, Canada M5S 3H6

Introduction: gramicidin, a model ion channel

B. A. Wallace

Department of Crystallography, Birkbeck College, University of London, Malet Street, London WC1E 7HX, UK

This symposium brings together a diverse group of people who are interested in gramicidin and related ion channels, with the goal of advancing our understanding of ion conduction across membranes. A large number of teams doing important work in this area are represented here, and as a result we can look forward to an exciting symposium.

This field has become increasingly interdisciplinary, and one of the reasons for holding such a symposium as this has been to bring together experts in the fields of biophysics, physiology, synthetic chemistry, theoretical chemistry, biology and engineering. I expect that the book that results from it will become a valuable reference for those working in the field of ion channels. I would like to start by outlining some of the topics we originally had in mind for discussion when developing the concept for this symposium. I expect we will cover all of these areas, and more, during the course of the next several days.

(1) Structural studies, including X-ray crystallography and diffraction, circular dichroism, and solution and solid-state NMR spectroscopy. Comparisons between results obtained on gramicidin by these different techniques and under different conditions should better define the polymorphic nature of this molecule and the relationships between its different conformational states.
(2) Structure/activity relationships. Gramicidin has been engineered (using synthetic peptide chemistry techniques rather than by cloning due to the presence of D-amino acids) more than any other ion channel. The resulting 'mutant' molecules have helped us to understand structure/function relationships by examining functional consequences of changes in the sequence of the molecule and correlating these with the structural studies.
(3) Theory and simulation analyses. The small size of gramicidin has made it particularly amenable to these types of studies, maximizing the utility of currently available computing resources to examine complex multi-component systems over realistic time-scales.

(4) Biological function. This may be a bit awkward for our community, which has focused on gramicidin's role as a model ion channel, because we don't know what its true biological role is, but it may be an interesting area to pursue.
(5) Lipid interactions. Gramicidin represents a good model system for examining interactions between polypeptides and lipids with different fatty acid chains and head groups.
(6) Applications. One important example of the use of gramicidin is its newly created role as part of an analytical sensor system, as well as its old use as an antibiotic, perhaps with some new twists.
(7) Other ion channels. Contrasts and comparisons will undoubtedly be made with the recently solved *Streptomyces lividans* potassium channel structure and with other polypeptide ion channels.

This symposium will be interdisciplinary with respect to the fields covered and the approaches taken, as well as truly international, with participants coming from Europe, the US, Canada, Australia and Russia. There should be many exciting discussions, resulting from the presence of experts in so many different areas. Most of us know each other's names because we read all the gramicidin papers, but we don't necessarily know everyone's faces, so it will be great to put faces to names and get to know each other and initiate in-depth dialogues. I expect this symposium and our discussions will turn out to have important consequences in the field for many years to come.

This symposium is particularly timely because of the many technical advances that have taken place in the last few years which have ultimately increased our understanding of ion conduction. Advances in X-ray crystallography and diffraction, such as new instrumentation for data collection (including synchrotrons and detector technology), and computational developments in phasing and refinement and graphics have allowed us to learn more about the structure of this intermediate-sized molecule. Advances in solution and solid-state NMR spectroscopy have included magnet technology and computing methods that have permitted more accurate and facile structure determinations. Computing advances have also played a major role in theoretical and dynamic studies and simulations. Advances in peptide synthesis, chemistry, characterization methods and conductance measurements have provided important new insights into the functioning of this molecule. All of these advances have had an important impact on our knowledge of gramicidin, and we hope that by extension, these advances in understanding gramicidin have had an important impact on our knowledge of the process of ion conduction across biological membranes.

This symposium is also timely due to the number of controversies that currently exist in the field. These include ones that have arisen as a result of new X-ray and

NMR structures, conductance calculations and comparisons with the potassium channel structure. Many more controversies will undoubtedly arise during the course of this symposium. I hope that we will be able to resolve some of them during the discussions, and that others may lead to further studies that will advance the field in the future.

Correlations of structure, dynamics and function in the gramicidin channel by solid-state NMR spectroscopy

T. A. Cross*, F. Tian, M. Cotten, J. Wang, F. Kovacs and R. Fu

*Department of Chemistry, Institute of Molecular Biophysics, and *National High Magnetic Field Laboratory, Florida State University, Tallahassee, FL 32310, USA*

Abstract. The high resolution structure of the gramicidin A channel has been determined in a lamellar phase environment using solid-state NMR spectroscopy. While the fold is similar to previous characterizations, channel function is exquisitely dependent on structural detail. There is essentially no structural change upon cation binding and no significant change in dynamics. The cations appear to be adequately solvated in their binding site by no more than two carbonyls and no fewer than three water molecules at any one time. The relatively large number of water molecules allows for geometric flexibility and little selectivity among monovalent cations. However, the dehydration energies of cations clearly explain the selectivity for monovalent versus divalent cations. Moreover, the binding site is shown to be delocalized, resulting in a shallow potential energy well so that efficient cation conductance can be realized. The potential energy barrier at the bilayer centre has been shown to be rate limiting under certain circumstances through a correlation between conductance and the electrostatic interactions between cations at the gramicidin monomer–monomer junction and the indole dipole moments at the lipid–water interface. The dynamics are functionally important. The time-scale for carbonyl fluctuations about the Cα-Cα axis and kinetic rates for cation movement in the channel are the same, suggesting a correlation between molecular dynamics and kinetics. These functional correlations will be described in light of the recent K^+ channel structure and the biological challenge to achieve both selectivity and efficiency.

1999 Gramicidin and related ion channel-forming peptides. Wiley, Chichester (Novartis Foundation Symposium 225) p 4–22

Gramicidin A has proven to be an exceptional model cation channel for exposing the principles by which cations can be transported across membranes by much more sophisticated proteinaceous channels. Although good models of the molecular fold have existed since 1971 (Urry 1971), it wasn't until high resolution structural and dynamic detail was available (Ketchem et al 1997, North & Cross 1995) that many of the functional insights have been achieved

(Fig. 1). This is not to belittle the knowledge from the fold that the channel supports a single file column of water molecules and that the polypeptide must provide much of the solvation environment for the cations. However, this does not explain the efficiency with which this channel conducts cations nor its selectivity, the issues of primary concern for the proteinaceous channels, such as the K^+ channel from *Streptomyces lividans* (Doyle et al 1998).

Solid-state NMR spectroscopy has made it possible to collect high resolution structural and dynamic data (Cross 1997) from membrane proteins. Uniformly

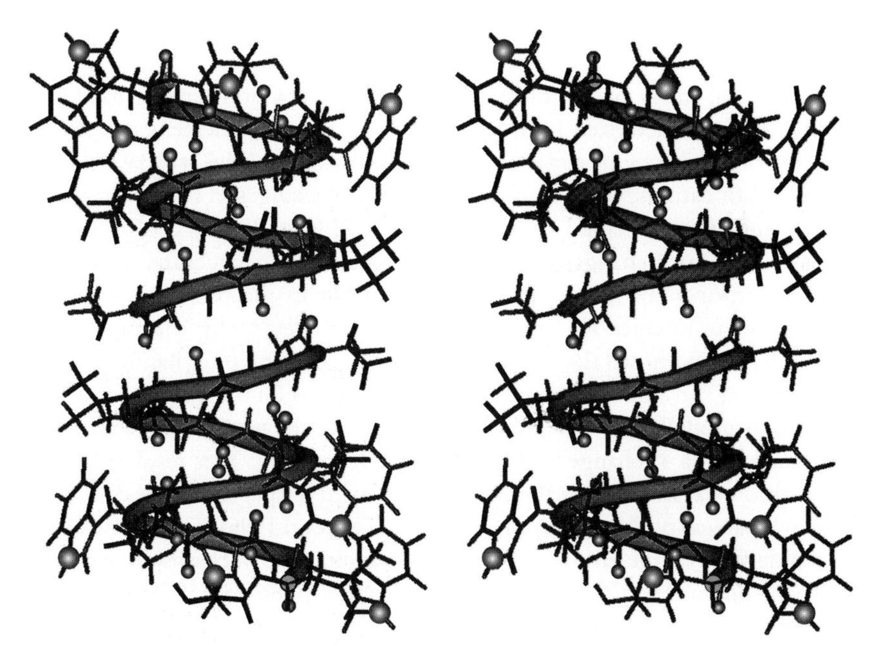

FIG. 1. Gramicidin A is a polypeptide of 15 amino acid residues having the sequence: formyl-Val1-Gly2-Ala3-D-Leu4-Ala5-D-Val6-Val7-D-Val8-Trp9-D-Leu10-Trp11-D-Leu12-Trp13-D-Leu14-Trp15-ethanolamine. This alternating sequence of L and D amino acids permits the formation of a β-strand-type structure with all of the side chains on one side, thereby encouraging the formation of a helix. This single-stranded helix of 6.5 residues per turn forms a pore of approximately 4 Å in diameter. The junction between the N-terminus of two monomers at the bilayer centre is supported by six intermolecular β-type hydrogen bonds. The monomer structure is defined by 120 precise orientational constraints from solid-state NMR of uniformly aligned samples in lamellar phase lipid bilayers (Ketchem et al 1997; protein data bank accession number 1MAG). The monomer–monomer geometry has been characterized by solution NMR in SDS micelles (Lomize et al 1992) and recently by solid-state NMR intermolecular distance measurements (R. Fu, M. Cotten & T. A. Cross, unpublished results 1999). The precision and accuracy of the structure is characterized by torsion angles defined to within an error of $\pm$ 5°.

aligned samples with a mosaic spread of less than 0.3° are possible with gramicidin in dimyristoyl phosphatidylcholine (DMPC) bilayers. The observation of anisotropic chemical shifts, dipolar and quadrupolar interactions present spectral resolution that is comparable to solution NMR spectra. Unlike the isotropic chemical shifts, which are difficult to interpret, the anisotropic interactions have a simple orientational dependence with respect to the magnetic field axis, leading to precise structural constraints. The accuracy of the constraints is dependent on how well characterized the spin interaction tensors are defined and on how well the molecular motions are characterized.

Cation binding

For gramicidin, many of the most variable spin interaction tensors have been experimentally characterized both in terms of their tensor element magnitudes and their orientations with respect to the molecular frame (Mai et al 1993). Upon the addition of cations, some significant (>1 ppm) changes in the observed ^{15}N chemical shifts from oriented samples are observed (Fig. 2; Tian et al 1996, Tian & Cross 1999). Such changes could be due to a change in structure, a change in dynamics and/or a change in chemical shift tensor in the presence of cations. Because dipolar interactions are not significantly affected by the presence of cations it has been argued that there is little to no structural or dynamic influence by the cations on the polypeptide backbone. Indeed, the largest structural deviations as modelled by a change in orientation of the peptide plane about the Cα-Cα axis is just a 4° change in time-averaged orientation. This result suggested that the cation was influencing the electron density in the peptide plane containing a carbonyl that was providing cation solvation (Tian et al 1996). The chemical shift tensor is defined by the electron density surrounding the nuclear spin and it has since been shown that ^{15}N tensors are significantly modified by the presence of cations in the channel (Tian & Cross 1998). Therefore, the changes in chemical shift reported in Fig. 2 define those peptide planes involved in cation solvation. The three carbonyls significantly involved are from Leu10, 12 and 14, whereas the carbonyls from Trp11, 13 and 15 and all others are not involved on a time-averaged basis to a great enough extent so that the average ^{15}N chemical shift tensors are affected.

These results lead to several fundamental conclusions. First, there is no significant structural change upon binding a cation to this channel in a low dielectric environment. This might appear to be a surprising result, and many of the early molecular dynamics calculations had suggested large deformations leading to ideal or near-ideal solvation environments for the cation (Åqvist & Warshel 1989, Roux & Karplus 1993). However, we have argued that such changes in conformation could lead to a more substantial potential energy well

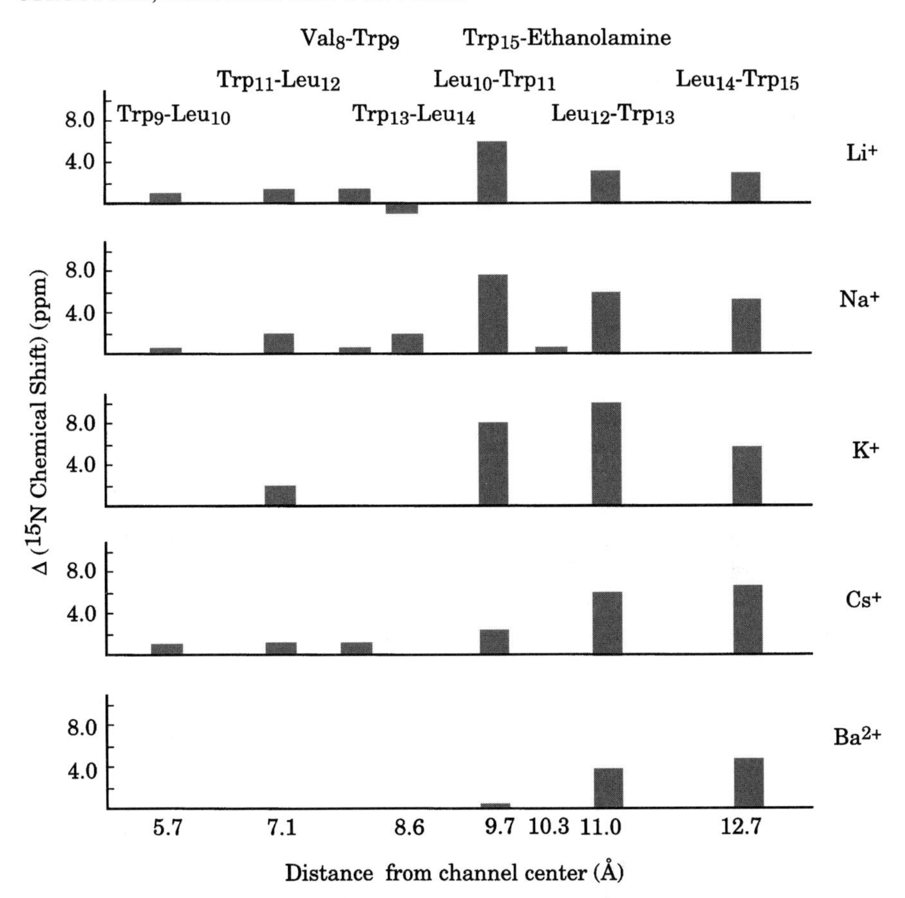

FIG. 2. Anisotropic ^{15}N chemical shift changes induced by 80% double occupancy by various cations in the gramicidin channel. Although the structure is a symmetrical dimer, when a cation binds at one end of the channel the symmetry is broken, and a second cation will bind with a weaker binding constant (Hinton et al 1986, Jing et al 1995, Urry 1987). The changes in chemical shift (with and without cations present) are plotted for the affected peptide planes as a function of the carbonyl oxygen distance from the bilayer centre.

and hence barrier to cation translocation (Tian et al 1996). What is desired for the efficiency of the channel is adequate rather than ideal solvation of the cation in the channel. Here, adequate means that the equilibrium binding constant must be strong enough to attract cations into the channel away from the bulk aqueous solvent where six waters solvate the cation in the primary hydration sphere. However, the question of why the structure of gramicidin is not distorted by the presence of cations is not answered by these arguments that define why it is good to have little distortion.

Delocalized binding

Part of the answer comes from a realization that the cation-binding site is a series of subsites, i.e. cation binding is delocalized. This conclusion is necessitated because favourable interactions with the Leu10, 12 and 14 carbonyls cannot be achieved simultaneously (Fig. 3). Not only is there too great a separation of the carbonyl oxygens in the plane of the bilayers (end view, Fig. 3B), but the carbonyl groups are separated substantially along the helical axis (~3 Å). Therefore, to induce the chemical shift changes observed in Fig. 2 it must be necessary for the cation to spend significant amounts of time in close vicinity to the individual carbonyls of Leu10, 12 and 14. This represents a delocalized binding site where the time-averaged influence of the cation is distributed over a significant spatial volume. The probability distribution is approximated by the histograms in Fig. 2 showing the influence of cations on the chemical shift tensors. Clearly, this probability distribution or time-averaged location for the cations in the channel is different for the various cations. The larger cations bind closer to the channel mouth and the divalent cation, Ba^{2+}, which is not conducted by the channel, shows no significant interaction with the Leu10 carbonyl, and induces only modest chemical shift changes in Leu12 and Leu14 carbonyl peptide planes despite its divalency. Not only does delocalized binding reduce the influence of the cation on the polypeptide structure, but it also generates a broad and shallow potential energy well for cation binding, ideal for displacing the cation from the binding site and moving it across the membrane. In addition, the entropic penalty associated with cation binding is minimized by allowing the cation to move within its binding site. Therefore, by avoiding such structural changes fast cation association rates can be achieved, as observed (Becker et al 1992).

Another question arises from this lack of structural change, and that is how does the channel provide adequate solvation for the cation without forming a constricted cluster of carbonyls about the cation? The dehydration energies for cations in the gaseous state are known (Dzidic & Kebarle 1970), and the energy required to remove each successive water is substantially greater than the removal of the previous water. Consequently, the removal of the last water prior to entry into the single file region of the channel (two waters per cation) has the largest energy barrier. This step appears to define the inner boundary of the cation-binding site. At the channel entrance the Leu14 carbonyl oxygen is the first site of interaction, displacing the first water molecule from the primary hydration sphere (Fig. 4). As the cation moves into the channel, more waters are stripped off through interactions with Leu10 and 12. Leu10 is the furthest into the channel that Na^{+} can penetrate while maintaining at least three waters in the hydration sphere. These waters provide considerable flexibility for solvating cations and hence provide a lack of selectivity among monovalent cations

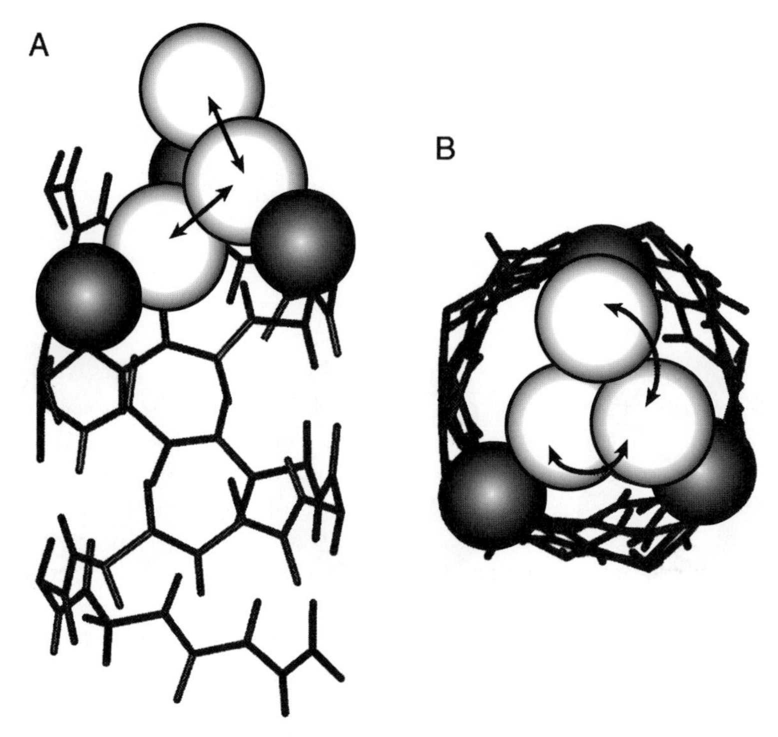

FIG. 3. (A) Side view and (B) end view of the channel backbone in which the three carbonyl oxygens of Leu10, 12 and 14 are presented in van der Waals radii. These are three sites that significantly interact with cations. K^+ (pale circles) is schematically shown interacting with the carbonyl oxygens (dark circles) through the peptide plane dipole moment dominated by a component parallel to the C-O bond direction. The three locations for the K^+ ions are intended to illustrate the span of the delocalized cation-binding site through which the cation becomes significantly dehydrated. Note that this delocalized volume involves significant translation in both the radial and axial directions, suggesting a helical path for the cations.

(Christensen et al 1975, Cox & Schneider 1992). Such flexibility will not be anticipated in the K^+ channel where specificity is exquisite. The stepwise dehydration, which had been previously suggested by molecular dynamics calculations (Åqvist & Warshel 1989, Jordan 1990, Roux & Karplus 1993), is important for generating an incremental pathway over large energy barriers associated with removing three waters from the cation. In this way, the association and dissociation rates are enhanced (Lehn 1973). To interact with the Val8 carbonyl in the second turn of the helix the hydration of even Li^+ has to be reduced to two waters. However, carbonyls from the first turn of the helix, which have formerly not participated in solvation, such as the Trp15 carbonyl, now compensate for the fourth water which has been stripped from the cation.

Throughout the rest of the channel this solvation environment remains relatively uniform with no local large potential energy steps until reaching the symmetry-related, cation-binding site in the adjacent monomer.

Although the lack of selectivity among monovalent cations is explained by having three mobile ligands in the cation-binding site, divalent cations clearly do not penetrate even to the Leu10 site in the first turn of the channel structure. Dehydration of divalent cations requires substantially more energy than monovalent cations (Blades et al 1990), and there is insufficient compensatory

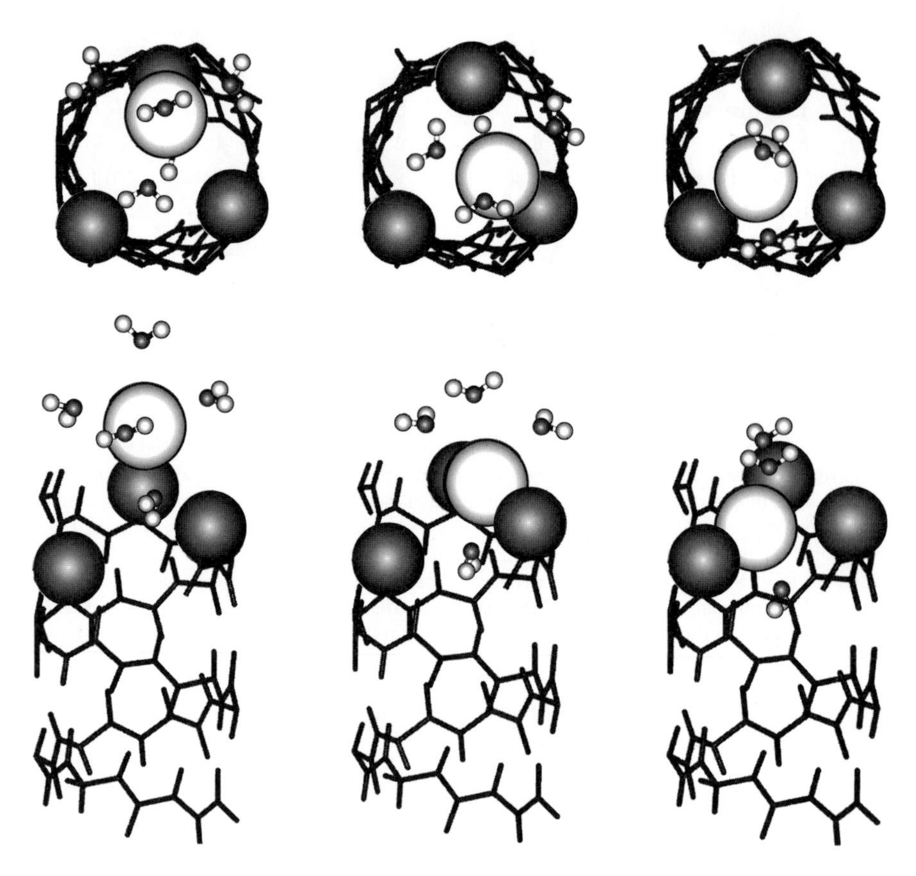

FIG. 4. End views (top) of the channel as in Fig. 3 showing the three individual carbonyl interactions with cations and a model of cation hydration for each configuration. The corresponding side views (bottom) are also shown. When the cation is interacting with Leu14 (left) as many as five waters can be found in the primary solvation sphere, whereas four are typically found when K^+ is in the vicinity of Leu12 (middle) and only three when in the vicinity of Leu10 (right). When monovalent cations are in close contact with the Val8 carbonyl (not shown) additional stabilization is provided by the Trp15 carbonyl and only two waters solvate the cation.

interaction energy available from the polypeptide backbone. Divalent cations block the channel, and the interactions with the Leu14 and Leu12 carbonyls demonstrate that while enough solvent waters are not stripped off for the cation to pass through the channel the interactions with Leu12 and Leu14 are significant enough to initiate dehydration of the divalent cation resulting in binding of such cations. Anions show no influence on the spectra documented by monitoring the chemical shift changes for Na^+ and Ba^{2+} salts of chloride and nitrate consistent with the known lack of anion binding to the channel. Although partial negative and positive charges are balanced in the peptide plane, the partial negative charge is focused on the carbonyl oxygen, whereas the partial positive charge is more distributed with much less charge density on the amide proton than oxygen, making the structure much less attractive to anions than cations. Furthermore, the refined structure has almost all peptide planes oriented with the carbonyl oxygens tipped in towards the channel axis and, therefore, the amide protons are oriented away from the channel and toward the lipid environment, making them inaccessible to anions (Ketchem et al 1997).

Functional role for dipoles

Solid-state NMR has also shed light on cation translocation through the channel. By accurately defining the orientation of the indole rings, Hu et al (1995) have determined the orientation of the indole dipoles with respect to the channel. Indeed, these dipoles are oriented such that the negative end is directed toward the channel axis at the bilayer centre. The influence of these dipoles has been assessed by comparing conductance with and without dipoles present (Becker et al 1991) to the sum of the monopole–dipole electrostatic interaction energy with the cation monopole located on the channel axis at the bilayer centre (Hu & Cross 1995). The elimination of dipoles was achieved by incrementally replacing the indoles with Phe. As shown in Fig. 5, a remarkable linear correlation is achieved between the natural log of conductance and the sum of these interaction energies. Such a correlation strongly suggests that the potential energy barrier at the bilayer centre is the rate-limiting step for cation conductance by gramicidin under these sample preparation conditions. Furthermore, the correlation also suggests that the structure of gramicidin in DMPC bilayers is similar to the structure in diphytanoyl phosphatidylcholine (DPhPC) bilayers used for the conductance measurements (Becker et al 1991, Busath et al 1998).

If the reduction in conductance by a factor of 20 (all Trp to all Phe) is equated with an increased potential energy barrier from an Arrhenius analysis then a dielectric constant for the 10–13 Å of intervening protein–lipid environment can be approximated as 5.1 (Hu & Cross 1995). However, the most important finding is that dipole–monopole electrostatic interactions are used by this molecule to enhance the efficiency of channel conductance. It appears that in the K^+ channel, a set of short helices are oriented such that the helix dipoles stabilize cations at the bilayer centre (Doyle et al 1998). Electrostatic interactions will be important over much longer distances in membrane proteins than typically observed in water-soluble proteins.

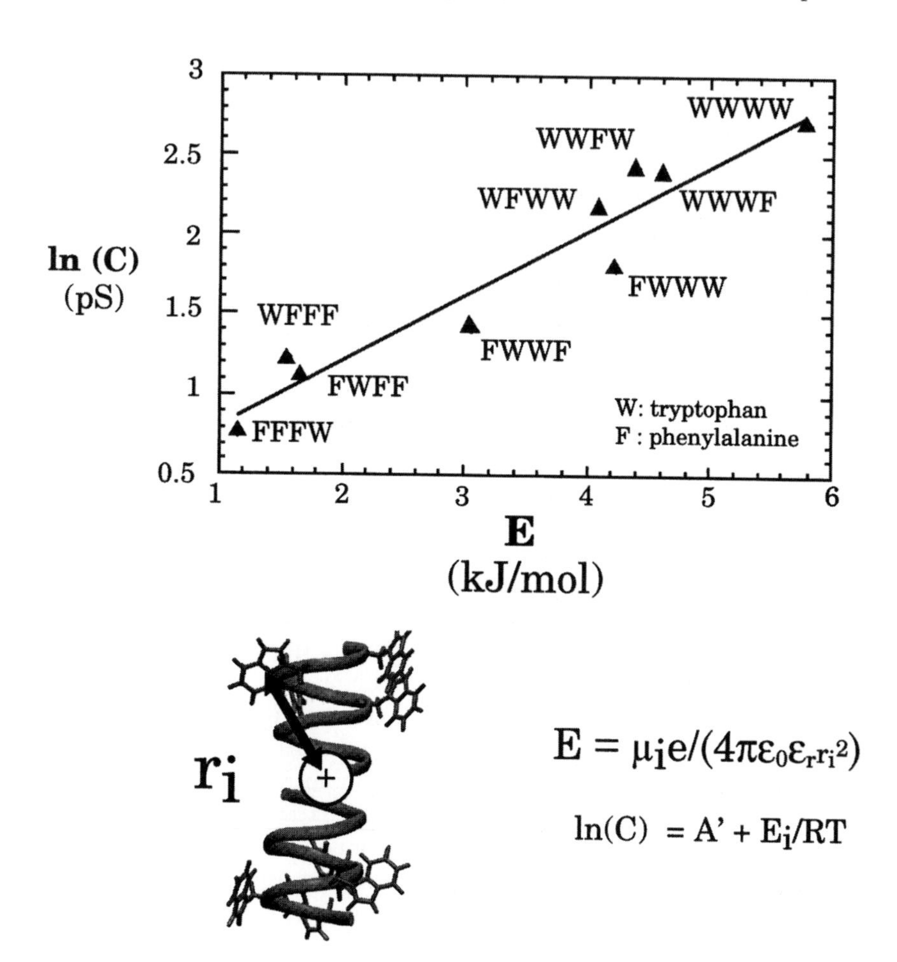

FIG. 5. Natural log of conductance for various gramicidin analogues were obtained by Becker et al (1991). The interaction energy scale represents the summation of monopole (cation at the monomer–monomer junction)–dipole (various indoles) interaction energy. This interaction energy is dependent on both distance and dipole strength in the direction of the distance vector, r_i, which are defined by the high resolution structure. This interaction is also dependent on the dielectric strength. Here, this scaling factor is estimated from the slope by first determining the energetic reduction in the potential energy barrier resulting in a factor of 20 loss in conductance. The ε_r value is thereby determined to be 5.1 and a linear correlation is demonstrated between lnC and the summed interaction energy.

Dynamics and transport rate

Understanding the fundamental determinants of rates of reaction and transport is a major challenge in structural biology. It is generally accepted that such an understanding will not come without a detailed characterization of molecular dynamics. Experimentally, dynamics are often characterized by order parameters or thermal factors that do not provide the high resolution characterization needed for correlating dynamics and function. Solid-state NMR has the advantage that isotropic motions are absent and spectral parameters are far more sensitive to the local motions than in solution NMR. Powder pattern spectra of the polypeptide backbone in hydrated lipid bilayers have been obtained as a function of temperature. The axial rotation rate of the channel about the bilayer normal is less than 1 Hz at 6°C and 10^6 Hz at 36°C (Lee et al 1993, North & Cross 1995). Below 200 K local librational motions of significant amplitude ($>5°$) cease. As the temperature is increased the tensor elements for single amide ^{15}N sites in the backbone are unequally averaged as molecular motions occur. By modelling the spectra between 200 and 283 K it is possible to determine both the axis about which the motions are occurring and the amplitude (Lazo et al 1995). Throughout gramicidin, the backbone motions appear to be occurring about the Cα-Cα axis, presumably the result of compensating motions about ψ_i and ϕ_{i+1} torsion angles. Unless the molecular motions are occurring in the millisecond and microsecond time-scale, the powder pattern spectra will not yield motional rates.

While powder patterns have characterized the motional axis and amplitude, relaxation parameters can now be used to determine motional frequencies in light of this experimentally defined motional model. The definitions of these frequencies require that either multiple relaxation parameters or relaxation parameters as a function of magnetic field strength are determined. For gramicidin T_1 relaxation at two field strengths were obtained, yielding frequencies of approximately 10^8 Hz and amplitudes of approximately 6° throughout the backbone (North & Cross 1995). This amplitude is significantly smaller than that determined from the powder patterns, which were 15–20°. The reason is that all motional frequencies faster than 10^4 Hz will average the powder patterns; however, only motions in the vicinity of the Larmor frequency (10^8 Hz) will induce T_1 relaxation. Relaxation parameters are highly non-linear detectors of motional frequencies, whereas powder patterns are linear detectors of motions greater the interaction strength represented by the powder pattern, in this case 10^4 Hz. Efficient relaxation rates clearly indicate the presence of motions near 10^8 Hz, and while these motions are of modest amplitude, motions in the picosecond time-scale are likely to be present and potentially account for the amplitude difference.

The 10^8 Hz rate is remarkably slow for a peptide plane motion. Even peptide plane motions one or two orders of magnitude faster are considered to be overdamped motions, the result of correlated motions (Usha et al 1991). Here, correlations between the motions of one peptide plane and its adjacent, next nearest neighbour, etc. planes could result in severe overdamping of the motion (Chiu et al 1991, Roux & Karplus 1991, Elber et al 1995). It is known that the motion of cations and water molecules in the channel are correlated; in other words, the column of water molecules moves as a unit while a cation is transported across the bilayer. The question for gramicidin was whether or not the correlations extended from the small molecules in the pore to the polypeptide backbone lining the pore. Apparently, motions of molecules and ions in the channel are coupled not only to each other, but to the peptide planes lining the channel and they, in turn, are coupled together. Based on single-channel conductance measurements (Andersen & Koeppe 1992) it has been estimated that cations spend approximately 10 nsec with each carbonyl cluster along the pathway between cation-binding sites. This is in remarkable agreement with the measured dynamic frequencies for the polypeptide backbone. Therefore, a correlation between the motion of the peptide planes and the kinetic translation of the cations is suggested. If such correlations exist it further suggests that the cations do not follow a one-dimensional random walk through the channel, but rather they follow a ballistic motion through the channel, thus spending as little time as possible in the low dielectric environs of the bilayer centre.

Gramicidin has proven to be a remarkable model channel from which great insights into cation solvation, selectivity and conductance efficiency have been achieved. Moreover, the need for high resolution structural and dynamic information has been demonstrated and the ability of solid-state NMR to uniquely provide it.

Acknowledgements

The authors are indebted to the staff of the National High Magnetic Field Laboratory and Florida State University NMR facilities; T. Gedris, J. Vaughn and A. Blue for their skilful maintenance and service of the NMR spectrometers; and H. Hendricks and U. Goli of the Bioanalytical Synthesis and Services Facility for their expertise and maintenance of the ABI 430A peptide synthesizer and HPLC equipment. This work has been supported by the National Institutes of Health grant number AI-23007 and the work was largely performed at the National High Magnetic Field Laboratory supported by the National Science Foundation Cooperative Agreement DMR-9527035 and the State of Florida.

References

Andersen OS, Koeppe RE II 1992 Molecular determinants of channel function. Physiol Rev 72 (suppl):S89–S157

Åqvist J, Warshel A 1989 Energetics of ion permeation through membrane channels. Solvation of Na^+ by gramicidin A. Biophys J 56:171–182

Becker MD, Greathouse DV, Koeppe RE II, Andersen OS 1991 Amino acid sequence modulation of gramicidin channel function: effects of tryptophan-to-phenylalanine substitution on the single-channel conductance and duration. Biochemistry 30:8830–8839
Becker MD, Koeppe RE II, Andersen OS 1992 Amino acid substitutions and ion channel function. Model-dependent conclusions. Biophys J 62:25–27
Blades AT, Jayaweera P, Ikonomou MG, Kebarle P 1990 Studies of alkaline earth and transition metal M^{2+} gas phase ion chemistry. J Chem Phys 92:5900–5906
Busath DD, Thulin CD, Hendershot RW et al 1998 Non-contact dipole effects on channel permeation. I. Experiments with (5F-indole)Trp^{13} gramicidin A channels. Biophys J 75:2830–2844
Chiu SW, Jakobsson E, Subramanian S, McCammon JA 1991 Time-correlation analysis of simulated water motion in flexible and rigid gramicidin channels. Biophys J 60:273–285
Christensen JJ, Eatough DJ, Izatt RM 1975 Handbook of metal ligand heats and related thermodynamic quantities. Marcel Dekker, New York
Cox BG, Schneider H 1992 Coordination and transport properties of macrocyclic compounds in solution. Elsevier Science, Amsterdam
Cross TA 1997 Solid-state nuclear magnetic resonance characterization of gramicidin channel structure. Methods Enzymol 289:672–696
Doyle DA, Cabral JM, Pfuetzner RA et al 1998 The structure of the potassium channel: molecular basis of K^+ conduction and selectivity. Science 280:69–77
Dzidic I, Kebarle P 1970 Hydration of the alkali ions in the gas phase. Enthalpies and entropies of reactions $M^+(H_2O)_{n-1}{}^+H_2O{=}M^+(H_2O)_n$. J Phys Chem 74:1466–1474
Elber R, Chen DP, Rojewska D, Eisenberg R 1995 Sodium in gramicidin: an example of a permion. Biophys J 68:906–924
Hinton JF, Whaley WL, Shungu D, Koeppe RE II, Millett FS 1986 Equilibrium binding constants for the group I metal cations with gramicidin-A determined by competition studies and Tl^+-205 nuclear magnetic resonance spectroscopy. Biophys J 50:539–544
Hu W, Cross TA 1995 Tryptophan hydrogen bonding and electric dipole moments: functional roles in the gramicidin channel and implications for membrane proteins. Biochemistry 34:14147–14155
Hu W, Lazo ND, Cross TA 1995 Tryptophan dynamics and structural refinement in a lipid bilayer environment: solid-state NMR of the gramicidin channel. Biochemistry 34:14138–14146
Jing N, Prasad KU, Urry DW 1995 The determination of binding constants of micellar-packaged gramicidin A by ^{13}C- and ^{23}Na-NMR. Biochim Biophys Acta 1238:1–11
Jordan PC 1990 Ion–water and ion–polypeptide correlations in a gramicidin-like channel. A molecular dynamics study. Biophys J 58:1133–1156
Ketchem RR, Roux B, Cross TA 1997 High-resolution polypeptide structure in a lamellar phase lipid environment from solid-state NMR derived orientational constraints. Structure 5:1655–1669
Lazo ND, Hu W, Cross TA 1995 Low-temperature solid-state ^{15}N NMR characterization of polypeptide backbone librations. J Magn Reson B 107:43–50
Lee K-C, Hu W, Cross TA 1993 ^2H NMR determination of the global correlation time of the gramicidin channel in a lipid bilayer. Biophys J 65:1162–1167
Lehn JM 1973 Design of organic complexing agents: strategies towards properties. Struct Bonding (Berlin) 11:1–70
Lomize AL, Orekhov VI, Arseniev AS 1992 Refinement of the spatial structure of the gramicidin A ion channel. Bioorg Khim 18:182–200
Mai W, Hu W, Wang C, Cross TA 1993 Three dimensional structural constraints in the form of orientational constraints from chemical shift anisotropy: the polypeptide backbone of gramicidin A in a lipid bilayer. Protein Sci 2:532–542

North CL, Cross TA 1995 Correlations between function and dynamics: time scale coincidence for ion translocation and molecular dynamics in the gramicidin channel backbone. Biochemistry 34:5883–5895

Roux B, Karplus M 1991 Ion transport in a model gramicidin channel. Structure and thermodynamics. Biophys J 59:961–981

Roux B, Karplus M 1993 Ion transport in the gramicidin channel: free energy of the solvated right-handed dimer in a model membrane. J Am Chem Soc 115:3250–3260

Tian F, Cross TA 1998 Cation binding induced changes in ^{15}N CSA in a membrane-bound polypeptide. J Magn Reson 135:535–540

Tian F, Cross TA 1999 Cation transport — an example of structural based selectivity. J Mol Biol 285:1993–2003

Tian F, Lee K-C, Hu W, Cross TA 1996 Monovalent cation transport: lack of structural deformation upon cation binding. Biochemistry 35:11959–11966

Urry DW 1971 The gramicidin A transmembrane channel: A proposed π(L,D) helix. Proc Natl Acad Sci USA 68:672–676

Urry DW 1987 NMR relaxation studies of alkali metal ion interaction with the gramicidin A transmembrane channel. Bull Magn Reson 9:109–131

Usha MG, Peticolas WL, Wittebort RJ 1991 Deuterium quadrupole coupling in N-acetylglycine and librational dynamics in solid poly(γ-benzyl-L-glutamate). Biochemistry 30:3955–3962

DISCUSSION

Roux: You mentioned that the structure is not uniquely determined because of the second-order Legendre polynomial $P_2[\cos(\theta)]$. However, I would stress that what is measured is not $P_2[\cos(\theta)]$ but rather its average. One might wish to interpret the data in terms of a single structure, or in terms of an ensemble of structures symmetrically distributed around an average, but it may not be possible. Once it goes into the non-linear function $P_2[\cos(\theta)]$, the distribution becomes skewed, i.e. the average of the function is not equal to the function of the average. Therefore, there is some unavoidable ambiguity in the interpretation of the solid-state NMR data in terms of a unique structure. Although the molecular dynamics simulation is not perfect, it provides a nice illustration of the problem I am describing. In practice, if I take a trajectory and compute the average that corresponds to a backbone ^{15}N solid-state NMR chemical shift, I may obtain the exact value that you measured, say 180 ppm. But if I look at the time course of the trajectory, I will observe fluctuations in the order of $\pm$30 ppm. Just because the average is 180 ppm does not mean that the molecular conformation is such that it corresponds to this value all the time — the structure fluctuates. On the other hand, the chemical shift calculated from the average structure may not correspond to a value of 180 ppm.

Cross: I should have mentioned that the structure I presented is a time-averaged structure.

Eisenberg: Over what time-scale?

Cross: We average over the time-scale of data acquisition, which is a matter of milliseconds.

Eisenberg: This is much longer than the permeation times.

Cross: We have also looked at the dynamics of the backbone, and in particular at how the tensor is averaged from the point at which the torsion angles are frozen out (below 200 K) to 280 K. From this we obtain a clear idea of the magnitude of tensor element averaging, and it is sensitive to picosecond motions.

Jakobsson: I have another time-averaging question. In our molecular dynamics simulations we see many more side chain transitions than Tim Cross. I would like to ask Benoît Roux if he sees any side chains undergoing torsion angles transitions.

Roux: We see isomerizations of leucines and valines, but not many.

Jordan: What temperature are these experiments performed at?

Cross: They are all run at about 30°C, which is above the phase transition temperature of the dimyristoylphosphatidylcholine (DMPC) lipids.

Davis: Regarding the question of time averaging by NMR spectroscopy, I would like to point out that Tim Cross has used deuterium NMR spectroscopy. He has measured deuterium quadrupolar splittings of hundreds of kilohertz, and he is also using dipolar couplings of a few hundred hertz. He is going from a time-scale of 10 μs to a time-scale of 10 ms, and he is therefore not measuring the same time average in all of these different experiments, so some parts of his structure are averaged over one time-scale and other parts by another.

Eisenberg: That's an important comment, but the currents are flat over that time-scale. We have direct experimental evidence that functionally there is only one potential mean force over that time-scale. When we're talking about ensembles of averaging, we have to remember that for most of us the main interest is the function of the channel, and that it is necessary to recover the potential energy profile and the underlying charge that produces it. We have the advantage that we know it exists, and it doesn't have to exist a priori for any molecular structure. The question is how to estimate it.

Davis: But you have to have a coherent view of what your structure is. We're talking about a time-averaged structure, but this is not consistent because it's averaged over several different time-scales.

Eisenberg: I don't disagree with the fact that they're different time-scales, but the observation that current is independent of time, and would change if there were a small change of potential barrier, proves that there is one time-independent structure from roughly a microsecond to seconds.

Cross: It is certainly true that the data are recorded with different time-scales. On the other hand, we have looked at the molecular motions occurring within those time-scales. We have observed global correlation times that occur within a 1 ms time-scale, and we have observed fluctuations of the peptide backbone that occur within a shorter time-scale. The only molecular motions that occur within the microsecond time-scale that we're talking about are some side chain motions for Val1 and Val7.

Koeppe: I would like to return to the topic of structure. I agree with Tim Cross on the indole function and the importance of the dipole. There are two ways to view Trp9. Figure 1 shows Trp9 next to Leu10 with an acyl chain that runs between them in acyl gramicidin. This alternative is similar to Alex Arseniev's structure in SDS (Arseniev et al 1986a) and to Jim Hinton's structure, but Tim Cross has Trp9 next to Trp15 (with the rings stacking). It turns out that both positions are allowed by Raman spectroscopy and deuterium NMR spectroscopy, so we have to use other considerations to distinguish the two possible orientations for Trp. There are two types of data that we have been using to determine this for gramicidin in DMPC. We showed that acylation gives rise to changes in the deuterium NMR spectra of side chains number 10 and number 9 (Koeppe et al 1995), and there are no other changes in the rest of the molecule. In another set of experiments using non-acylated gramicidin (Greathouse et al 1997), we mutated Leu10 and saw large changes in the deuterium spectrum of Trp. Therefore, we believe that the side chain of Trp9 is oriented toward that of Leu10 in DMPC as well as in SDS.

Arseniev: The most important point is to determine how well the structure can explain other observable characteristics of the channel. I should mention that we revealed some correlation between structural dynamics and properties of the gramicidin A-bound ions in dodecyl sulfate micelles (Arseniev et al 1986b). The dynamics of the structure is much stronger when a lithium ion is bound instead of a sodium ion, i.e. the channel becomes more flexible. However, when thallium ions replace sodium ions the channel becomes more rigid. In principle, this correlates with ion conductance: the conductance of Li^+ is lower, and Tl^+ is higher, than that of Na^+. Moreover, the ratios of Tl^+- and Na^+-binding constants to gramicidin A in dodecyl sulfate micelles are the same as those determined for the channel state in the lipid bilayers. The correlation between lifetimes (stabilities) of gramicidin ion channels in lipid bilayers and the thermostability of gramicidin structure in micelles is also remarkable. According to the NMR data melting temperatures of N-acetyl(desformyl)gramicidin A, gramicidin A and succinyl-bis(desformyl)gramicidin A incorporated into SDS micelles are 45°C, 65°C and >85°C, respectively (Barsukov et al 1987, Maslennikov et al 1988), i.e. the higher the thermostability of gramicidin structure in micelles the longer the lifetime of the corresponding channel in lipid bilayers. Thus, not only the spatial structure but also thermodynamic properties of gramicidin channel are reproduced in the micelles.

Cross: I would like to make two brief comments. The first is with regard to Roger Koeppe's side chain. I agree that the NMR data that I presented are consistent with the two Trp9 conformational possibilities. We have no NMR data that support the structure in which Trp15 stacks with Trp9. I would also

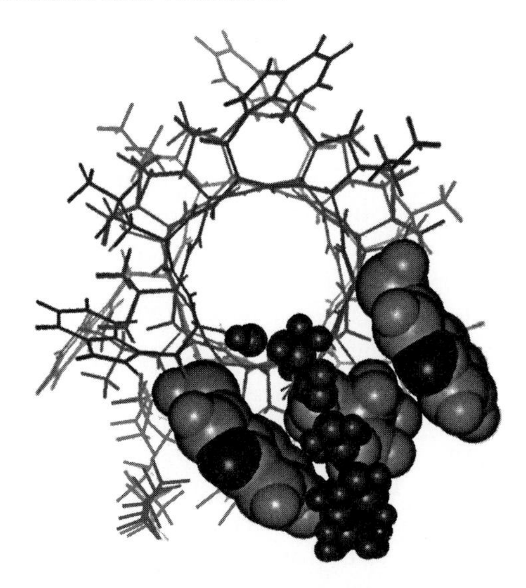

FIG. 1. End view of wire model of the single-stranded $\beta^{6.3}$-helical membrane-spanning gramicidin A channel, with the side chains of Trp9, Leu10 and Trp11 depicted as full space-filling atoms in positions determined by two-dimensional NMR in SDS (Arseniev et al 1986a). In dimyristoylphosphatidylcholine (DMPC) bilayers, Raman spectra (Takeuchi et al 1990) and deuterium NMR spectra (Koeppe et al 1994) are consistent with the illustrated orientation for Trp9, and the acyl chain of acyl gramicidin (smaller space-filling atoms) runs between the #9 and #10 side chains (Koeppe et al 1996).

like to point out that there are a number of side chain structural differences, i.e. differences in rotameric state between the SDS structure and the lipid bilayer structure, so just because it exists in SDS in one conformation does not necessarily mean that it is present in the same conformation in bilayers.

Koeppe: I would suggest that one should distinguish whether there are leucine changes or tryptophan changes. It is much easier to move a leucine residue than a tryptophan residue.

Cross: We're not talking about dynamics, or rotations, but rather the energetics associated with a side chain in one conformational state versus another.

Jakobsson: But there's not as much difference between leucine rotamers in terms of energy as there is for tryptophan rotamers.

Cross: It depends upon the packing interactions.

Sansom: Talking of packing interactions, could you remind me what the peptide to lipid ratio is in the solid state?

Cross: In our samples we kept it at 1:8. We have done a few experiments at 1:40 and 1:60 to see whether there are any changes in chemical shift, and there are not.

Sansom: So there's no evidence for changes in rotamer states when you dilute the peptide? If you have one gramicidin molecule and eight lipid molecules, there is just enough lipid to go around gramicidin. However, in terms of the ease of occupying different rotamer states with tryptophan, the restricted packing of the lipid may bias towards one structure. We don't know whether the differences between your solid-state NMR experiments and the earlier experiments are caused by dilution, such that there is a greater possibility for tryptophan reorientation when the peptide/lipid ratio is reduced.

Smart: Fluorescence energy transfer could answer this, because it should be possible to see the tryptophan signal.

Jakobsson: Mukherjee & Chattopadhyay (1994) have done tryptophan fluorescence in gramicidin. They saw that the tryptophan residues exist in a couple of different states.

Separovic: I have a comment in relation to Mark Sansom's remark about restriction of tryptophans. Our ^{13}C-labelled Trp aligned samples have a ratio of about 1:15 gramicidin:lipid. If we leave them at about 30°C for a day, we start to see a ^{13}C powder pattern. We have to heat the sample to 50°C to get the ^{13}C aligned signal back, which suggested to us that the tryptophan residues were interacting with each other. Our lipid signal still remained aligned, so we thought we had aligned membranes with aggregated gramicidin tryptophans.

I would also like to ask Jim Hinton whether there is a difference between the structure of gramicidin in SDS and the structure of gramicidin in dodecylphosphocholine (DPC), and what sort of root mean square deviations (RMSDs) he gets, because we obtain RMSDs of 2 Å for gramicidin in SDS.

Hinton: The two structures are essentially the same. The RMSDs in SDS are less than 1.8 Å, although we haven't finished refining the structure in DPC yet. We've also done the structure of seven other gramicidin analogues in SDS, and they also look essentially the same.

Wallace: I would like to ask Alex Arseniev if he has done RMSD calculations between his best-averaged structures?

Arseniev: No, but if we compare the data from different solid-state or solution NMR spectroscopy experiments, they give rise to similar structures.

Wallace: My impression was that this was one of the advantages of having two types of studies: for the first time both solid-state and solution NMR have been done on the same molecule, with the finding that, at least to an outsider, the structures they obtained look similar.

Cross: I agree. The solid-state NMR data help to validate the use of SDS micelles as the medium for doing additional solution-state NMR. The backbone fold and hydrogen bond pattern are the same. The only substantial differences are in the side chain rotameric states.

Davis: I would like to discuss the backbone. There is close agreement between Alex Arseniev's structure and Tim Cross' structure. Now that they are both here, I would like to ask about the publication in *Structure*, in which you refined the structure and removed all the ambiguities in the dihedral angles, and ended up with a single structure (Ketchem et al 1997). However, you had previously developed a nice way of deducing the dihedral angles from a couple of dipolar couplings — the ^{13}C-^{15}N and the ^{1}H-^{15}N dipolar interactions — but when you refined the structure, all your dihedral angles differed by about 30° from what you started. Could you explain this?

Cross: The solid-state NMR experiments define the orientation of the peptide plane with respect to a single axis, the bilayer normal. However, the normal to the peptide plane has two possible solutions, $\cos\theta$ and $\cos(180–\theta)$, each of which has the same bond orientations with respect to the bilayer normal. We refer to this ambiguity as a chirality, and there are two chirality solutions for each peptide plane. In the first paper (Ketchem et al 1993) the chiralities were assumed. In the second (Ketchem et al 1997) they were determined through refinement of the structure against all of the experimental data, not just the dipolar results, as well as CHARMM energy. If the chiralities change, the dihedral angles could change by 30°, but even then the bond orientations remain much the same.

Davis: But I maintain that the dihedral angles have changed, even though you insist that the dipolar couplings are preserved. Your coordinates agree exactly with all of your dipolar couplings, but the dihedral angles in that molecule are not close to any of the dihedral angles you have published.

Cross: That's not true, except where the chiralities have changed. If the orientation of the peptide plane is changed through a change in chirality, then here is a Ψ_i and $\Phi_i + 1$ compensating change of approximately equal magnitude and opposite sign.

Roux: We did some energy refinement studies on Tim Cross' initial structure using the NMR data as a restraint, and the configurational search we designed was based on a Metropolis Monte Carlo with small moves. The structure was not allowed to deviate away from the initial condition. Therefore, I am surprised at your claims that they did deviate, because most of the atoms moved by less than about 0.5 Å.

Cross: We designed that method to search all of the chirality ambiguities, hence the conformational space is consistent with the experimental data.

References

Arseniev AS, Lomize AL, Barsukov IL, Bystrov VF 1986a Gramicidin A transmembrane ion-channel. Three-dimensional structure reconstruction based on NMR spectroscopy and energy refinement. Biol Membr (USSR) 3:1077–1104

Arseniev AS, Barsukov IL, Bystrov VF, Ovchinnikov YA 1986b Spatial structure of gramicidin A transmembrane ion channel — NMR analysis in micelles. Biol Membr (USSR) 3:437–462

Barsukov IL, Lomize AL, Arseniev AS, Bystrov VF 1987 Spatial structure of succinyl-bis(desformyl)gramicidin A in micelles. NMR conformational analysis. Biol Membr (USSR) 4:171–193

Greathouse DV, Hatchett J, Jude AR, Koeppe RE, II, Providence LL, Andersen OS 1997 Neighboring aliphatic aromatic side chain interactions between residues 9 and 10 in gramicidin channels. Biophys J 72:396 (abstr)

Ketchem RR, Hu W, Cross TA 1993 High-resolution conformation of gramicidin A in a lipid bilayer by solid-state NMR. Science 261:1457–1460

Ketchem RR, Roux B, Cross TA 1997 High-resolution polypeptide structure in a lamellar phase lipid environment from solid-state NMR derived orientational constraints. Structure 5:1655–1669

Koeppe RE II, Killian JA, Greathouse DV 1994 Orientations of the tryptophan 9 and 11 side chains of the gramicidin channel based on deuterium NMR spectroscopy. Biophys J 66:14–24

Koeppe RE II, Killian JA, Vogt TCB et al 1995 Palmitoylation-induced conformational changes of specific side chains in the gramicidin transmembrane channel. Biochemistry 34:9299–9306

Koeppe RE II, Vogt TCB, Greathouse DV, Killian JA, de Kruijff B 1996 Conformation of the acylation site of palmitoylgramicidin in lipid bilayers of dimyristoylphosphatidylcholine. Biochemistry 35:3641–3648

Maslennikov IV, Arseniev AS, Bystrov VF 1988 Spatial structure of N-acetyl(desformyl)gramicidin A in micelles. Biol Membr (USSR) 5:459–474

Mukherjee S, Chattopadhyay A 1994 Motionally restricted tryptophan environments at the peptide–lipid interface of gramicidin channels. Biochemistry 33:5089–5097

Takeuchi H, Nemoto Y, Harada I 1990 Environments and conformations of tryptophan side chains of gramicidin A in phospholipid bilayers studied by Raman spectroscopy. Biochemistry 29:1572–1579

X-ray crystallographic structures of gramicidin and their relation to the *Streptomyces lividans* potassium channel structure

B. A. Wallace

Department of Crystallography, Birkbeck College, University of London, Malet Street, London WC1E 7HX, UK

Abstract. Gramicidin has been used extensively as a model system for structure/function studies of ion channels. Long before crystals of other ion channel proteins were produced, crystals of gramicidin had been prepared, even though it was many years before the first forms of those crystals were solved. There now exist a large number of crystal structures of both uncomplexed and ion-complexed forms of gramicidin crystallized from organic solvents. In all these crystals, the molecules are double helices, although they differ in helical pitch, handedness and side chain orientations, depending on the conditions used for crystallization. Since many of these structures have been discussed in detail in a recent review (Wallace 1998), this chapter concentrates on recently reported structures and how they relate to previously described X-ray and NMR structures. It also discusses how the crystal structure of a K^+ complex of gramicidin relates to the recently solved structure of a K^+ complex of the potassium channel from *Streptomyces lividans* and argues that this demonstrates that gramicidin is indeed a good model structure for biological ion channels, despite the presence of D-amino acids in its sequence.

1999 Gramicidin and related ion channel-forming peptides. Wiley, Chichester (Novartis Foundation Symposium 225) p 23–37

Gramicidin crystal structures

Crystals of gramicidin were first reported in 1949 (Hodgkin 1949), and since then many studies have attempted to characterize various types of gramicidin crystals and determine their structures. However, this molecule posed a difficult problem for X-ray structure determination and it wasn't until nearly 40 years later that the first crystal structures of it were solved (Wallace & Ravikumar 1988, Langs 1988). This was because gramicidin falls in the awkward intermediate-size range for crystal structure determinations: too small for multiple isomorphous replacement methods typically used for proteins, and too large for traditional direct methods used for small molecules. Hence, it required the use of a variety of novel approaches for structure solution, including single wavelength anomalous scattering from incorporated caesium ions (Wallace et al 1990), new types of direct methods (Langs 1988) and molecular replacement studies with various types of models (Langs et al 1991, Doyle & Wallace 1997, Burkhart et al 1997, 1998a). None of these approaches were straightforward. Heroic attempts were even made using neutron diffraction of crystals prepared with protonated and deuterated solvents for phasing (Koeppe & Schoenborn 1984). A variety of different refinement methods have been employed for the gramicidin crystal structures, including PROLSQ (Wallace & Ravikumar 1988, Langs et al 1991), SHELX-97 (Burkhart et al 1997), XPLOR (Doyle & Wallace 1997) and RESLSQ (Langs 1988), which may account for some of the variations in the details of the reported geometries. The structures range in resolution from high (0.86 Å; Langs 1988) to only moderate (2.5 Å; Doyle & Wallace 1997). In all of the crystal forms the structure of the molecule is different. In some cases (i.e. two different alcoholic solvents without ions), the differences are relatively minor, involving mostly side chain conformations, but in others, the differences are drastic, involving changes of handedness of the helices, as well as differences in helical pitch and hydrogen-bonding patterns. The resulting structures differ considerably in the size and regularity of their central hole (Smart et al 1993, Smart et al 1997). The widely varying crystal structures are a clear demonstration of the polymorphism of this flexible polypeptide (Wallace 1991).

Uncomplexed forms

The first X-ray structure of an uncomplexed form of gramicidin to be solved was that of crystals prepared from ethanol (Langs 1988). In this structure gramicidin forms a left-handed antiparallel double helix with 5.6 residues/turn ($\pi\pi^{5.6}$) and has its tryptophan side chains oriented principally in a direction parallel to the helix axis. Its backbone fold is somewhat irregular, resulting in a central hole whose diameter varies dramatically along its length (Smart et al 1993). Later, the structure of a crystal form prepared from methanol was solved (Langs et al 1991). In it gramicidin had a similar fold, except that its side chains were oriented more or less normal to the helix axis, and the backbone and central hole were much more uniform. It was argued that these differences were primarily the result of different packing constraints in the two crystal forms, one orthorhombic and the other monoclinic.

The structures of the uncomplexed forms from ethanol (Fig. 1A) and methanol (Fig. 1B) have recently been re-refined (Burkhart et al 1997). Both are at high resolution and differ in mostly minor ways from the structures originally reported in these solvents (Langs 1988, Langs et al 1991). For example, the rms difference between the original structure in ethanol (Brookhaven Protein Data Bank code 1GMA) and the re-refined structure (Brookhaven Protein Data Bank code 1ALZ) is only 0.06 $Å^2$ for all atoms. The major differences appear to be that different side chains are assigned to have multiple positions in the original and re-refined structures. Curiously, however, for the ethanol form, the R-factor increased by 9% (from 0.07 to 0.16) upon re-refinement.

The structure of an uncomplexed form prepared from n-propanol (Fig. 1C) has also recently been determined (Burkhart et al 1997); it, too, is from orthorhombic crystals. This structure, like the other uncomplexed structures, is a left-handed antiparallel double helix with 5.6 residues/turn. As in the ethanol structure, its tryptophan side chains are oriented with their planes approximately parallel to the helix axis. The tryptophans at position 11 are disordered. The propanol form is much more similar to the ethanol form than the methanol form, with an rms difference of 1.8 $Å^2$ for all atoms, and an rms of 0.26 $Å^2$ for backbone atoms only.

Finally, Burkhart et al (1998b) have recently reported the structure of an uncomplexed form prepared from glacial acetic acid, which is a right-handed, antiparallel, double-stranded helix. This is a particularly interesting structure in

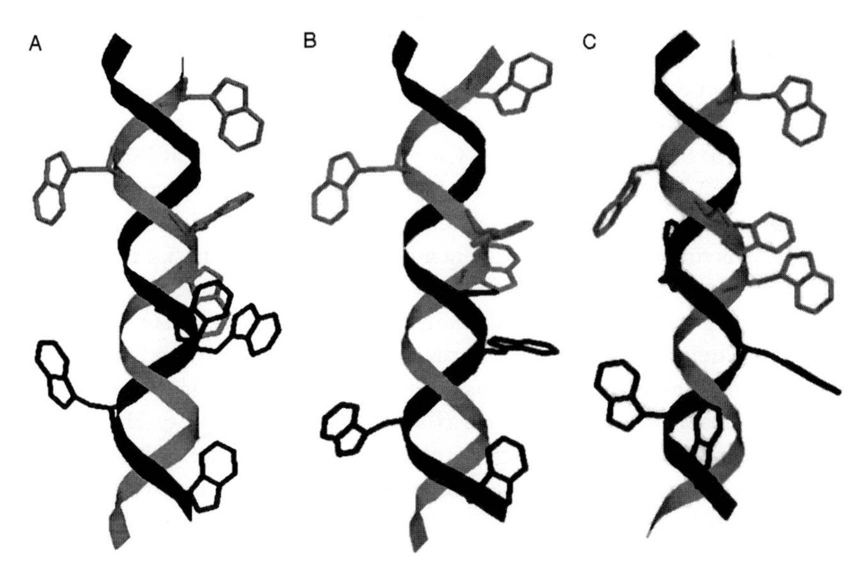

FIG. 1. X-ray structures of uncomplexed forms of gramicidin crystallized from: (A) ethanol (Burkhart et al 1997); (B) methanol (Burkhart et al 1997); and (C) propanol (Burkhart et al 1997). These are drawn, using Rasmol (Sayle & Milner-White 1995), as ribbon diagrams with only the tryptophan side chains shown. The two polypeptide chains in each dimer are shown in grey and black, respectively.

that it is reported to have the $\pi\pi^{7.2}$ hydrogen-bonding pattern previously found only in ion complexes (see next section). The authors compared their structure with the ^{15}N NMR data of Nicholson et al (1987) and suggested that their structure is consistent with the chemical shift anisotropy data of gramicidin in a planar bilayer. They have used this to claim that the acetic acid structure is the 'membrane-active' form. However, not only was this comparison made with an early preliminary NMR data set, but the interpretation is contrary to a wide range of chemical, conductance and physical measurements that have shown that the principal membrane-active form is the N-to-N helical dimer structure. The crystal structure is undoubtedly correct, but the interpretation appears to be misguided. The acetic acid form appears to be yet another polymorphic form present in organic solvents, and again demonstrates the versatility of the molecule in adopting structures of different hands, helical pitch and relative chain orientations.

Ion-complexed forms

Crystals of an ion complex of gramicidin (CsSCN) were first reported by Koeppe et al (1978), who showed by Patterson analyses that the dimensions of this structure were considerably different from those of gramicidin in an uncomplexed form, but they did not solve the structure. Preparations of crystals of a number of other ion-complexed forms have subsequently been reported (Koeppe et al 1979, Kimball & Wallace 1984, Doyle & Wallace 1994, 1995, Burkhart et al 1998c). The first crystal structure of an ion-complexed form of gramicidin to be solved was that of CsCl (Wallace & Ravikumar 1988, Wallace et al 1990; Fig. 2A). As was found for most of the uncomplexed forms, it is a left-handed double helix, but in contrast to the $\pi\pi^{5.6}$ structure found in the absence of ions, this structure had a $\pi\pi^{7.2}$ hydrogen-bonding pattern and 6.4 residues per turn. Like most of the ion-free forms, in the ion-complexed form, the tryptophan side chains are oriented nearly parallel to the helix axis. Recently, a right-handed double helix with a $\pi\pi^{7.2}$ hydrogen-bonding pattern and different ion stoichiometries and positions as well as a different stagger between the polypeptide chains has been reported (Burkhart et al 1998a; Fig. 2B). It has a more regular structure than the left-handed form, without the deformations of the backbone hydrogen bonds in the regions of the ions, but also has tryptophan side chains that lie relatively parallel to the helix axis. It was suggested that this structure was derived from the same crystal form reported earlier (Wallace & Ravikumar 1988), and that the earlier crystal structure was in error. However, the new crystals were prepared under significantly different conditions (including a fourfold higher ion concentration) and clearly diffract to higher resolution than the previous crystals. The two types of crystals are non-isomorphous, having slightly different unit cell dimensions. As can be seen from other gramicidin structures, virtually identical unit cells can lead to considerably different molecular structures (Doyle & Wallace 1997) so the small differences in unit cell dimensions reported for the two CsCl forms are significant. Furthermore, the sizes of the Bijvoet differences and partial structure signals in the earlier data set (Wallace et al 1990) indicate that there must be the equivalent of at least 1.5 caesium ions per dimer (more if the B factors are high); in the Burkhart et al structure, the caesium/gramicidin dimer stoichiometry is only 1.0. In addition, the ion positions in the Burkhart et al structure are inconsistent with the Patterson and anomalous Patterson maps calculated for the earlier crystals: four of the caesium sites in the asymmetric units of each structure are very similar (rms=1.83 Å^2), although they lie in different positions with respect to the polypeptide chains. The remaining caesium sites in the two structures are different, as are all of the chloride sites. However, it is interesting to note that two of the other caesium sites in the Burkhart et al structure are near to what have been identified as chloride sites in the Wallace and Ravikumar structure; they cannot be caesiums in the Wallace et al data due to the absence of peaks at these locations in the anomalous map (although they could be mistaken for partially occupied caesiums in a native map). Thus, from all of these differences, the two crystal forms must be different, and thus they appear to be a further example of the polymorphism of the molecule.

Other ion complexes of gramicidin have been reported, although not all have been solved. Smaller ions, such as potassium, form crystals with slightly shorter unit cell lengths in the longest dimension, which is a consequence of a distortion

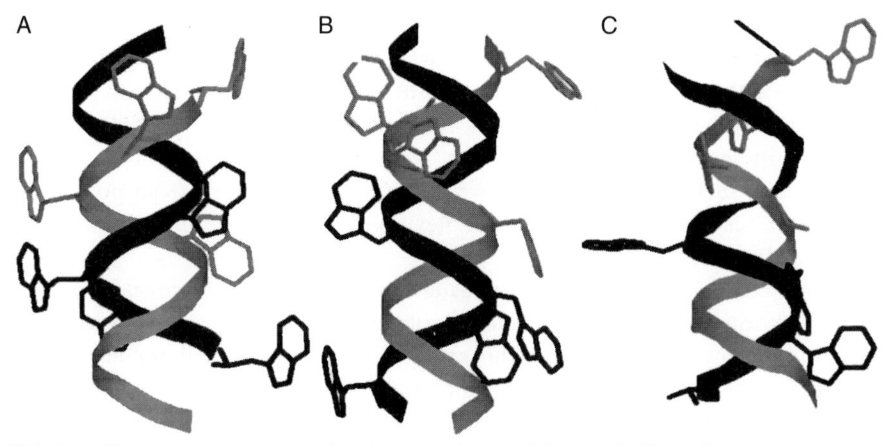

FIG. 2. X-ray structures of complexed forms of gramicidin: (A) CsCl (Wallace & Ravikumar 1988); (B) CsCl (Burkhart et al 1998a); and (C) KSCN (Doyle & Wallace 1997). Drawn as in figures.

in the polypeptide backbone that occurs in the region adjacent to the ions (Doyle & Wallace 1997; Fig. 2C). Crystals containing rubidium chloride (Doyle & Wallace 1995) are of a different space group, but also have similar sub-cell dimensions, with the long axis again foreshortened relative to either of the CsCl crystals.

Lipid-complexed forms

To date, all the crystal structures of gramicidin solved have been of double helical forms, which are not the principal conducting form of the molecule. Crystals have been reported of a lipid complex, containing gramicidin and dipalmitoyl phosphatidylcholine (DPPC), that may be of a helical dimer form as found in membranes, but that structure has not yet been solved (Wallace & Janes 1991). Raman spectroscopic studies (Short et al 1987) suggested the gramicidin structure in these crystals is the same as is found in bilayers, chemical analyses showed that the crystals contain both lipid and protein, and modelling studies (Crouzy et al 1992) have suggested that a complex containing a 6.3 residue/turn helical dimer plus two lipid molecules per gramicidin monomer would be consistent with the unit cell and space group. However, these crystals are highly non-isomorphous from individual crystal to individual crystal, a characteristic, which when combined with a high sensitivity to radiation damage, has prevented collection of a complete native data set and solution of its structure. Thus, future X-ray studies on the helical dimer will depend on finding a lipid-complexed crystal form with modified physical characteristics more suitable for X-ray structure determination.

Comparison of gramicidin crystal and NMR structures

It is of considerable interest to examine whether the solution (NMR) and solid-state (X-ray) structures of gramicidin are the same, and if they are not, what the differences in the structures can tell us about dynamics and environmental effects on this molecule. Such comparisons are especially important for a small polypeptide such as gramicidin, in which crystal-packing forces can potentially distort the crystal structure, and the flexibility of the molecule can result in poorly defined solution structures.

Direct comparisons can now be made for several of the gramicidin forms, bearing in mind that the crystals and solutions were not prepared under identical conditions, so that some of the differences seen could be attributable to these variables. The structures of the uncomplexed forms from 'ethanol' are the most facile to compare, although the crystals are from an ethanol/benzene azeotrop (essentially all ethanol) and contain ethanol and water solvent molecules in the crystal, but no detectable benzene, while the NMR structure was determined in a 1:4 mixture of ethanol and benzene, so its solvent environment will be considerably more hydrophobic. In ethanol solution, it has been shown (Veatch et al 1974) that gramicidin adopts a mixture of four interconvertable conformers, and indeed, the structure found in the crystals (species 3) represents, at room temperature, only ~57% of the total population present in solution (Chen & Wallace 1997). The ethanol/benzene mixture was chosen for the NMR experiments to bias the population towards a single conformer, the same one present in the crystals, the left-handed antiparallel $\pi\pi^{5.6}$ double helix. Both the solution and solid-state structures of gramicidin are of similar forms, but there are notable differences between them, especially in the side chain conformations and their degree of disorder (Pascal & Cross 1994). This may be a consequence of distortions due to packing in the crystals, which mostly affect the side chains at the periphery of the molecules, or it may be an environmental effect resulting from the different solvents used.

The other forms whose NMR and X-ray structures can be compared are the caesium-containing ion complexes. Again, the solvent conditions necessary to produce the crystals and for the NMR studies are not identical, and these must be considered in any conclusions derived for this polymorphic molecule. The X-ray structures (Wallace & Ravikumar 1988, Burkhart et al 1998a) were determined for crystals obtained from methanol solutions, whereas the NMR studies (Arseniev et al 1985) were done in a 1:1 methanol/chloroform mixture (again considerably more hydrophobic than the crystallization solvent); the crystals contained CsCl while the solution contained CsSCN. In addition, the ion concentrations were vastly different and the temperatures used for data collection were significantly different (higher for the solution study than for either of the crystal studies). Despite these differences, one of the X-ray structures (Burkhart et al 1998a) is of the same general fold (a right-handed antiparallel $\pi\pi^{7.2}$ double helix) as the NMR structure, although the X-ray and NMR structures differ considerably in detail. The other crystal structure (Wallace & Ravikumar 1988) is of the opposite hand, although again with the same sort of helical pitch, and so is quite different from the solution structure. Burkhart et al (1998b) argue that this is evidence that the true ion-complexed structure is right-handed. However, once again, a mixture of conformers is present in solution for the ion complexes, the composition of which depends on the ion concentration, temperature, etc. (Chen & Wallace 1996), so conclusions about which component will crystallize under these conditions are not straightforward. Since the NMR spectrum (Chen & Wallace 1996) of a CsCl chloride complex in methanol—as in the crystallization mixture—is distinctly different from that of the CsSCN complex in methanol/chloroform, the mixture used for the structure solution by NMR, the NMR structure reported may not reflect the conformation found in the crystal environment. The NMR structure of the CsCl/methanol solution remains to be solved. Therefore, any differences seen in the ion complexes may be another reflection of the flexibility of this molecule with environment.

Thus, it is clear that there are real solid state/solution structural differences, not a terribly surprising result given the changes in structure seen with virtually every other environmental variation. However, this clearly demonstrates that the NMR and X-ray techniques can provide us with important complementary information on the structures and dynamics of this polymorphic molecule.

Comparison of gramicidin and the potassium channel structure

Although gramicidin has been used as a model ion channel for many years, it had been argued that ion-binding sites in 'real' channels were likely to be formed from side chain rather than the polypeptide backbone atoms, as the side chains would be more accessible in proteins with typical secondary structures. As a result, a gramicidin-type structure which binds ions via its backbone carbonyl oxygens could be irrelevant to biological channels. With the recent solution of the crystal structure of the potassium channel from *Streptomyces lividans* (Doyle et al 1998), it is now possible to compare the types of binding motifs used in this biological channel and in gramicidin.

While there are now a number of gramicidin X-ray and NMR structures potentially available for such a comparison, the most direct comparison can be made between the potassium ion complex of gramicidin (Doyle & Wallace 1997) and the potassium ion complex of the *S. lividans* potassium channel. Although the gramicidin in the potassium complex is of the double helical pore structure and not the helical dimer channel structure, the general principles adopted by this structure are likely to extend to other gramicidin structures. The advantage of using this crystal structure for the comparison is that the details of the actual binding sites of the cations can be seen, whereas in all of the NMR structures of the helical dimer forms, the ion-binding sites are only implicated by extrapolation from other structural and chemical data.

The cation-binding sites in the K^+/gramicidin structure and the K^+/potassium channel structure are remarkably similar, despite the fact that gramicidin contains D-amino acids. Both molecules produce an ion conduction pathway using carbonyl groups of their polypeptide backbones (Fig. 3). Each creates a hole for the ions by the absence of steric interference from side chains of the adjacent amino acids. In the case of gramicidin, this is accomplished by having alternating L- and D-amino acids that place all the side chains on one face of the polypeptide away from the pore lumen (Fig. 3A); in the case of the potassium channel, this is possible due to the presence of alternating glycine residues, which have no side chains to block the lumen, as the sequence around the binding site region is VGYG (Fig. 3B). Furthermore, in both structures the side chains in the region

A

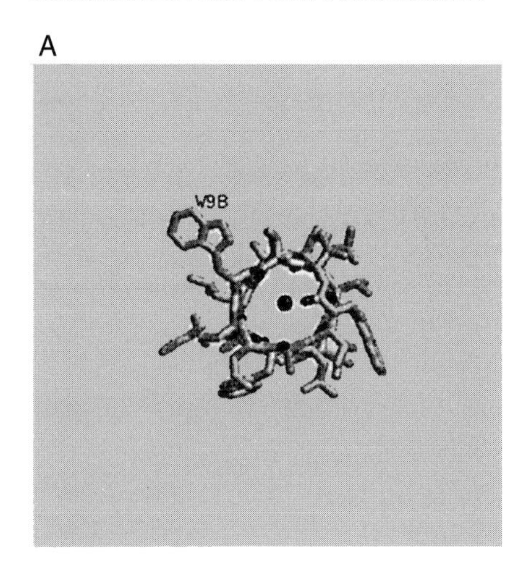

B

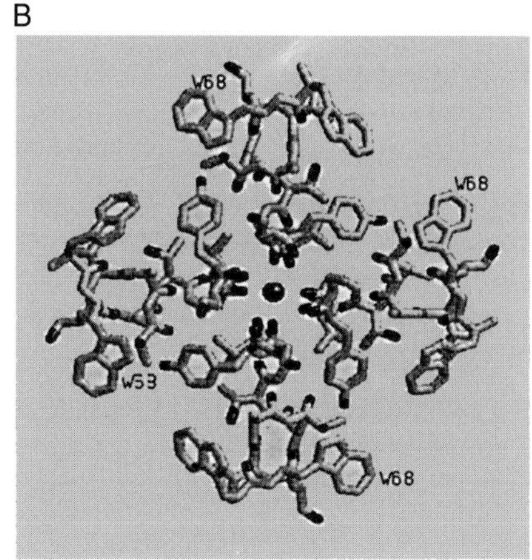

FIG. 3. Comparison of the K^+-binding sites: (A) in KSCN-gramicidin (Doyle & Wallace 1997); and (B) in the *Streptomyces lividans* potassium channel (Doyle et al 1998). Both views looking down the ion conduction pathways.

of the ion-binding sites seem to be predominantly aromatic in nature and may have a role in stabilizing the surrounding structure. Thus, it is clear that nature has used the same sort of ion-binding motif in both gramicidin and in the potassium channel, and, consequently this confirms (much to the relief of people working in the field) that gramicidin is indeed a good model system for biological ion channel studies.

References

Arseniev AS, Bystrov VF, Barsukov IL 1985 NMR solution structure of gramicidin A complex with cesium cations. FEBS Lett 180:33–39

Burkhart BM, Gassman RM, Pangborn W, Duax WL 1997 Gramicidin structure: influence of a lipid-like solvent. Biophys J 72:395 (abstr)

Burkhart BM, Li N, Langs DA, Duax WL 1998a Will the real gramicidin channel please stand up: the right-handed double-stranded double-helical dimer complexed with Cs^+. Biophys J 74:232 (abstr)

Burkhart BM, Li N, Langs DA, Duax WL 1998b The membrane active forms of gramicidin A, at last. American Crystallographic Association Abstracts, p 67

Burkhart BM, Pangborn WA, Duax WL 1998c Cation complexes with the right-handed double-stranded double-helical ($DSDH_R$) dimer of gramicidin. Biophys J 74:392 (abstr)

Chen Y, Wallace BA 1996 Binding of alkaline cations to the double-helical form of gramicidin. Biophys J 71:163–170

Chen Y, Wallace BA 1997 Solvent effects on the conformation and far ultraviolet circular dichroism spectra of gramicidin A. Biopolymers 42:771–781

Crouzy S, Janes RW, Roux B, Wallace BA 1992 Modeling gramicidin/lipid interactions in a crystalline complex. In: Roux B (ed) Report of the CECAM Workshop on Molecular

Modeling of Membrane Proteins and Lipids. Centre European de calcal atomique et moléculaire, Lyon, France, p 1–4

Doyle DA, Wallace BA 1994 Caesium-binding sites in the gramicidin pore. Biochem Soc Trans 22:1043–1045

Doyle DA, Wallace BA 1995 Crystallization and characterisation of a rubidium chloride complex of gramicidin A. Protein Peptide Letts 2:371–376

Doyle DA, Wallace BA 1997 The crystal structure of a gramicidin/potassium thiocyanate complex. J Mol Biol 266:963–977

Doyle DA, Cabral JM, Pfuetzner RA et al 1998 The structure of the potassium channel: molecular basis of K^+ conduction and selectivity. Science 280:69–77

Hodgkin DC 1949 X-ray analysis and protein structure. Cold Spring Harbor Symp Quant Biol 14:65–78

Kimball MR, Wallace BA 1984 Crystalline ion complexes of gramicidin A. Ann NY Acad Sci 435:551–554

Koeppe RE II, Schoenborn BP 1984 5 Å Fourier map of gramicidin A phased by deuterium-hydrogen solvent difference neutron diffraction. Biophys J 45:503–507

Koeppe RE II, Hodgson KO, Stryer L 1978 Helical channels in crystals of gramicidin A and a cesium–gramicidin A complex: an X-ray diffraction study. J Mol Biol 121:41–54

Koeppe RE II, Berg JM, Hodgson KO, Stryer L 1979 Gramicidin A crystals contain two cation binding sites per channel. Nature 279:723–725

Langs DA 1988 Three dimensional structure at 0.86 Å of the uncomplexed form of the transmembrane ion channel peptide gramicidin A. Science 241:188–191

Langs DA, Smith GD, Courseille C, Précigoux G, Hospital M 1991 Monoclinic uncomplexed double-stranded, antiparallel, left-handed $\beta^{5.6}$-helix ($\uparrow\downarrow \beta^{5.6}$) structure of gramicidin A: alternate patterns of helical association and deformation. Proc Natl Acad Sci USA 88:5345–5349

Nicholson LK, Moll F, Mixon TE, LoGrasso PV, Lay JC, Cross TA 1987 Solid-state ^{15}N NMR of oriented lipid bilayer bound gramicidin A. Biochemistry 26:6621–6626

Pascal SM, Cross TA 1994 Polypeptide conformational space. Dynamics by solution NMR, disorder by X-ray crystallography. J Mol Biol 241:431–439

Sayle RA, Milner-White EJ 1995 RASMOL: biomolecular graphics for all. Trends Biochem Sci 20:374–376

Short KW, Wallace BA, Myers RA, Fodor SPA, Dunker AK 1987 Comparison of lipid/gramicidin dispersions and cocrystals by Raman scattering. Biochemistry 26:557–562

Smart OS, Goodfellow JM, Wallace BA 1993 The pore dimensions of gramicidin A. Biophys J 65:2455–2460

Smart OS, Neduvelil JG, Wang S, Wallace BA, Sansom MSP 1997 HOLE: a program for the analysis of the pore dimensions of ion channel structural models. J Mol Graph 14:1–7

Veatch WR, Fossel ET, Blout ER 1974 The conformation of gramicidin A. Biochemistry 13:5249–5256

Wallace BA 1991 Alternate folding motifs for gramicidin: crystallographic and spectroscopic analyses of polymorphism. In: Nall BT, Dill KA (eds) Conformations and forces in protein folding. AAAS Publications, Washington, DC, p 188–195

Wallace BA 1998 Recent advances in the high resolution structures of bacterial channels: gramicidin A. J Struct Biol 121:123–141

Wallace BA, Janes RW 1991 Co-crystals of gramicidin A and phospholipid. A system for studying the structure of a transmembrane channel. J Mol Biol 217:625–627

Wallace BA, Ravikumar K 1988 The gramicidin pore: crystal structure of a cesium complex. Science 241:182–187

Wallace BA, Hendrickson WA, Ravikumar K 1990 The use of single-wavelength anomalous scattering to solve the crystal structure of a gramicidin A/caesium chloride complex. Acta Crystallogr B 46:440–446

DISCUSSION

Cross: Where are the ion-binding sites in potassium crystal structure?

Wallace: There are three potassium sites within the double helical lumen. They involve the backbone carbonyls of residues 6 and 8 (for the fully occupied K^+ site near the centre of the pore), and residue 13 and the formyl carbonyl for the partially occupied sites near the ends of the pore. The two closest potassiums are 11.2 Å apart, whereas the sites further apart are separated by 14.7 Å.

Cross: This is quite different from the gramicidin channel and from the gramicidin/caesium structure.

Wallace: Yes, there are only two caesium sites in the pore lumen in our CsCl crystals. They are separated by 11.6 Å but are located about 7 Å from the end of the pore, and so involve carbonyl groups from completely different residues in the binding site. I should note, however, that whilst I have used the potassium crystal structure as my example for comparison with the potassium channel, the same principles should also be true for any of the gramicidin structures, whether double helical or helical dimer, because they all utilize their backbone carbonyls to bind the cations, as does the potassium channel. I specifically used the potassium/gramicidin complex for my comparison simply because it is the one structure that allows us to directly look at a potassium in its binding site.

Jakobsson: I'm sure we are going to talk a lot about the analogies between the K^+ channel and gramicidin, so it may be worth pointing out that although we talk about gramicidin as being a non-selective channel, it does have a much higher affinity for potassium than it does for sodium. The ratio of those affinities is about the same as the selectivity of the *Streptomyces* K^+ channel that Schrempf et al (1995) determined, so we could call gramicidin a potassium channel. Based on change in reversal potential, Schrempf et al said the KcsA channel is only threefold more selective for K^+ than Na^+. On the other hand, by the method of rubidium uptake in vesicles, different experiments give 10/1 to many/one, according to Cuello et al (1998) and Heginbotham et al (1998). Although most members of the g/aromatic/G (i.e. potassium) family are selective for potassium, besides KcsA there are other members where the selectivity has been seen to be relatively weak, i.e. in the 3:1 to 5:1 range (for example Gauss et al 1998, Ludwig et al 1998). So it's fair to say that gramicidin is as K^+ selective as the more weakly selective members of the K channel family, and likely for the same reason, because the basic structural motif in both the K channel selectivity filter and the gramicidin channel give a polarized 'cage' for holding the ion that is about the same size in both cases (for an insightful published comment that is pertinent to this, see Phillips & Luisi 1998).

Jordan: In your gramicidin structure how many carbonyls are associated with the potassium-binding site? In MacKinnon's potassium channel there are about eight (Doyle et al 1998).

Wallace: There are only two closely associated carbonyls in the case of the potassium thiocyanate complex structure.

Roux: It's also true that the gramicidin channel is a β-helix that hydrogen bonds to itself, so that the carbonyls are not freely available for ion co-ordination. In contrast, there may be some uncertainties in that region in the case of the KcsA potassium channel.

Koeppe: For the benefit of those peripherally involved with this field, I would like to suggest a change in terminology, i.e. that we change the 'pore structure' designations to 'double-stranded' channels. This would help to distinguish them from the single-stranded channel structure. I would also like to mention that it is possible under certain conditions to observe some double-stranded transport activity, but these conditions are so difficult that I question whether it is a useful model to compare with the potassium channel from the transport point of view. I would prefer to make comparisons with single-stranded gramicidin channels. I will give more details on this in my presentation this afternoon.

Busath: Why do the crystals that contain gramicidin and the lipid dipalmitoyl phosphatidylcholine (DPPC) dissolve or melt upon radiation exposure? Does tryptophan absorb the irradiation?

Wallace: There are many possibilities, and tryptophans may be involved. Radiation damage involves hitting one molecule with an X-ray, followed by the transfer of free radicals throughout the crystal, but we don't know why these crystals are more sensitive than the other crystal forms of gramicidin. We do know that they are also more temperature sensitive, and this may be related to the state of their lipid molecules because the temperatures at which they melt differ for crystal complexes with dimyristoyl phosphatidylcholine (DMPC) and DPPC lipids.

Busath: What is the lipid:peptide ratio?

Wallace: The actual lipid:peptide ratio in the crystals is small, i.e. four phospholipids per dimer. The diffraction patterns suggest it is in a bilayer-like motif, although it's clearly not a bilayer.

Jakobsson: Are other ions present in the crystals, and if so does their presence affect the crystal?

Wallace: We have not yet been able to create lipid complex crystals that have ions because ions precipitate out the lipids before they complex with gramicidin. We're working on that.

Davis: Have you tried diether lipids as well as diester lipids?

Wallace: Yes, but we had no luck in forming crystals from them. We just haven't found the trick for making the lipid complexes any better, despite setting up many different conditions. Unfortunately, finding the trick or having some luck is what successful crystallization is all about.

Cross: What is the evidence that gramicidin A can form a double helix in lipid bilayers?

Koeppe: I'll elaborate on this in my presentation, but suffice to say that a 22 carbon membrane is required for a hint of double-stranded channel activity in the case of gramicidin A homodimers (Mobashery et al 1997).

Wallace: We used the same lipids as Olaf Andersen used in his conductance studies (Mobashery et al 1997) to produce small unilamellar vesicles, and by circular dichroism we also found that anything less than 22 carbons did not form a double helix (Galbraith & Wallace 1999).

Koeppe: The latest estimate from Olaf Andersen (personal communication 1998) is that in dioleoyl phosphatidylcholine or diphytanoyl phosphatidylcholine membranes the single-stranded channel is about 15–20 kJ/mol more stable than the occasional double-stranded channel that one might observe in these membranes.

Killian: In order to obtain the channel conformation of gramicidin it seems that the peptide needs to be surrounded by lipids, because of the interactions between the tryptophans and the interfacial region of the lipids. Wouldn't this make it virtually impossible to obtain crystals of gramicidin in the channel conformation because such a small amount of lipid is present?

Wallace: That is what we originally thought. We set up hundreds of crystallizations with different lipid:gramicidin ratios, and at the higher lipid ratios we obtained what appeared to be rolled up sheets (i.e. cigar-like structures), possibly two-dimensional crystals. In all cases examined, the crystals that grew didn't have the same final lipid:gramicidin ratio that was in the initial crystallization mixture. However, if we tried to set up the crystals with the final lipid:gramicidin ratio, we didn't get any crystals growing at all.

Killian: Have you tried using glycolipids? Because there may be a specific interaction between the tryptophan side chains and the glycosyl moiety.

Wallace: No, but it's a good suggestion.

Jakobsson: Have you tried different lengths of hydrocarbon chain in the lipids? I would have thought that shorter-chain hydrocarbons would have a better length for hydrophobic matching and would have faired better.

Wallace: For a year or so we thought DMPC would be the correct chain length, based on what we know about the structures, so we tried all the variations we could think of, but the crystals we obtained were hopeless. Then we finally tried DPPC even though we thought it was the wrong fatty acid chain length, and we obtained crystals!

Hinton: Have you tried to crystallize covalently linked monomers?

Wallace: We were given a sample of some cross-linked monomers produced by Stuart Schreiber's group (Stankovic et al 1989), and we tried to obtain crystals, but without success. However, we didn't follow up the initial experiments, so perhaps we should try again.

I would also like to ask Roger Koeppe a question. In your early neutron diffraction studies of gramicidin crystals (Koeppe & Schoenborn 1984) did you use pure gramicidin A? I ask this because there is a recent *Biophysical Journal* paper (Burkhart et al 1998) which suggests that it is not possible to crystallize pure gramicidin A, rather a certain amount of sequence heterogeneity is required to produce crystals. However, it was my recollection that you were able to produce crystals from pure gramicidin A.

Koeppe: We have separated and crystallized pure valine gramicidin A and isoleucine gramicidin A, and the unit cell dimensions are slightly different (Koeppe & Weiss 1981).

Jakobsson: Is it possible to do this in the cubic phase?

Wallace: We did some preliminary work on this a short time ago, but we haven't yet succeeded.

Rosenbusch: But there is lots of variation between various parameters such as lipid geometry, temperature and pressure.

Wallace: We only tried conditions similar to those you used for bacteriorhodopsin (Pebay-Peyroula et al 1997), but we know that there are many variables so we should try some different conditions.

Killian: It may not work in this system because gramicidin can induce non-lamellar structures and may not segregate out from a cubic phase to form crystals.

References

Burkhart BM, Gassman RM, Langs DA, Pangborn WA, Duax WL 1998 Heterodimer formation and crystal nucleation of gramicidin D. Biophys J 75:2135–2146

Cuello LG, Romero JG, Cortes DM, Perozo E 1998 pH-dependent gating in the *Streptomyces lividans* K^+ channel. Biochemistry 37:3229–3236

Doyle DA, Cabral JM, Pfuetzner RA et al 1998 The structure of the potassium channel: molecular basis of K^+ conduction and selectivity. Science 280:69–77

Galbraith TP, Wallace BA 1999 Phospholipid chain length alters the equilibrium between pore and channel forms of gramicidin. Faraday Discusions 111:159–164

Gauss R, Seifert R, Kaupp UB 1998 Molecular identification of a hyperpolarization-activated channel in sea urchin sperm. Nature 393:583–587

Heginbotham L, Kolmakova-Partensky L, Miller C 1998 Functional reconstitution of a prokaryotic K^+ channel. J Gen Physiol 111:741–749

Koeppe RE II, Schoenborn BP 1984 5 Å Fourier map of gramicidin A phased by deuterium-hydrogen solvent difference neutron diffraction. Biophys J 45:503–507

Koeppe RE II, Weiss LB 1981 Resolution of linear gramicidins by preparative reversed-phase high-performance liquid chromatography. J Chromatogr 208:414–418

Ludwig A, Zong X, Jeglitsch M, Hofmann F, Biel M 1998 A family of hyperpolarization-activated mammalian cation channels. Nature 393:587–591

Mobashery N, Nielsen C, Andersen OS 1997 The conformational preference of gramicidin channels is a function of lipid bilayer thickness. FEBS Lett 412:15–20

Pebay-Peyroula E, Rummel G, Rosenbusch JP, Landau E 1997 X-ray structure of bacteriorhodopsin at 2.5 Å from microcrystals grown in lipidic cubic phases. Science 277:1676–1681

Phillips K, Luisi B 1998 Ion discrimination in proteins and DNA. Science 281:883 (abstr)

Schrempf H, Schmidt O, Kummerlen R et al 1995 A prokaryotic potassium channel with two predicted transmembrane segments from *Streptomyces lividans*. EMBO J 14:5170–5178
Stankovic CJ, Heinemann SH, Delfino JM, Sigworth FJ, Schreiber SL 1989 Transmembrane channels based on tartaric acid–gramicidin A hybrids. Science 244:813–817

General discussion I

The effects of ion binding

Koeppe: There are several ways to interpret the observation that in the presence of an ion the spectrum of the backbone, but not that of the side chain, changes. One possibility is that the side chain is in the same orientation, but because of the swivel at $\chi 1$ and $\chi 2$, the backbone moves a little bit without a corresponding movement of the side chain.

Cross: Yes, the constancy of the side chain suggests that the backbone doesn't move, but the backbone dipolar data clearly demonstrate that the peptide planes do not structurally change.

Hinton: In your ion-binding studies, what's the thermodynamic concentration of the sodium ions in the hydrated lipid system? The reason I ask is, are you seeing free ions or ion pairs in the channel?

Cross: We don't believe we're seeing ion pairs, because when we change the anion we don't see any changes in the influence the ion has on the structure, so there do not appear to be any observable interactions between the anion and the structure. This is a reasonable suggestion in the light of the electrostatics surrounding the end of the channel with the exposed carbonyls.

Hinton: Is it possible that some free ions are binding to some channels, and that the rest are sampling ion pairs?

Cross: We've done some preliminary work with titrations, where we observe a levelling off at single-site saturations and then some double occupancy saturation.

Roux: The crystal structure of the double helical gramicidin structure may also give us some clues about ion binding. The NMR spectroscopy results suggest that there are two potential ion-binding sites. At a low ion concentration a single site is probably occupied, and as the concentration increases both sites may be occupied. The final observed structures would be a mixture of non-occupied, singly occupied and doubly occupied channels.

It is possible that for Leu10 and Trp11 the orientation of the carbonyl which binds most strongly to the ion could differ from the orientation of the NH, because to tilt the NH plane you have to change ϕ and ψ, and we also calculated that ω must change by about 5°. In fact, ω fluctuates spontaneously by about 8°, so the peptide plane is not a rigid unit. Also, what's measured as a function of ion position is an average of a function, and that's not necessarily equal to the function of the average. It is possible that in the doubly occupied channel, the two same sites are not occupied, but rather one site might be shifted slightly. If this is the case, then it introduces a certain amount of uncertainty, and the data have to be interpreted carefully. I would like to ask Bonnie Wallace whether she sees any deviations of carbonyls in her crystal structures due to the presence of ions.

Wallace: In our CsCl crystal structure, the carbonyls closest to the ions deviate from the helix axis by up to $\sim 30°$, but have comparable B factors to the other carbonyls.

Arseniev: I would like to mention our work on ion interactions with gramicidin. In organic solvents such as methanol or ethanol there is a mixture of four different gramicidin species. All of these are double helices—parallel and antiparallel, right-handed and left-handed—and all have 5.6 residues per turn and the maximum number of hydrogen bonds (28) allowable for gramicidin dimers (Bystrov & Arseniev 1988). Saturated concentrations of Li^+ and Na^+ salts induce complete unfolding of the double helices. K^+, Tl^+ and Cs^+ salts shift the equilibrium from the double helices with 5.6 residues per turn to the right-handed antiparallel double helix with 7.2 residues per turn (Arseniev et al 1985). In the case of micelles, Li^+ and, to a lesser degree, Na^+ ions provide some flexibility of the single-stranded structure of the channel, whereas Tl^+ ions lead to more rigid structure (Arseniev et al 1986). It turns out that degree of the flexibility of the double- and single-stranded helices is determined by the size of the monovalent cations. ^1H NMR spectroscopy doesn't provide unambiguous data on the position of monovalent cations in their complexes with gramicidin. However, the position of the divalent cation Mn^{2+} in the single-stranded structure of gramicidin in micelles has been determined (Bystrov et al 1990). The chemical shift data on the binding of monovalent and divalent cations are puzzling, so we used the relaxation data instead. We showed that Mn^{2+} ions block the gramicidin channel in lipid bilayers, and then we measured distances between the Mn^{2+}-binding site and gramicidin protons in SDS micelles. The Mn^{2+} was found to be near the carbonyl groups of D-Leu10,12 and 14, oriented into bulk solvent. The distance between the nearest oxygen atom of the D-Leu12 carbonyl group and Mn^{2+} is 6.4 ± 2.1 Å. Thus, Mn^{2+} ions preserve their first hydration shell upon binding to the mouths of the gramicidin channel.

Cross: I would also like to mention some of our results on the ω torsion angle. We have looked at the sodium double occupancy-induced chemical shifts of the Leu10-Trp11 peptide plane. We looked at how much the peptide plane needs to be rotated about the $C\alpha$-$C\alpha$ axis in order to achieve these changes in chemical shift (Fig. 1). This is about a 10° change for the ^{13}C carbonyl chemical shift. These data are extrapolated from Smith et al (1990). The change in ^{15}N chemical shift suggests an 8° change in orientation, whereas the change in N-H dipoles suggests a 2° change in peptide plane orientation. This could be accommodated for by a change in the ω torsion angle, although this doesn't address the problem of what has happened to the ^{15}N chemical shift, which is on the same side of ω as the N-H dipolar interaction. I believe that this plane is relatively fixed in geometry, and that electrostatic effects affect both the ^{13}C and ^{15}N chemical shift tensors.

Sansom: It doesn't hold to assume that the peptide is rigid. The S.D. on ω in atomic resolution X-ray structures that have been refined without any constraints is about 10°, which fits with Benoît Roux's results.

Cross: Our time-averaged structure shows a number of different values for ω torsion ranging from 165° to −173°. I agree that plus or minus 10° is fine. What I'm questioning is whether or not there is a change in ω upon cation binding.

Sansom: In the crystal structures there are variations in ω depending on the environment of the particular peptide, and this is averaged over an entire set of environments. The environmental factors changing ω values are probably somewhat softer than the effect of ion binding. In the crystal structures, you are talking about putting something with a relatively low charge density next to a peptide bond and you observe deviations of the order of 10°. You're putting something (i.e. an ion) that's creating a stronger electrostatic field than that experienced by peptide bonds in a crystal and you're saying that this doesn't change the ω value.

Cross: But let me ask, is a rotation about ϕ and ψ associated with a softer potential than a rotation about ω? I would say the answer is yes. We're not seeing a change in ϕ and ψ upon cation binding, so I don't understand why you are suggesting that there is a change in ω.

Jakobsson: The issue is that the force acting to create the ω distortion is high because there is a large fractional difference in the proximity of the sodium to the

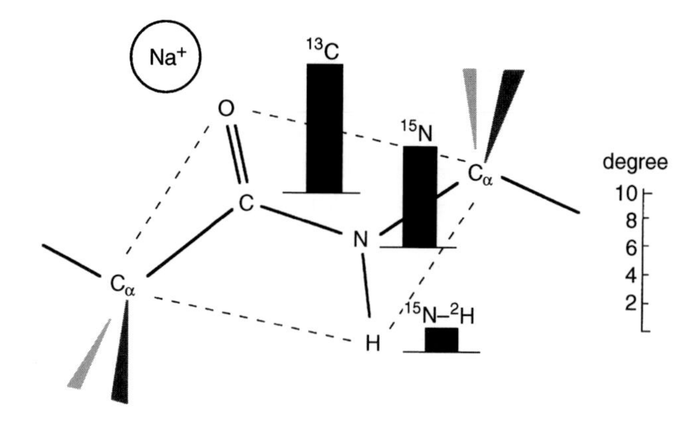

FIG. 1. Changes in NMR observables are interpreted as a possible change in orientation about the Cα-Cα axis due to Na^+ binding in the gramicidin channel. The ^{13}C chemical shift data are from Smith et al (1990). See also Tian et al (1996).

carbonyl, as opposed to the NH, so ϕ and ψ appear to be floppier, but they just don't have as much force acting on them.

Cross: But there would be just as much force on ψ as there would be on ω.

Roux: NMR does not measure ϕ and ψ directly.

Cross: No, but we do not see a significant change in the orientation of N-H on one side of ω.

Roux: Despite the fact that molecular dynamics is not perfect, we see a 10° NH deviation, and a deviation of 15° for ϕ and ψ. Therefore, these deviations are not linearly related and you can't add them together.

Hinton: Is the binding site really at Leu10? I recollect that Urry et al (1982) showed that it was between residues 11 and 13, so where is the binding site?

Cross: The three carbonyls affected by the presence of sodium, potassium or lithium are the Leu10 carbonyl attached to Trp11, the Leu12 carbonyl and the Leu14 carbonyl. We believe that there's a delocalized binding site, so there isn't a single, fixed site.

Hinton: I agree. We did some NOESY11 experiments in which we looked at the cation-inhibited exchange of protons between backbone amide protons and water protons in the channels, and it looks like binding occurs in that region.

Cross: You brought up a good point, which is that Dan Urry has done a lot of work on the changes in isotropic chemical shift in the presence of cations. We have tried to interpret that data and we are having difficulty sorting it out because much of the early data were interpreted in terms of a left-handed helix and not a right-handed helix.

Jordan: In a doubly occupied channel do you observe a delocalization at both sites? Because if you do, this suggests that the water chain is like a transducer, transmitting information between the well-separated sites.

Cross: No. What I'm saying is that our time-averaged experiments suggest that the cations interact with these three sites. I'm not saying anything about how that occurs at both ends. We see one signal for the Trp11 nitrogen and we're looking at both ends.

Jordan: So you can't compare what's happening on the right- and left-hand side of the molecule?

Cross: No. We're not looking at them individually.

Separovic: I would like to make a comment. We observe a change of −10 ppm for the Leu10 carbonyl and −6 ppm for both Leu12 and Leu14. We didn't expect to see any changes in the chemical shift anisotropies (CSAs) of the tryptophan carbonyls, but we measured them and saw no changes.

Bechinger: How does the intramolecular hydrogen bonding between the amides and the carboxylic residues one helical turn up affect the flexibility of the ϕ and ψ angles of the peptide bonds of Leu10, Leu 12 and Leu14?

Cross: We find that the dynamics of these nitrogen sites throughout the polypeptide backbone are substantial. They are greater than what is observed for an α-helix, for instance. We estimate that the peptide plane fluctuates by $\pm 15°$ or so, and that changes in ϕ and ψ account for those changes. However, we are saying that in the presence of cations the time-averaged structure does not appear to change.

Eisenberg: I would like to mention that we have direct experimental evidence of the ensemble that Benoît Roux has proposed, although our evidence is based on a different channel, the calcium release channel of cardiac muscle, and it may be an isolated phenomenon. We have looked at the effects of lithium, sodium, potassium, rubidium and caesium, and mixtures thereof, from 20 mM to 2 M salt. We found, to our amazement, that the same fixed charge for the channel fits all of the data. One extra constant per ion, namely a diffusion coefficient, and one charge are sufficient to fit all of the data. Therefore, this is an extreme example of an ensemble in which only a simple property of that ensemble, i.e. the charge of the protein, is involved.

Smart: Is it possible to resolve the difference between the Cross and Roux views of the ion-binding position by performing a back calculation of what NMR observables could be expected to be measured based on the molecular dynamics trajectories Roux has performed?

Roux: It is possible to address the difficulties in interpreting the solid-state NMR in terms of structure with molecular dynamics simulations. Although the simulations are not perfect, they provide a detailed view of the magnitude of the fluctuations and local average structure. Let us consider the notion that the sodium ion binds over a range of sites along the channel axis, and analyse the consequences on the solid-state NMR observables. For example, it is possible that when the Na^+ ion is at one position, it deflects a carbonyl oxygen more than when it is at another position. What is observed is the weighted average with the ion at all positions in the channel mouth. If it binds over a region of 2 Å wide and deflects a carbonyl significantly only if it is bound within 0.25 Å of that carbonyl, then you have to divide the deflection by eight to get the apparent deflection that corresponds to the average that would be observed by solid-state NMR (the fraction of length that causes deflection, i.e. 0.25 Å divided by 2.0 Å). Furthermore, when the concentration is increased so that there are doubly occupied channels, the interpretation is based upon the assumption that both ions contribute to structural distortions that are similar to those of the singly occupied channel (in other words, double occupancy is just an amplification of the same channel distortions). However, it is possible that the binding site position is shifted in the doubly occupied channel and that the high concentration measurements do not correspond to the maximum channel distortion. These factors should be kept in mind when interpreting the solid-state NMR in terms of a structure. My feeling is that it is difficult to make a clear statement about carbonyl deflections in terms of degrees upon ion binding at this point.

References

Arseniev AS, Bystrov VF, Barsukov IL 1985 NMR solution structure of gramicidin A complex with caesium cations. FEBS Lett 180:33–39

Arseniev AS, Barsukov IL, Bystrov VF, Ovchinnikov YA 1986 Spatial structure of gramicidin A transmembrane ion channel — NMR analysis in micelles. Biol Membr (USSR) 3:437–462

Bystrov VF, Arseniev AS 1988 Diversity of the gramicidin A spatial structure: two-dimensional ^{1}H NMR study in solution. Tetrahedron 44:925–940

Bystrov VF, Arseniev AS, Barsukov IL, Golovanov AP, Maslennikov IV 1990 The structure of the transmembrane channel of gramicidin A: NMR study of its conformational stability and interaction with divalent cations. Gaz Chim Ital 120:485–491

Smith R, Thomas DE, Atkins AR, Separovic F, Cornell BA 1990 Solid-state ^{13}C-NMR studies of the effects of sodium ions on the gramicidin A ion channel. Biochim Biophys Acta 1026:161–166

Tian F, Lee K-C, Hu W, Cross TA 1996 Monovalent cation transport: lack of structural deformation upon cation binding. Biochemistry 35:11959–11966

Urry DW, Prasad KU, Trapane TL 1982 Location of monovalent cation binding sites in the gramicidin channel. Proc Natl Acad Sci USA 79:390–394

Design and characterization of gramicidin channels with side chain or backbone mutations

Roger E. Koeppe II, Denise V. Greathouse, Lyndon L. Providence*, S. Shobana* and Olaf S. Andersen*

*Department of Chemistry and Biochemistry, University of Arkansas, Fayetteville, AR 72701, and *Department of Physiology and Biophysics, Cornell University Medical College, New York, NY 10021, USA*

Abstract. Mutations and chemical substitutions of amino acid side chains and backbone atoms have proved vital for understanding the folding, structure and function of gramicidin channels in phospholipid membranes. The channel's pore is lined by peptide backbone groups; their importance for channel structure and function is shown by a single amide-to-ester replacement within the backbone, which greatly reduces the resulting channel conductance and lifetime. The four tryptophans and the intervening leucines together govern the formation and dissociation of conducting channels from single-stranded subunits. Conducting double-stranded gramicidin conformations (channels) occur rarely in membranes — except when the sequence has been altered to permit special arrangements of tryptophans or (infrequently) in unusually thick membranes. The tryptophans anchor the single-stranded channels to the membrane/solution interface, and the indole dipoles promote cation transport through the channels. Removal of any indole dipole reduces ion conductance; whereas 5-fluorination of an indole, which increases its dipole moment, enhances ion conductance. Some sequence changes at the formyl-NH-terminus (in the membrane interior, away from the tryptophans), including fluorination of the formyl-NH-terminal valine, introduce voltage-dependent channel gating. Gramicidin channels are not just static conductors, but also dynamic entities whose structure and function can be manipulated by backbone and side chain modifications.

1999 Gramicidin and related ion channel-forming peptides. Wiley, Chichester (Novartis Foundation Symposium 225) p 44–61

Gramicidin channels provide a remarkable combination of properties — tuneable chemistry, known structure and exquisitely well-characterized function — crucial for detailed understanding and manipulations that alter the characteristics of the transmembrane channels. The recent development of chemical sensors based on the gramicidins (Cornell et al 1997) represents a culmination of diverse research from many perspectives, as well as a true opening of new horizons in molecular design.

Most of the salient information about gramicidin channels has come from: (i) electrophysiological single-channel analysis, which provides information about both function (Hladky & Haydon 1984) and structure (Durkin et al 1990); (ii) spectroscopic methods, such as NMR, infra-red, circular dichroism and fluorescence spectroscopy (reviewed in Busath 1993 and Cross 1994); (iii) other physical methods such as size-exclusion chromatography (Salom et al 1998); and (iv) chemical modifications of the amino acid sequence and end groups (Koeppe & Andersen 1996, Greathouse et al 1999). Ironically, X-ray crystallography of linear gramicidins — despite its long history dating back to 1948, with pioneering structures in 1988 (Wallace & Ravikumar 1988, Langs 1988) and additional structures continuing to emerge — has contributed little to the understanding of the functional transmembrane channels. This situation has arisen because all of the crystal structures to date derive from lipid-free solvent environments that fail to mimic the phospholipid membrane. Given this state of affairs, it is fortunate that the channel structure could be determined by multi-dimensional NMR (Arseniev et al 1986) and solid-state NMR (Ketchem et al 1993, 1997).

The sequence of the parent molecule, gramicidin A (gA) from *Bacillus brevis*, is formyl-$\text{VGA}\underline{\text{L}}\text{A}^{5}\underline{\text{V}}\text{V}\underline{\text{V}}\text{W}\underline{\text{L}}^{10}\text{W}\underline{\text{L}}\text{W}\underline{\text{L}}\text{W}^{15}$-ethanolamine, in which the underlined amino acids are of D-chirality (Sarges & Witkop 1965). Several hundred modifications of this sequence have been reported and have contributed to the understanding of the structure and function of the transmembrane channel. Recently, the first backbone modification was reported (Jude et al 1998). This chapter will not deal with solvent structures, but will limit itself to a discussion of selected subsets of mutations that address particular issues concerning (conducting) gramicidins within membranes.

Experimental procedures

A variety of methods for the solid-phase synthesis of side chain- and end-modified gramicidins were reviewed recently (Greathouse et al 1999). The electrophysiological methods have been reported by Andersen (1983). The methods for the ester modification in the backbone will be described (Jude et al 1998).

Gramicidin channel structure, helix sense and end modifications

At equilibrium, gramicidin A adopts a single, unique folded conformation in bilayers whose hydrophobic thickness matches fairly closely the gramicidin A dimer length. The same structure is observed when gramicidin A is incorporated into selected detergents such as SDS. Other molecular conformations may have preceded this unique (single-stranded) ‘channel’ conformation in time (e.g. prior to membrane insertion or peptide refolding), but these conformations do not contribute to the NMR spectra or to the characteristic circular dichroism spectrum of the gramicidin A channel, which shows positive ellipticity maxima at 218 nm and 230 nm (Wallace et al 1981; Fig. 1A). This single-stranded channel conformation has been determined at atomic resolution by multi-dimensional and by solid-state NMR (Arseniev et al 1986, Ketchem et al 1993); these studies confirmed the essential features predicted by Urry (1971). End views of the structure are shown in Fig. 2, which depicts a channel with and without side chain modifications at positions 10, 12 and 14 (see below).

The gramicidin A sequence consists of alternating L and D amino acids, but left-handed gramicidin A channels are not observed. The helix sense of the gramicidin A channel is right-handed, as determined by both 2D and solid-state NMR, as well as by a combination of circular dichroism spectroscopy and single-channel analysis. The enantiomer, gramicidin A^-, however, forms left-handed channels. Heterodimer formation (hybrid channel) experiments, which are sensitive to the presence of minor conformations, show that neither left-handed gramicidin A channels nor left-handed gramicidin A monomeric subunits form to a measurable extent (Koeppe et al 1992). Even in lipids of different chirality, gramicidins retain their unique helix sense, which must be determined by the amino acid sequence (Providence et al 1995). The tryptophans largely determine the channel's preference to be single-stranded (Durkin et al 1992, Salom et al 1998), as well as channel's right-handed helix sense; but when all Trps are replaced by Phe (see below), the single-stranded channels that form are nevertheless right-handed (Providence et al 1995).

Within the context of the increasing numbers of double-stranded gramicidin crystal structures that are appearing, it is worthwhile to review again the chemical evidence that gramicidin A channels form by the pairwise association of single-stranded subunits (see Andersen & Koeppe 1992 for a more comprehensive review). Early convincing evidence for single-stranded gramicidin A channels came from the findings that non-perturbing ^{13}C and ^{19}F labels at the channel's N-terminal are shielded from paramagnetic ions in the aqueous solution, whereas labels at the C-terminal are accessible (Weinstein et al 1979). Removal or modification of the formyl-NH end groups abolishes, or greatly diminishes, the analogue's channel-forming ability (Bamberg et al 1977), whereas cross-linking of the formyl groups is permitted and increases the channel lifetime (Stankovic et al 1989), which indicates that the opposing formyl groups are near each other in the channel structure. By contrast, the C-terminal ethanolamine can be modified in a variety of chemical reactions with retention of channel activity. Hybrid channel experiments using many varieties of mutants (e.g. Durkin et al 1990, Cifu et al 1992) also support fully the channel structure that has been seen by NMR. Notably, if only one single-stranded monomer misses a residue (or has an extra

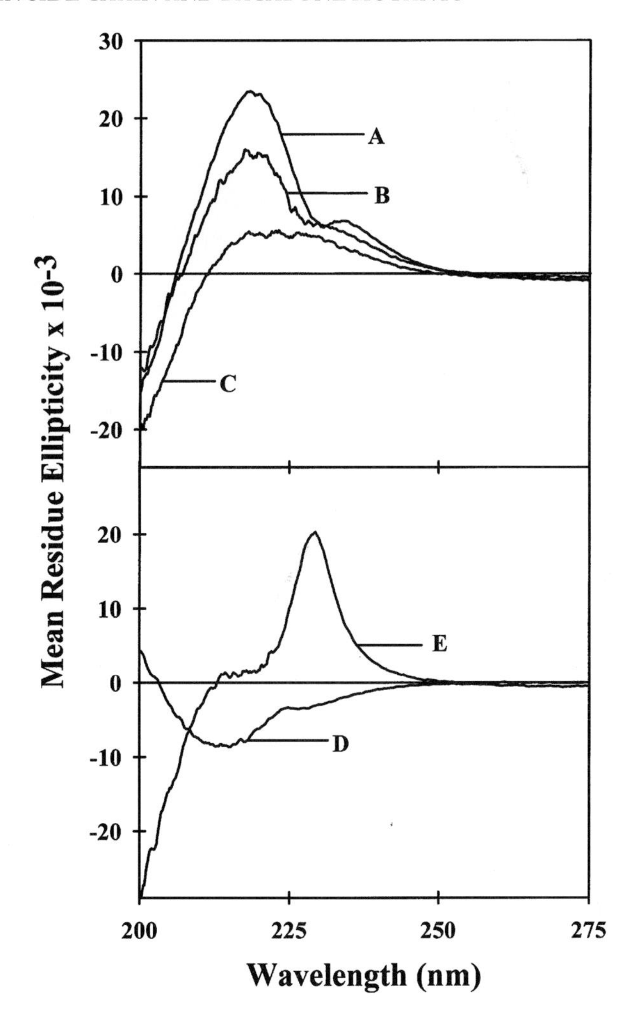

FIG. 1. Variations in the mixture of gramicidin conformations in dimyristoyl phosphatidylcholine (DMPC) membranes when the C-terminal $(\text{Trp-D-Leu})_n$ sequence is changed—as shown by circular dichroism spectroscopy. (A) The spectrum of native gramicidin A (gA) provides a fingerprint for the right-handed single-stranded $\beta^{6.3}$ channel conformation (Wallace et al 1981). (B) The spectrum of [D-Ala10]gA (see model in Fig. 2) shows little change from that of gramicidin A. (C) The spectrum for the triply substituted [D-Ala10,12,14]gA shows extensive loss of the single-stranded channel conformation. (D) The spectrum of the 'interchange' molecule [L-Leu9,11,13,15, D-Trp10,12,14]gA (gLW) shows that the predominant conformation is left handed (and independently is known to be single-stranded; see Greathouse et al 1999). (E) [Val5]gLW adopts primarily an inert right-handed double-stranded conformation. These conclusions have been confirmed by size-exclusion chromatography (Greathouse et al 1999, Jude et al 1999).

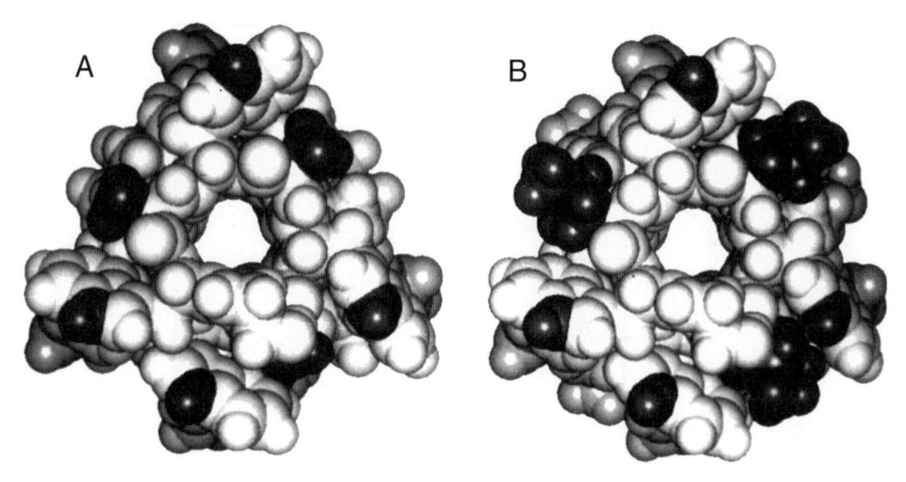

FIG. 2. End view of space-filling single-stranded molecular models for (A) [D-Ala10,12,14]gA and (B) native gramicidin A which has D-Leu at positions 10, 12 and 14. The D-Ala10,12,14 and D-Leu10,12,14 side chains are dark, as are the indole NH groups of the neighbouring tryptophans. The less bulky alanines alter the packing around the Trp residues, which causes a reduction in the gramicidin single-channel conductance and lifetime, and promotes the formation of non-conducting double-stranded structures in phospholipid membranes (see text and Jude et al 1999).

residue) at the formyl-NH-terminal, then the subunit junction is perturbed and hybrid channels are destabilized by ~10 kJ/mol (Durkin et al 1993); such a modification should not affect double-stranded dimers.

Insights into the structural basis for the single-stranded preference in lipid bilayers can be gleaned from Fig. 3, which shows the different distribution patterns of the four Trp residues in single-stranded channels and double-stranded dimers. In the double-stranded structure, the Trp residues are distributed all along the exterior molecular surface—in stark contrast to their organization in single-stranded channels. The energetic cost of burying the Trp residues in the non-polar bilayer core is likely to account for the single-stranded/double-stranded preference (Durkin et al 1992).

Modification of the backbone: implications for transport

The topology of the gramicidin A channel is such that the pore through which the ions pass is lined exclusively by peptide backbone groups; the side chains interact with membrane lipids and the membrane/water interface. To assess the importance of the peptide backbone for channel properties, we synthesized a modified gramicidin A with a single backbone amide replaced by an ester bond. The ester analogue [f-Val-O-Gly]gA has a single NH→O substitution between Val1 and

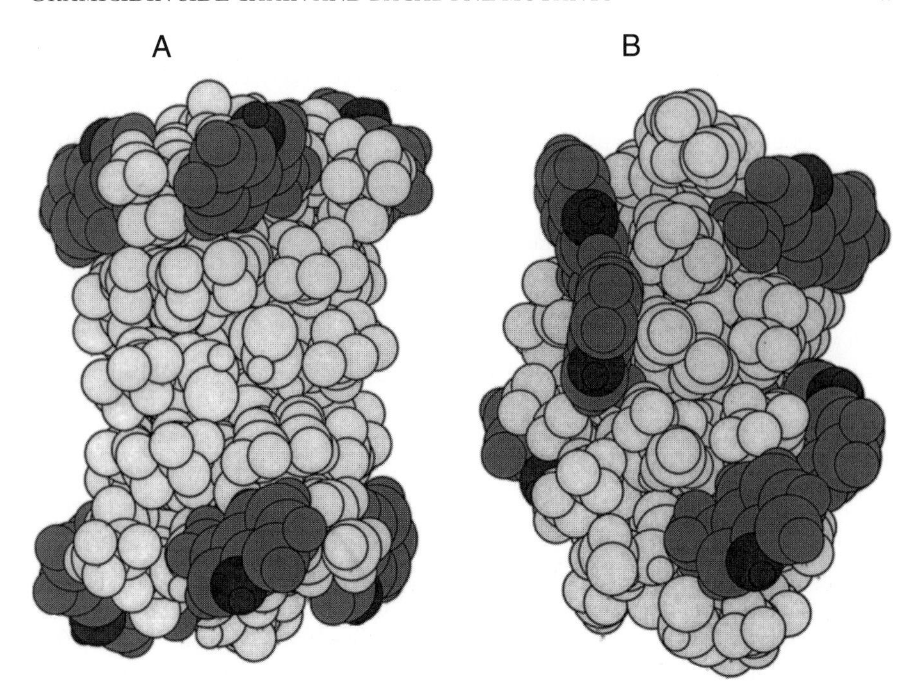

FIG. 3. Side view of space-filling models of a single-stranded right-handed $\beta^{6.3}$-helical gramicidin A dimer (A) and a double-stranded left-handed $\pi\pi^{6.4}$-helical gramicidin A dimer (B). The single-stranded $\beta^{6.3}$-helical dimer was drawn using coordinates that were refined from Arseniev et al (1986); the double-stranded $\pi\pi^{6.4}$-helical dimer was drawn using the coordinates of structure 1AV2, deposited by Burkhart et al (1998) in the Protein Data Bank (http://www.pdb.bnl.gov). The side chains of Trp 9, 11, 13 and 15 are shown in grey, and the indole NH groups are dark grey. Note the different distributions of the Trp residues along the molecular axis in the two structures.

Gly2 in the backbone. [f-Val-O-Gly]gA forms both homodimeric channels and hybrid channels with gramicidin A in planar bilayers. The homodimeric channels have greatly reduced conductance and lifetime, which demonstrates the importance of the backbone amide carbonyl groups for channel folding and function (Jude et al 1998).

Modifications in the Trp-Leu region: implications for folding and transport

Tryptophans

Gramicidin A has L-Trp at positions 9, 11, 13 and 15 in the sequence. These tryptophans cluster at the membrane/water interface (Fig. 2 and Fig. 3A) and drive the formation of the single-stranded right-handed channels (O'Connell et al 1990). As individual Trps in the sequence are successively changed to Phe, the propensity toward single-stranded channel formation is diminished incrementally, to the point where [Phe9,11,13,15]gA forms >90% double-stranded structures and correspondingly few single-stranded channels in dimyristoyl phosphatidylcholine (DMPC) and other membranes (Salom et al 1998). Notably, these double-stranded conformations do not conduct ions; the conducting, homodimeric channels formed by [Phe9,11,13,15]gA channels are 'short'-lived single-stranded channels (Durkin et al 1992, Girshman et al 1997).

In a special case, when the usual hybrid channels cannot form between a chain-shortened des-Val1 gramicidin and enantiomeric [D-Phe9,11,13,15]gA$^-$, ion-conducting double-stranded channels are observed (Durkin et al 1992). These unusual double-stranded channels are characterized by their exceptionally long lifetimes (order of minutes).

Successive Trp→Phe substitutions diminish the single-channel conductance of the single-stranded gramicidin channels that do form (Becker et al 1991). The effect is due to sequential loss of the indole ring dipoles, each of which promotes both cation entry and cation permeation due to electrostatic interactions between the permeating ion and the side chain dipole. This dipolar mechanism was verified by 5-fluorination of the indole ring, which enhances the dipole moment of the ring and thereby increases the single-channel conductance (Fig. 4; Andersen et al 1998).

Residues adjacent to Trp

The 'spacer' leucines between the Trps in the gramicidin sequence modulate the folding and function of gramicidin A channels—possibly through alterations in the average orientation and dynamics of the Trp residues. The side chains of Trp9 and Leu10, for example, interact closely because a Leu10→Ala10 replacement, which does not alter the basic channel fold (Fig. 1B), can be detected through a change in the solid-state NMR spectrum of deuterated Trp9 in oriented channels (Fig. 5).

More extensive modifications, at positions 12 and 14 in addition to position 10, decrease the proportion of single-stranded channels with a corresponding increase in the proportion of double-stranded conformers. Whereas [D-Ala10]gA largely retains the single-stranded channel conformation (Fig. 1B), the triply substituted [D-Ala10,12,14]gA exists as a mixture of single-stranded and double-stranded conformations in DMPC (Fig. 1C); [D-Val10,12,14]gA and [D-Ile10,12,14]gA behave similarly (Jude et al 1999). These conformational mixtures are observed even though all of the Trps in the sequence remain intact. Thus, although the Trps provide a major driving force for the formation of single-stranded channels,

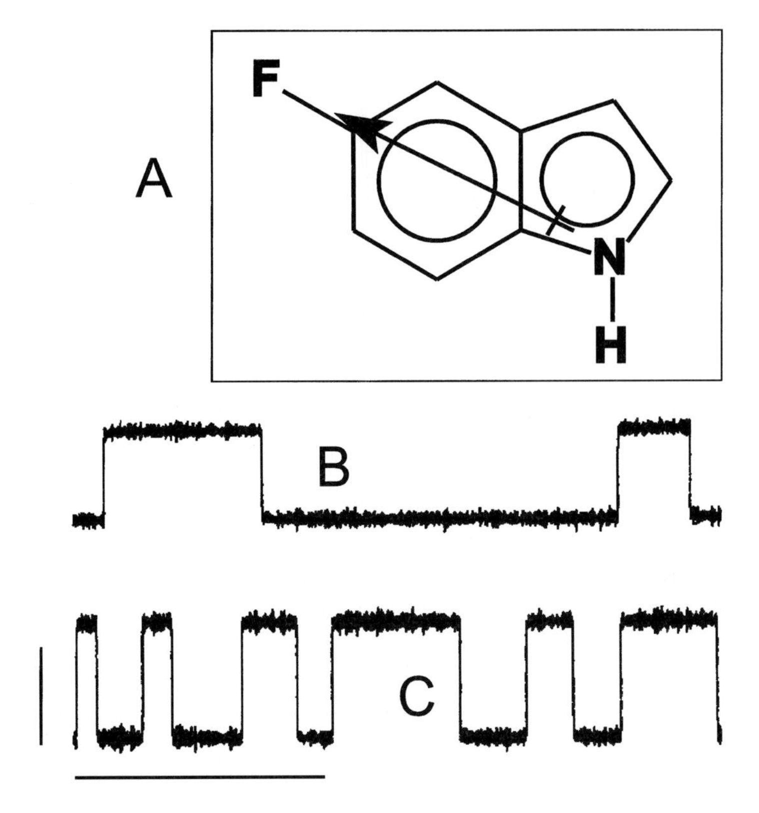

FIG. 4. Illustration of the 5-fluoro-indole ring and of enhanced single-channel conductance when Trp9 of gramicidin A (gA) is 5-fluorinated. (A) 5-F-indole, which has a dipole moment of 3.6 D, compared to 2.1 D for indole. (B) Single-channel current trace for [Phe11]gA. (C) Single-channel current trace for [5-F-Trp9, Phe11]gA. The calibration bars denote 2 pA (vertically), and 5 s (horizontally). For details, see Andersen et al (1998).

the Trps alone are not sufficient; the neighbouring 'spacer' residues are also important, with leucines being the most favourable for channel formation (Jude et al 1999).

Sequence modifications that confer voltage dependence

Gramicidin heterodimers usually have properties intermediate between those of the two symmetrical channels (Durkin et al 1990). Selected amino acid substitutions at position one, however, impart qualitatively new functions to hybrid gramicidin channels. When hexafluorovaline is substituted at position 1 on only one subunit, the hybrid channel conductance and average duration are

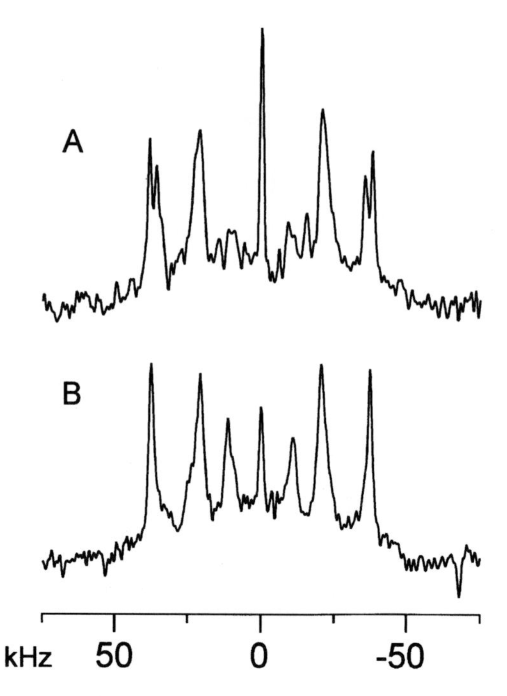

FIG. 5. ^{2}H NMR spectra at 46 MHz of oriented samples of (A) [D-Ala10, d_5-Trp9]gA and (B) [d_5-Trp9]gA, in hydrated dimyristoyl phosphatidylcholine (DMPC) with the membrane normal oriented at β=90° with respect to the magnetic field H_o, 15:1 lipid:peptide and a temperature of 47°C. The twinning of the outer resonances into two peaks in (A) indicates that the outer resonances should be assigned to two geometrically similar deuterons, i.e. those at positions C4 and C7 (Cδ2 and Cε3) of the Trp9 indole ring (for details see Koeppe et al 1994, Hu et al 1995). In the folded gramicidin A channel, Trp9 is situated close to the side chain of residue 10 (see Fig. 2; Arseniev et al 1986, Koeppe et al 1996, Jude et al 1999).

lower than those of either corresponding symmetric channel, and the hybrid channels exhibit voltage-dependent gating (Oiki et al 1994, 1995). In combination with either [Val1]gA, [Ala1]gA or [Gly1]gA, [F_6Val1]gA forms hybrid channels that show voltage-dependent transitions between at least two stable conductance states. The 'gating' behaviour depends on the presence of the F_6Val1 residue in only one monomer. Changes at position one (Val, Ala or Gly) in the *trans* monomer influence the quantitative description of the voltage dependence; therefore, both subunits participate in a voltage-dependent conformational change from a low conductance (closed) to a high conductance (open) state.

When changing the permeant ion from Cs^+ to H^+, the voltage-dependent gating is not affected (Oiki et al 1995). The gating is therefore a property of the channel itself (independent of the passing ions), and these voltage-dependent gramicidins offer an attractive model system for a deeper understanding of the molecular workings of a (voltage-dependent) molecular switch.

Summary and conclusions

Chemical changes in the amino acid sequence, side chains and backbone of gramicidin A have been used to help understand the folding and function of membrane-spanning gramicidin channels, and to introduce new structural and functional properties. A single ester bond in the polyamide backbone dramatically reduces channel conductance and lifetime. The helix sense and backbone interactions (subunit intertwining) are sensitive to changes in not only the tryptophans but also the leucines of the (Trp-Leu)-rich C-terminal half of the gramicidin molecule. Selected single amino acid substitutions, which break the channel symmetry and introduce a side chain dipole, confer voltage-dependent channel gating. These findings provide general insights into the folding of membrane proteins, the critical role of the peptide backbone chemical functional groups, the importance of interfacial tryptophan indole rings and the mechanism of channel gating.

Acknowledgement

The authors are grateful for financial support for research in our laboratories from the National Institutes of Health (grants GM34968 and GM21342).

References

Andersen OS 1983 Ion movement through gramicidin A channels. Single-channel measurements at very high potentials. Biophys J 41:119–133

Andersen OS, Koeppe RE II 1992 Molecular determinants of channel function. Physiol Rev 72:S89–S158

Andersen OS, Greathouse DV, Providence LL, Becker MD, Koeppe RE II 1998 Importance of tryptophan dipoles for protein function: 5-fluorination of tryptophans in gramicidin A channels. J Am Chem Soc 120:5142–5146

Arseniev AS, Lomize AL, Barsukov IL, Bystrov VF 1986 Gramicidin A transmembrane ion-channel. Three-dimensional structure reconstruction based on NMR spectroscopy and energy refinement. Biol Membr (USSR) 3:1077–1104

Bamberg E, Apell HJ, Alpes HJ 1977 Structure of the gramicidin A channel: discrimination between the π-L,D and the beta-helix by electrical measurements with lipid bilayer membranes. Proc Natl Acad Sci USA 74:2402–2406

Becker MD, Greathouse DV, Koeppe RE II, Andersen OS 1991 Amino acid sequence modulation of gramicidin channel function: effects of tryptophan-to-phenylalanine substitutions on the single-channel conductance and duration. Biochemistry 30:8830–8839

Burkhart BM, Li N, Langs DA, Pangborn WA, Duax WL 1998 The conducting form of gramicidin A is a right-handed double-stranded double helix. Proc Natl Acad Sci USA 95:12950–12955

Busath DD 1993 The use of physical methods in determining gramicidin channel structure and function. Annu Rev Physiol 55:473–501
Cifu A, Koeppe RE II, Andersen OS 1992 On the supramolecular structure of gramicidin channels. The elementary conducting unit is a dimer. Biophys J 61:189–203
Cornell BA, Braach-Maksvytis V, King L et al 1997 A biosensor that uses ion-channel switches. Nature 387:580–583
Cross TA 1994 Structural biology of peptides and proteins in synthetic membrane environments by solid-state NMR spectroscopy. Annu Rep NMR Spectrosc 29:124–158
Durkin JT, Koeppe RE II, Andersen OS 1990 Energetics of gramicidin hybrid channel formation as a test for structural equivalence. Side-chain substitutions in the native sequence. J Mol Biol 211:221–234
Durkin JT, Providence LL, Koeppe RE II, Andersen OS 1992 Formation of non-$\beta^{6.3}$-helical gramicidin channels between sequence-substituted gramicidin analogues. Biophys J 62:145–159
Durkin JT, Providence LL, Koeppe RE II, Andersen OS 1993 Energetics of heterodimer formation among gramicidin analogues with an NH_2-terminal addition or deletion: consequences of missing a residue at the join in the channel. J Mol Biol 231:1102–1121
Girshman J, Greathouse DV, Koeppe RE II, Andersen OS 1997 Gramicidin channels in phospholipid bilayers with unsaturated acyl chains. Biophys J 73:1310–1319
Greathouse DV, Koeppe RE II, Providence LL, Shobana S, Andersen OS 1999 Design and characterization of gramicidin channels. Methods Enzymol 294:525–550
Hladky SB, Haydon DA 1984 Ion movements in gramicidin channels. Curr Top Membr Transp 21:327–372
Hu W, Lazo ND, Cross TA 1995 Tryptophan dynamics and structural refinement in a lipid bilayer environment: solid state NMR of the gramicidin channel. Biochemistry 34:14138–14146
Jude AR, Providence L, Andersen O, Greathouse D, Koeppe RE II 1998 Influence of an amide-to-ester replacement in the gramicidin channel backbone on ion permeability and channel stability. Biophys J 74:A387
Jude AR, Greathouse DV, Koeppe RE II, Providence LL, Andersen OS 1999 Modulation of gramicidin channel structure and function by the aliphatic 'spacer' residues 10, 12 and 14 between the tryptophans. Biochemistry 38:1030–1039
Ketchem RR, Hu W, Cross TA 1993 High-resolution conformation of gramicidin A in a lipid bilayer by solid-state NMR. Science 261:1457–1460
Ketchem RR, Roux B, Cross T 1997 High resolution polypeptide structure in a lamellar phase lipid environment from solid state NMR-derived orientational constraints. Structure 5:1655–1669
Koeppe RE II, Andersen OS 1996 Engineering the gramicidin channel. Annu Rev Biophys Biomol Struct 25:231–258
Koeppe RE II, Providence LL, Greathouse DV et al 1992 On the helix sense of gramicidin A single channels. Proteins 12:49–62
Koeppe RE II, Killian JA, Greathouse DV 1994 Orientations of the tryptophan 9 and 11 side chains of the gramicidin channel based on deuterium NMR spectroscopy. Biophys J 66:14–24
Koeppe RE II, Vogt TCB, Greathouse DV, Killian JA, de Kruijff B 1996 Conformation of the acylation site of palmitoylgramicidin in lipid bilayers of dimyristoylphosphatidylcholine. Biochemistry 35:3641–3648
Langs DA 1988 Three-dimensional structure at 0.86 Å of the uncomplexed form of the transmembrane ion channel peptide gramicidin A. Science 241:188–191
O'Connell AM, Koeppe RE II, Andersen OS 1990 Kinetics of gramicidin channel formation in lipid bilayers: transmembrane monomer association. Science 250:1256–1259
Oiki S, Koeppe RE II, Andersen OS 1994 Asymmetric gramicidin channels: heterodimeric channels with a single F6-Val-1 residue. Biophys J 66:1823–1832

Oiki S, Koeppe RE II, Andersen OS 1995 Voltage-dependent gating of an asymmetric gramicidin channel. Proc Natl Acad Sci USA 92:2121–2125
Providence LL, Andersen OS, Greathouse DV, Koeppe RE II, Bittman R 1995 Gramicidin channel function does not depend on phospholipid chirality. Biochemistry 34:16404–16411
Salom D, Pérez-payá E, Pascal J, Abad C 1998 Environment- and sequence-dependent modulation of the double-stranded to single-stranded conformational transition of gramicidin A in membranes. Biochemistry 37:14279–14291
Sarges R, Witkop B 1965 Gramicidin A. V. The structure of valine- and isoleucine-gramicidin A. J Am Chem Soc 87:2011–2020
Stankovic CJ, Heinemann SH, Delfino JM, Sigworth FJ, Schreiber SL 1989 Transmembrane channels based on tartaric acid–gramicidin A hybrids. Science 244:813–817
Urry DW 1971 The gramicidin A transmembrane channel: a proposed π(L,D) helix. Proc Natl Acad Sci USA 68:672–676
Wallace BA, Ravikumar K 1988 The gramicidin pore: crystal structure of a cesium complex. Science 241:182–187
Wallace BA, Veatch WR, Blout ER 1981 Conformation of gramicidin A in phospholipid vesicles: circular dichroism studies of effects of ion binding, chemical modification, and lipid structure. Biochemistry 20:5754–5760
Weinstein S, Wallace BA, Blout ER, Morrow JS, Veatch WR 1979 Conformation of gramicidin A channel in phospholipid vesicles: a ^{13}C and ^{19}F nuclear magnetic resonance study. Proc Natl Acad Sci USA 76:4230–4234

DISCUSSION

Woolley: Did you calculate what the voltage dependence would be if the only effect of the voltage was to cause reorientation of the dipole of the fluorinated residue?

Koeppe: The rotation of the dipole itself accounts for only ~20% of the voltage dependence (Oiki et al 1995). We think that the side chain has to pull some of the rest of structure with it, and that this may alter the peptide bond orientation.

Eisenberg: I would like to point out that this calculation, although honoured by many years of history, has no fundamental basis. It is necessary to have an explicit model of gating before you can compute its energetics. This is relatively straightforward. If you consider a switch that turns a light on and off, you know absolutely nothing about the physics of the switch. It can operate via any number of physical principles. If the mechanism is dissipated, i.e. if heat is generated, then it's not enough just to have the energy in the initial and final state.

Sansom: If you know the difference in energy of a dipole between two orientations in an electrostatic field, and you also know that the difference in energy between those two orientations is too small to account for the voltage dependence of a process, surely you can conclude that the change in dipole orientation is not the switch?

Eisenberg: You can conclude that if it is an isolated system in which there are no other energy sources, and in this kind of system it is anything but isolated. There are all sorts of other energy sources.

Hladky: Roger Koeppe said that just rotating the dipole is not enough, and you're saying essentially the same thing. Roger is saying that something pulls the backbone around, and you're asking us to consider the whole system at once, but you are both concluding that the dipole alone isn't enough.

Eisenberg: What I'm saying is that we should talk about explicit models of gating, not semi-thermodynamic arguments, which are not useful. We should say how we think gating occurs and then try to test the model.

Roux: If the effect was caused by the dipole orientation, presumably the channel would still be intact and open. Do you believe that the barrier to permeation is suddenly increased, or do you believe that the channel is just not conducting because it is physically blocked?

Koeppe: It is possible to have physical closing by sterically blocking the channel or electrostatic closing by making the barrier so high that the ion cannot go through at a detectable rate. In the case of glycine^1/hexafluorovaline^1 hybrid channels, there are clear low conducting and high conducting states. However, the second set of data for valine^1/hexafluorovaline^1 hybrid channels suggests there is an open state and a low conducting state that is indistinguishable from closed. For me, closed and low conducting are equivalent. The dimer stays together, but it goes from high to low, or it may even be closed.

Roux: If you increased the voltage, would you be able to detect small conductance levels that may appear as closed channels?

Koeppe: In the glycine^1/hexafluorovaline^1 hybrid channel we always see a low conducting state. The low conducting state is not zero.

Roux: This tells us something about the structure. If ions can still go through but at a much lower rate, presumably the dimer structure is still intact and the pore is open.

Busath: I have a question along the same lines. I see your data on hydrogen switching in a new light after reading Benoît Roux's recent paper (Pomès & Roux 1998). Benoît reflects on the observation that the hydrogen conductance is rate limited by the water re-orientation. He says that the water re-orientation begins at one end of the channel most of the time because of the proximity of the end of the channel to the bulk water. The implication of his theory is that the water re-orientations near the centre of the channel (where the hexafluorovaline residue is located) become inconsequential because the first couple of water re-orientations represent the rate-limiting step. It seems to me that to interpret the hydrogen passage data, we have to think about how the hexafluorovaline dipole at the centre of the channel affects the water re-orientations at the centre of the channel, and whether they affect it more than the water re-orientations at the rate-limiting step.

Koeppe: I would say that those are two different issues. You are talking about the permeation mechanism for protons. I'm not talking about how the protons get through, rather I'm asking when the voltage-dependent channel is open to caesium, is it also open to protons?

Busath: Yes, that is true. I'm keen to extrapolate further on and addressing the implications of your results. For example, could we obtain some information about that dipole–dipole interaction from your results?

Koeppe: Not from this experiment. We would have to do more proton permeation experiments.

Eisenberg: Let's imagine that you could get a closed state with a measurable current, then it might be interesting to look at current–voltage (I–V) relationships in asymmetrical solutions, both of the open and closed states. With a physical model such as Poisson–Nernst–Planck you may then be able to determine the difference in the underlying charge (Eisenberg 1998a,b).

Koeppe: That would be nice, although these are technically difficult experiments.

Hladky: I have a point concerning the nomenclature. If we are going to refer to a lesser conducting state that is still conducting, to avoid confusion with the rest of the literature I suggest we call it a sub-state rather than a closed state.

Roux: Is it possible to do equilibrium measurements on these modified channels of the type that Jing et al (1995) did to determine whether some of the single ion and double ion association constants are modified by this?

Koeppe: It's unlikely. The problem with doing spectroscopy with hybrid channels is that it competes with the parent channels, which are also present. The voltage-dependent analogue that we want to study is not simple. We have a new homodimer that is voltage dependent. It has 14 residues, beginning with L-alanine 2 then L-alanine 3, but it has a chirality problem. By itself it is voltage dependent as a homodimer. We hope to do some deuterium labelling with this voltage-dependent homodimer, because up until now the problem has been how to study the hybrid channel by itself, especially when it has a low lifetime.

Hinton: Do you have any idea about the length of the double helix in the Duax crystal structure (Burkhart et al 1998)? If you start with a double helix, then you have to be able to pull it apart to some extent to insert it into the membrane to form a transporting channel. Do you have any idea what the stagger is, and how it pulls apart? You can pull it so far, but then you would have a dimer. If you shorten it, however, you will have problems with the lipid itself.

Wallace: Duax's caesium structure and our caesium structure differ in their stagger — ours has a stagger of three residues, and theirs has a stagger of one residue. Consequently, the two structures could be considered to be two different steps in the process of pulling the double helix apart.

Koeppe: None of us know for sure what the structure of the conducting units that present a double-stranded channel signature might be. You're suggesting that it might be a double helix with quite a severe stagger between the two strands, which is a good idea. A hybrid structure like that is proposed by Fredrick Heitz for a different analogue that is partly double helical and partly single stranded (Lelievre et al 1989).

Hinton: If it pulls apart and then somehow fits into the membrane, each of the two molecules has to be extended, and that decreases the diameter of the channel. In a transport configuration (i.e. decreased diameter) the larger caesium ions might have difficulty fitting into the channel. Did he do this experiment with caesium or sodium?

Koeppe: Actually, the experiments in thick membranes were based on proton conductance. In order to observe a double-stranded structure, you first have to destabilize the standard channel. What I believe is happening in the thick membrane is that it becomes less favourable for hydrophobic matching for both the standard single-stranded channel and for double-stranded structures, but that the relative change for the single-stranded channel is more than for the double-stranded channel, because the double strands can form a greater variety of structures (see Mobashery et al 1997).

Heitz: Are you saying that the non-conducting state is a double helix?

Koeppe: No, my model for the hybrid channels with hexafluorovaline is that it is a single-stranded dimer. The model is similar to Tim Cross' model (Ketchem et al 1993), but the hexafluorovaline has done something to temporarily go to a low conducting sub-state or a closed state. Those two monomers are still together, and they can re-open before they dissociate. The opening and closing of the voltage-dependent channel is too rapid to be explained by the conversion from a double-stranded to a single-stranded structure.

Smart: For normal gramicidin, what is the model for the closed conformation? Is there any evidence that it is monomeric?

Huang: We have some evidence indicating that the closed conformation is monomeric. We did some X-ray in-plane scattering experiments, which I will elaborate in my presentation. We obtained some evidence that monomers are floating in the monolayers.

Smart: What is the equilibrium?

Huang: A Russian group has measured the association constant (Rokitskaya et al 1996).

Hladky: Haydon & Hladky (1972) showed that the increased conductance was quadratic in time, under conditions where one would expect the rate of arrival of gramicidin at the membrane to be linear in time. Veatch et al (1975) did some measurements that may not be quantitatively as good as we would hope for because they were pushing their techniques to the limit, but they showed that the conductance produced by a fluorescent derivative of gramicidin was proportional to the square of fluorescence. Both of these sets of results are consistent with the idea that there is a surplus of monomer in the membrane, and that there is an 'equilibrium' between the monomer and the conducting dimer conformation. It was emphasized much later by Stark et al (1986) that all we (and Bamberg & Läuger 1973) had shown was that there was a species in the membrane and the conducting form was a dimer of that species. Therefore, all of the early work would have been consistent with a tetramer being the conducting unit, but since there are good models for the dimer, we have no reason to worry about a tetrameric form.

Separovic: Antoinette Killian, Tim Cross and Jim Hinton have shown that if the double-stranded form is heated, it is converted into the single-stranded form (Killian et al 1988, Arumugam et al 1996, Buster et al 1988). What happens, therefore, if you heat a double-stranded form in a 22-carbon lipid? Is it converted to a single-stranded form?

Wallace: We recently did a circular dichroism experiment in which we heated gramicidin for 16 h at 55°C in the presence of a 22 carbon lipid (Galbraith & Wallace 1999), and we observed a mixture of double-stranded and single-stranded forms.

Separovic: Does the equilibrium favour the double-stranded form?

Wallace: No. We observed a maximum of only 40% of the conformers to be double helices. The helical dimer form still predominates.

Cross: I would like to mention a couple of points about double strands in bilayers. The first is that the structure of a parallel gramicidin intertwined double helix is stable in a lipid bilayer as a double strand. If you heat it up to 60–65°C for several days, you will see that this double-stranded structure is converted to the channel state. Presumably, this is because all the tryptophans are on one side, and to convert it to a head-to-head dimer you have to ensure that those tryptophans are transported across the bilayer. Secondly, we've been doing some experiments recently on gramicidin M in lipid bilayers. We are convinced that we're looking at an intertwined double helix that has no significant pore. There is no NH exchange for the NHs that are near the centre of the lipid bilayer, so I believe that the only way to make the double-stranded form conduct is to make sure that it does not have a 5.6 residue per turn structure.

Ring: In the tryptophan substitution experiments, we know that hydrophobic mismatch is important, and we also know that the boundary lipids are important. Therefore, when the 'bulky' tryptophans are removed, there is a profound influence on the effects of boundary lipids. I would expect that the resulting decreased 'inner' radius of the deformation region (the dimple) as well as removing the 'spare' empty space under the tryptophans would result in increased compression energy from those boundary lipids, which would then destabilize the channel. Since hydrophobic mismatch is important, then that could be as important for destabilizing the channel as the change in interaction of tryptophans with the permeating ion. Is this a likely scenario?

Huang: The mismatch energy can be comparable to the compression energy.

Jakobsson: My feeling is that the specific affinity that the tryptophans have for the interfacial region of the membrane is more important than their bulk. It also occurred to me that when Roger Koeppe compared the two structures (i.e. the Duax structure with the head-to-head helices) it would be useful to highlight the tryptophans in those structures and also in a visualization of the K channel and other membrane proteins of known structure. In the high resolution known structures you would see rows of aromatic residues, tryptophans and tyrosines, in that region, and they are not there in the Duax structure. Steve White's lab has thermodynamic evidence for the special propensity of tryptophans to partition into the interfacial region of phospholipid membranes, in a way that is different from partitioning into other organic solvents (Wimley & White 1996).

Koeppe: I would like to mention that when the tryptophans are substituted a particular channel lifetime pattern is observed that cannot as yet be explained (Becker et al 1991), i.e. there is a series of non-linear changes in lifetimes involving both increases and decreases of channel lifetime as particular tryptophans at different positions in the sequence are substituted. Perhaps this relates to some of your ideas about the lipids.

References

Arumugam S, Pascal S, North CL et al 1996 Conformational trapping in a membrane environment: a regulatory mechanism for protein activity? Proc Natl Acad Sci USA 93:5872–5876

Bamberg E, Läuger P 1973 Channel formation kinetics of gramicidin A in lipid bilayer membranes. J Membr Biol 11:177–194

Becker MD, Greathouse DV, Koeppe RE II, Andersen OS 1991 Amino acid sequence modulation of gramicidin channel function: effects of tryptophan-to-phenylalanine substitutions on the single-channel conductance and duration. Biochemistry 30:8830–8839

Burkhart BM, Li N, Langs DA, Pangborn WA, Duax WL 1998 The conducting form of gramicidin A is a right-handed double-stranded double helix. Proc Natl Acad Sci USA 95:12950–12955

Buster DC, Hinton JF, Millett FS, Shungu DC 1988 ^{23}Na-nuclear magnetic resonance investigation of gramicidin-induced ion transport through membranes under equilibrium conditions. Biophys J 53:145–152

Eisenberg B 1998a Ionic channels in biological membranes. Electrostatic analysis of a natural nanotube. Contemp Phys 39:447–466

Eisenberg B 1998b lonic channels in biological membranes: natural nanotubes. Acc Chem Res 31:117–125

Galbraith TP, Wallace BA 1999 Phospholipid chain length alters the equilibrium between pore and channel forms of gramicidin. Faraday Discusions 111:159–164

Haydon DA, Hladky SB 1972 Ion transport across thin lipid membranes: a critical discussion of mechanisms in selected systems. Q Rev Biophys 5:187–282

Jing N, Prasad KU, Urry DW 1995 The determination of binding constants of micellar-packaged gramicidin A by ^{13}C-and ^{23}Na-NMR. Biochim Biophys Acta 1238:1–11

Ketchem RR, Hu W, Cross TA 1993 High-resolution of gramicidin A in a lipid bilayer by solid-state NMR. Science 261:1457–1460

Killian JA, Prasad KU, Hains D, Urry DW 1988 The membrane as an environment of minimal interconversion. A circular dichroism study on the solvent dependence of the conformational behavior of gramicidin in diacylphosphatidylcholine model membranes. Biochemistry 27:4848–4855

Lelievre D, Trudelle Y, Heitz F, Spach G 1989 Synthesis and characterization of retro gramicidin A-DAla-gramicidin A, a 31-residue-long gramicidin analog. Int J Pept Protein Res 33:379–385

Mobashery N, Nielsen C, Andersen OS 1997 The conformational preference of gramicidin channels is a function of lipid bilayer thickness. FEBS Lett 412:15–20

Oiki S, Koeppe RE II, Andersen OS 1995 Voltage-dependent gating of an asymmetric gramicidin channel. Proc Natl Acad Sci USA 92:2121–2125

Pomès R, Roux B 1998 Free energy profiles for H^+ conduction along hydrogen-bonded chains of water molecules. Biophys J 75:33–40

Rokitskaya TI, Antonenko YN, Kotova EA 1996 Photodynamic inactivation of gramicidin channels: a flash-photolysis study. Biochim Biophys Acta 1275:221–226

Stark G, Strässle M, Takácz Z 1986 Temperature-jump and voltage-jump experiments at planar lipid membranes support an aggregational (micellar) model of the gramicidin A ion channel. J Membr Biol 89:23–37

Veatch WR, Mathies R, Eisenberg M, Stryer L 1975 Simultaneous fluorescence and conductance studies of planar bilayer membranes containing a highly active and fluorescent analog of gramicidin A. J Mol Biol 99:75–92

Wimley WC, White SH 1996 Experimentally determined hydrophobicity scale for proteins at membrane interfaces. Nat Struct Biol 3:842–848

Engineering charge selectivity in alamethicin channels

G. Andrew Woolley, Andrei V. Starostin, Radu Butan, D. Andrew James, Holger Wenschuh* and Mark S. P. Sansom†

*Department of Chemistry, University of Toronto, 80 St. George Street, Toronto, Canada M5S 3H6, *Jerini BioTools, 12489 Berlin, Germany, and †Laboratory of Molecular Biophysics, The Rex Richards Building, Department of Biochemistry, University of Oxford, South Parks Road, Oxford OX1 3QU, UK*

Abstract. The peptide alamethicin provides a system for engineering ion channel charge selectivity. To define alamethicin charge selectivity experimentally, we measured single-channel current–voltage relationships in KCl gradients using covalently linked peptide dimers. Two factors were found to contribute to the charge selectivity of these channels: (i) the ionic strength of the surrounding solutions; and (ii) the distribution of fixed charge on the peptide. Native alamethicin channels exhibited either cation selectivity or anion selectivity depending on which end of the channel was at the low salt side of the membrane. When the glutamine residue at position 18 in the sequence was replaced with a lysine residue, an anion-selective channel was obtained regardless of which end of the channel was at the low salt side of the membrane.

1999 Gramicidin and related ion channel-forming peptides. Wiley, Chichester (Novartis Foundation Symposium 225) p 62–73

Simple, robust, cation-selective ionophores, such as valinomycin, and channels, such as gramicidin, are often-used tools in cell biology. These compounds find application in controlling cation flux and membrane potential in a wide range of systems (Woolley et al 1995). There are, however, no anion-selective counterparts to these compounds (Chao et al 1989, Woolley et al 1995). Anion-selective ionophores could be useful for studying chloride-dependent processes, anion channel-related disease states, as well as for controlling membrane potential in a manner complementary to the cation ionophores.

In view of this situation, we initiated a research programme to design an anion-selective channel. We opted to begin with the well-studied peptide ion channel alamethicin (Fig. 1). The alamethicin system is attractive for a number of reasons: (i) the peptide is small, robust (cannot be denatured) and will spontaneously insert into membranes; (ii) it produces well-resolved single-channel currents that are reliably measured with standard electrophysiological techniques; (iii) it can be

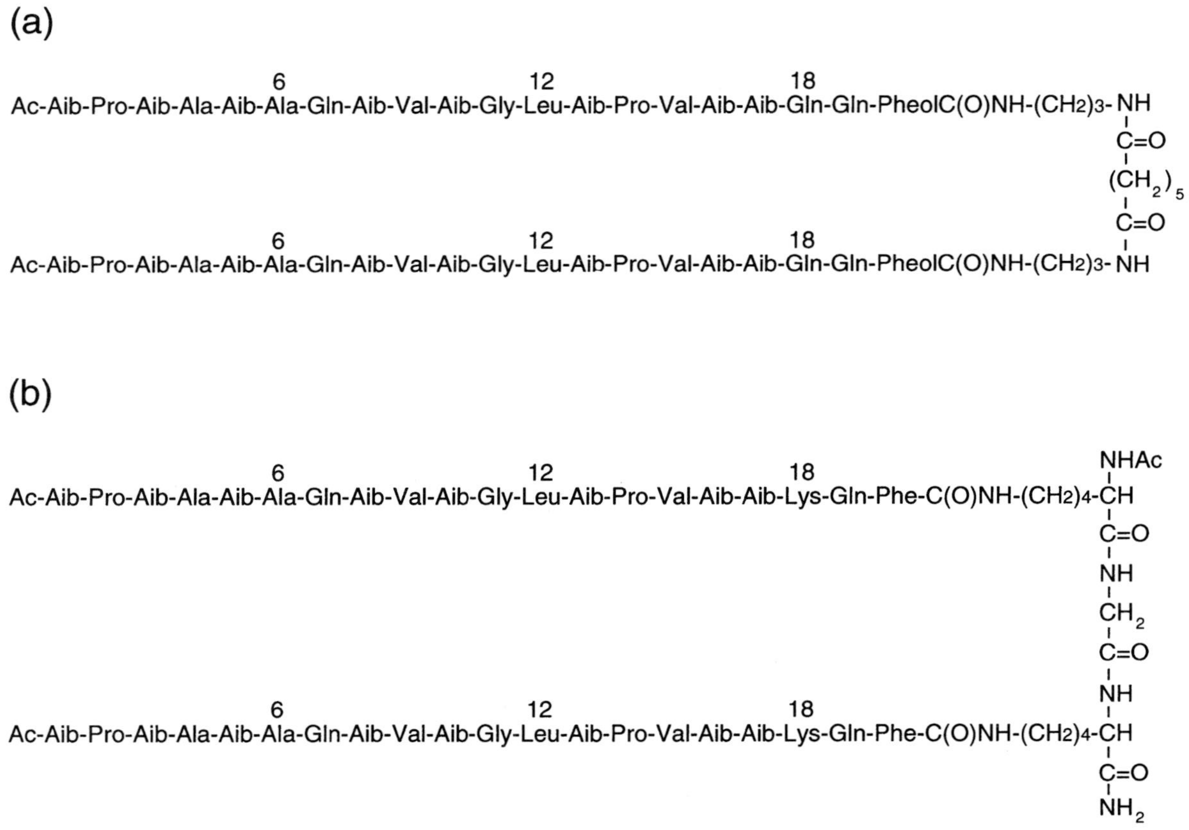

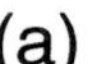

(a)

Ac-Aib-Pro-Aib-Ala-Aib-Ala-Gln-Aib-Val-Aib-Gly-Leu-Aib-Pro-Val-Aib-Aib-Gln-Gln-PheolC(O)NH-$(CH_2)_3$-NH-C(=O)-$(CH_2)_5$-C(=O)-NH-$(CH_2)_3$-NHC(O)Pheol-Gln-Gln-Aib-Aib-Val-Pro-Aib-Leu-Gly-Aib-Val-Aib-Gln-Ala-Aib-Ala-Aib-Pro-Aib-Ac

(residue numbers 6, 12 and 18 are marked above each chain)

(b)

Ac-Aib-Pro-Aib-Ala-Aib-Ala-Gln-Aib-Val-Aib-Gly-Leu-Aib-Pro-Val-Aib-Aib-Lys-Gln-Phe-C(O)NH-$(CH_2)_4$-CH(NHAc)-C(=O)-NH-CH_2-C(=O)-NH-CH($C(=O)NH_2$)-$(CH_2)_4$-NHC(O)-Phe-Gln-Lys-Aib-Aib-Val-Pro-Aib-Leu-Gly-Aib-Val-Aib-Gln-Ala-Aib-Ala-Aib-Pro-Aib-Ac

(residue numbers 6, 12 and 18 are marked above each chain)

FIG. 1. Primary structure of (a) alamethicin-bis(N-3-aminopropyl)-1,7-heptanediamide, and (b) alamethicin-K18 (alm-K18).

chemically synthesized so that any required modification can be made; and (iv) a considerable body of knowledge exists concerning the properties of the alamethicin system (for reviews see Breed et al 1997, Cafiso 1994, Woolley & Wallace 1992).

Alamethicin forms channels via self-association of membrane-bound helices. These helices can assemble into conducting bundles of different sizes depending on the number of monomers per bundle. Thus, macroscopic recordings of alamethicin channels reflect properties of this ensemble of helix bundles. For instance, the voltage dependence of the macroscopic conductance arises from a voltage dependence of the rate of channel formation rather than an intrinsic voltage dependence of the single-channel conductance (Latorre & Alvarez 1981). Likewise, measurements of alamethicin ion selectivity have been based on macroscopic current measurements and so reflect the selectivity of the ensemble of channels. Measurements with alm-Rf30 (Glu18) and alm-Rf50 (Gln18) indicated the channel passed both cations and anions with cations preferred, ($P_{K^+}/P_{Cl^-} = 2.7$ for alm-Rf30; Eisenberg et al 1973). In order to relate ion selectivity to specific conducting structures, we have focused exclusively on microscopic (single-channel) measurements. Measurement of single-channel selectivity has been made possible by the covalent linking of alamethicin peptides into dimers. Dimers form channels with extended lifetimes—sufficient for a characterization of the single-channel current–voltage (I–V) relationship (Woolley et al 1997, You et al 1996).

Alongside the experimental work, we developed detailed atomic models of alamethicin bundles. Figure 2 shows representative space-filling structures of hexameric bundles formed by three alamethicin dimers. The process of building these models has been described in detail elsewhere (Breed et al 1997, You et al 1996). The models were used to calculate electrostatic energy profiles for a cation (or anion) along a trajectory through the pore by numerically solving the linearized Poisson–Boltzmann equation (Woolley et al 1997). Although these electrostatic profiles apply strictly to the equilibrium case (Chen et al 1992), it was found that if they were used to describe the transmembrane potential gradient (when summed with a linear applied-field term), they successfully predicted the shapes of experimentally determined I–V curves. In particular, this treatment successfully predicted the ionic strength dependence of current rectification in alamethicin channels (Woolley et al 1997).

These electrostatic calculations show a broad zone of favourable electrostatic interactions between a cation and the pore in the region from the glycine at position 11 to the C-terminus. These favourable interactions should act as a source of cation selectivity. The electrostatic calculations further imply that an anion-selective channel should result if a positively charged site were placed in this zone. This prediction led us to synthesize alamethicin with lysine at

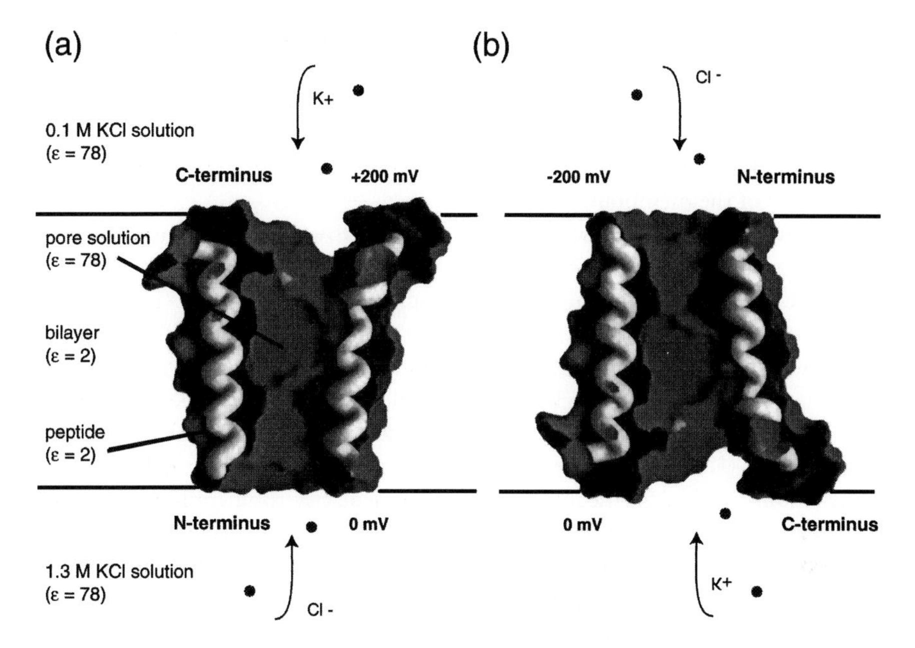

FIG. 2. Models of hexameric alamethicin-bis(N-3-aminopropyl)-1,7-heptanediamide channels showing the channel orientation with respect to the salt gradient and applied field. The dielectric regions used for electrostatic calculations are indicated. (a) For a voltage ramp starting from +200 mV, the channel is oriented with the C-terminal end at the low salt side of the membrane. At the beginning of the voltage ramp, cations enter the C-terminal end of the channel and anions enter the N-terminal end. (b) For a voltage ramp starting from −200 mV, the channel is oriented with the N-terminal end at the low salt side of the membrane. Figure prepared using GRASP (Nicholls et al 1991).

position 18 (alm-K18, Fig. 1b)—a pore-lining site that should not interfere with helix–helix packing. The selectivity properties of the channels formed by alm-K18 dimers appear consistent with expectations derived using the models.

Results and discussion

Initially, we investigated the selectivity properties of native (Gln18, Fig. 1a) dimeric alamethicin channels. We performed single-channel I–V measurements for particular conducting states in the presence of a KCl gradient. Figure 3a,b shows results obtained for the putative (You et al 1996) hexamer level. As discussed previously (Woolley et al 1997), channels form by N-terminal insertion from the side of the membrane made positive by the applied field. In symmetrical salt solutions, +200 → 200 mV and −200 → +200 mV voltage ramps produce I–V curves that are related by reflection through the origin (see Woolley et al 1997; and compare 1.3/1.3 M curves in Fig. 3a,b). In the presence of a salt gradient, however, a +200 → −200 mV voltage ramp results in the C-terminal end of the channel being at the low salt side of the membrane (the high salt is at ground; see Fig. 2a). A −200 → +200 mV ramp results in the N-terminal end of the channel being at the low salt side of the membrane (see Fig. 2b).

When the C-terminal end of the peptide is at the low salt side of the membrane (Fig. 3a), zero current is observed at a positive voltage. This is indicative of a cation-selective channel. Using the GHK equation (Lear et al 1997) a value for $P_{K^+}/P_{Cl^-} = 4.2$ is obtained. In contrast, when the N-terminal end of the peptide is at the low salt side of the membrane (Fig. 3b), zero current is observed at slightly negative voltages, indicating some anion selectivity ($P_{Cl^-}/P_{K^+} = 2.7$). Similar behaviour has been reported for the Leu-Ser series of peptides studied by Lear and colleagues (Kienker & Lear 1995).

This finding highlights the importance of the ionic environment in dictating the functional properties of the channel. The shape of the alamethicin-bis(N-3-aminopropyl)-1,7-heptanediamide (alm-BAPHDA) I–V curve in symmetrical salt solutions is affected by ionic strength in a way that can be explained by shielding of fixed channel charge (Woolley et al 1997). One is inclined to explain the observed dependence of charge selectivity on ionic strength in the same terms, i.e. differential shielding of the each end of the channel. Qualitatively, the observed behaviour is what one might predict. The N-terminal end of the peptide has partial positive charge (due to the alignment of dipoles). When it is at the low salt side of the membrane, the charge is less shielded and an anion-selective channel results. The converse applies to the C-terminal end.

To test the prediction that fixed positive charge in the region of the channel from position 11 to the C-terminus could reverse the charge selectivity, we synthesized alamethicin dimers with lysine at position 18 (alm-K18, Fig. 1b). Single-channel I–V curves of alm-K18 in a KCl gradient are shown in Fig. 3c,d. In contrast to alm-BAPHDA, alm-K18 shows anion selectivity regardless of whether the N-terminus or the C-terminus is at the low salt side of the membrane ($P_{Cl^-}/P_{K^+} = 3.0$ and $P_{Cl^-}/P_{K^+} = 1.8$, respectively). We assume that the N-terminus still inserts first from the positive side of the membrane since it is formally uncharged and has not been modified; however, C-terminal insertion cannot be ruled out. The degree of selectivity depends, as before, on the orientation of the salt gradient, but for either orientation the charge selectivity is anionic.

A more striking demonstration of the anion selectivity of alm-K18 is obtained in a larger KCl gradient. Figure 4 shows a single-channel I–V curve of the putative octamer level (four dimers) in a 1.3 M/0.01 M KCl gradient. The permeability ratio calculated in this case ($P_{Cl^-}/P_{K^+} = 3.6$) is when the C-terminus is at the low salt side of the membrane. In this case, the electrostatic field of the lysine residues is

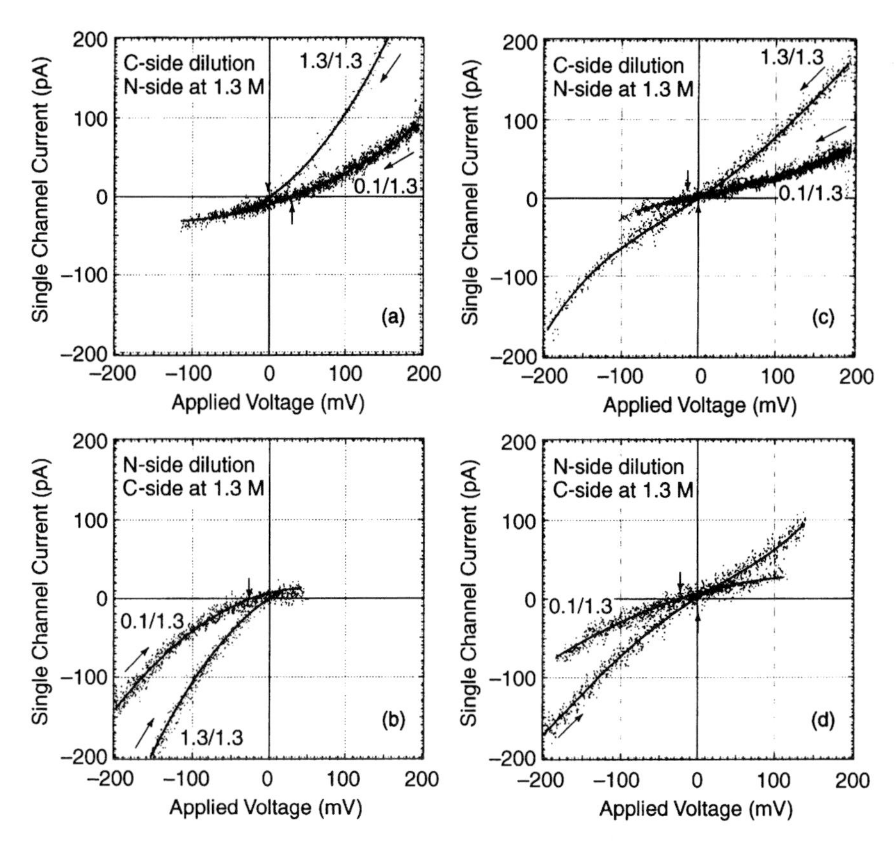

FIG. 3. Single-channel current–voltage (I–V) curves for alamethicin-bis(N-3-aminopropyl)-1,7-heptanediamide (alm-BAPHDA; panels a, b) and alamethicin-K18 (alm-K18; panels c, d). In each case, the high salt side of the membrane (1.3 M KCl, 5 mM BES, pH 6.8) is at ground. In panels (a) and (c), +200 mV is applied, the channel opens (oriented as shown in Fig. 2a) and the voltage is ramped to -200 mV (indicated by arrows). Symmetrical salt solutions give zero voltage reversal potentials. When the C-terminus is at the low salt side of the membrane (0.1 M KCl, 5 mM BES, pH 6.8) non-zero reversal potentials are observed (indicated by arrows). E_{rev} (alm-BAPHDA)=+28 mV (panel a); E_{rev} (alm-K18)=-12 mV (panel c). In panels (b) and (d), -200 mV is applied and channels open, this time with the N-terminal end at the low salt side of the membrane. Again, non-zero reversal potentials are observed (E_{rev} (alm-BAPHDA)=-20 mV (panel b); E_{rev} (alm-K18)=-22 mV (panel d).

presumably less shielded than with a 1.3 M/0.1 M KCl gradient. Anion selectivity is apparently maintained despite the increase in pore diameter expected for an 8-mer channel. The extra two lysine residues contributed by an extra alm-K18 dimer may be sufficient to offset the effects of a larger pore diameter.

The degree of anion selectivity is presumably affected by the magnitude of positive charge in the channel lumen. The Lys18 residues in the helix bundles

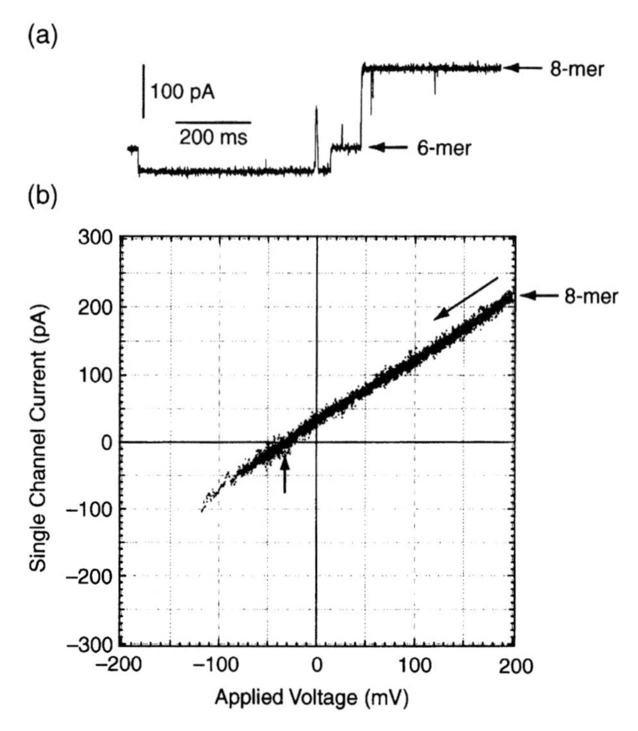

FIG. 4. Single channel properties of alamethicin-K18 (alm-K18) in the presence of a large KCl gradient (N-terminal side is at 1.3 M KCl, C-terminal side is at 0.01 M KCl; both sides pH 6.8, 5 mM BES). (a) Constant voltage (+200 mV) recording showing 6-mer and 8-mer levels stabilized. (b) Current–voltage (I–V) curve from +200 mV (direction of voltage gradient indicated). A reversal potential of −31 mV is observed.

may not all be positively charged. Calculations of effective pK_a values in analogous situations are consistent with this proposal (Adcock et al 1998). A systematic study of the effect of pH and side-chain pK_a on alamethicin channel selectivity is underway.

Acknowledgements

We thank the Natural Sciences and Engineering Research Council of Canada and the Canadian Cystic Fibrosis Foundation for funding this project.

References

Adcock C, Smith GR, Sansom MSP 1998 Electrostatics and the ion selectivity of ligand-gated channels. Biophys J 75:1211–1222

Breed J, Biggin PC, Kerr ID, Smart OS, Sansom MSP 1997 Alamethicin channels — modelling via restrained molecular dynamics simulations. Biochim Biophys Acta 1325:235–249

Cafiso DS 1994 Alamethicin: a peptide model for voltage gating and protein–membrane interactions. Annu Rev Biophys Biomol Struct 23:141–165
Chao AC, Dix JA, Sellers MC, Verkman AS 1989 Fluorescence measurement of chloride transport in monolayer cultured cells. Mechanisms of chloride transport in fibroblasts. Biophys J 56:1071–1082
Chen DP, Barcilon V, Eisenberg RS 1992 Constant fields and constant gradients in open ionic channels. Biophys J 61:1372–1393
Eisenberg M, Hall JE, Mead CA 1973 The nature of the voltage-dependent conductance induced by alamethicin in black lipid membranes. J Membr Biol 14:143–176
Kienker PK, Lear JD 1995 Charge selectivity of the designed uncharged peptide ion channel Ac-(LSSLLSL)3-CONH2. Biophys J 68:1347–1358
Latorre R, Alvarez O 1981 Voltage-dependent channels in planar lipid bilayer membranes. Physiol Rev 61:77–150
Lear JD, Schneider JP, Kienker PK, DeGrado WF 1997 Electrostatic effects on ion selectivity and rectification in designed ion channel peptides. J Am Chem Soc 119:3212–3217
Nicholls A, Sharp KA, Honig B 1991 Protein folding and association: insights from the interfacial and thermodynamic properties of hydrocarbons. Proteins 11:281–296
Woolley GA, Wallace BA 1992 Model ion channels: gramicidin and alamethicin. J Membr Biol 129:109–136
Woolley GA, Pfeiffer DR, Deber CM 1995 Use of ionophores for manipulating intracellular ion concentrations. In: Conn PM, Kraicer J, Dixon SJ (eds) Measurement and manipulation of intracellular ions. Academic Press, San Diego, CA (Methods in Neurosciences 27) p 52–68
Woolley GA, Biggin PC, Schultz A et al 1997 Intrinsic rectification of ion flux in alamethicin channels: studies with an alamethicin dimer. Biophys J 73:770–778
You S, Peng S, Lien L, Breed J, Sansom MSP, Woolley GA 1996 Engineering stabilized ion channels: covalent dimers of alamethicin. Biochemistry 35:6225–6232

DISCUSSION

Eisenberg: In asymmetrical solutions Andrew Woolley's current–voltage (I–V) relationships were linear. I would like to point out that many people have calculated I–V relationships at different concentrations in, for example, the calcium release channel using barrier models, and they have been unable to demonstrate a linear relationship.

Jakobsson: I'm not sure I agree with that. We've done Brownian dynamics simulations in which we created a well in the interior of the channel (Bek & Jakobsson 1994). This favours multiple occupancy and we observed linear I–V relationships.

Eisenberg: But the checking has to be done over a wide range of concentrations with the same set of parameters. I'm suggesting people should try to do this.

Ring: I have a question and a remark. First, are you saying that in principle, even with 10 barriers, for example, an approximately linear relationship cannot be achieved over the relevant range? And second, it is necessary to distinguish between barriers and voltage dependencies. This is important to remember because if the barrier is steep enough, then there will not be a voltage dependency. Therefore, the I–V curve reflects the voltage dependency, and it doesn't reflect the fact that there are multiple barriers.

Eisenberg: We have to define what we're talking about by barrier models. I define them in the way they are defined in Hille (1992).

Hladky: That's not the way that most people now define them.

Eisenberg: I disagree. People have used that equation to try to fit I–V relationships with one set of parameters over a range of voltages and over a reasonable set of concentrations, and they have reported that they can't do it.

Hladky: I agree that we must look at data over a wide range of concentrations. That range has to be wide enough so that any effect of the charge present to buffer concentrations can be obviated.

I also wondered whether Andrew Woolley is now in a position to address an old puzzle in the field of alamethicin research, i.e. the size cut-off. Gordon & Haydon (1975) showed in the early 1970s that the apparent size cut-off is independent of the conductance level. This is contrary to what one would expect from the addition model in which the size of the pore becomes larger and larger accounting for the increase in conductance.

Huang: Can you explain why it is cut off?

Hladky: In Gordon and Haydon's study using a series of cations with chloride as the anion, the point in the series at which the cation appears not to be conducted, and just the chloride appears to contribute, lies between dimethylammonium and tetramethylammonium for all conductance levels, even though the sequential addition model in which additional molecules of alamethicin are added each time the conductance level increases suggests that larger ions should be able to pass when the conductance level is higher. Gordon & Haydon (1976) took this as evidence against an expanding pore and proposed instead that higher conductance states meant more open pores in a fixed size co-operative aggregate of pores. However, as far as I know the observation that the aggregate changes size is no longer disputed. The only remaining discrepancy is the question of the size cut-offs.

Smart: Bezrukov & Vodyanoy (1993) describe an experiment relevant to this point. Rather than analysing the conductivity of ions of different sizes, polyethylene glycol was added to the conducting medium. By varying the molecular weight of the polymer the channels can be sized, and in particular the end radius can be found. Although the original interpretation of the data was against the barrel stave model, a reinterpretation (Smart et al 1997) using the HOLE program shows the data are compatible with the models of alamethicin produced by Sansom and co-workers based on the barrel stave hypothesis. In particular the end radii revealed by access resistance of the first three significant conductance states show a progressive increase which suggests that the barrel becomes larger (Smart et al 1997).

Sansom: One possible rationalization is to assume that for the highest conductance levels (which correspond to 10 or 12 monomers) the bundle deviates from circular symmetry in cross-section, i.e. the bundles start to flatten. If you do a careful analysis of the increased conductance from level to level (N), you find that the increases are not parabolic in their dependence on N.

Hladky: In the calculations supporting the idea that the conductance increased in proportion to pore area (Hanke & Boheim 1980, Stein 1986) the area available for diffusion was calculated without correction for the finite ionic radius.

Sansom: But even if you re-do that calculation and subtract the ionic radius, the result never quite matches what you would expect. If it's switching from a circular hole towards a slit, the cut-off point depends on where the switch starts.

Hladky: Eisenberg et al (1977) confirmed Gordon and Haydon's qualitative results that an expanding circular pore could not explain the low conductance of large cations but found, as did Hanke & Boheim (1980), that larger ions could still pass through the channel. They proposed that the shape is instead a slit. If so, the calculations based on the assumption of a circular cross section are irrelevant. It appears to me that the shape of the pore, the exact number of alamethicin monomers present in each conductance state, and the selectivity are still open questions. The point I am trying to make is that it would be interesting if Andrew Woolley and his colleagues could develop their approach to establish the number of monomers in each state and thus look at conductances and selectivity for pore states consisting of a known rather than an assumed number of monomer units.

Bechinger: I would like to comment on the entropic loss in energy during the association of several peptide monomers into larger transmembrane polymers. I always assumed that this term is mainly compensated by favourable van der Waals interactions between the peptide helices. The model you presented includes the formation of aggregates that have an ellipsoidal cross-section. One would expect that the van der Waals interactions in such an aggregate are not optimal for all peptide contacts. Also, the tethering of monomers into dimers reduces the possibilities to assume the best spatial arrangements in particular in cases where a channel size corresponding to an odd number of monomers is observed.

Sansom: In Andrew Woolley's dimers there are several different tethers, and they are all sufficiently flexible such that they do not perturb the pattern. What may be peculiar about alamethicin is that the α-aminoisobutyric acid (Aib) residues seem to generate a smoother helix, although we don't have solid calculations to support this. If you take a pure poly-Ala helix or a pure poly-Leu helix and try to pack them together, the poly-Ala and poly-Leu generate a tightly packed structure, whereas the surface of the poly-Aib helix is much smoother, giving rise to a greater degree of flexibility in the packing arrangement (Breed et al 1995). Further, the bundle may not be held together simply by van der Waal's interactions. We suspect that it may also be held together by electrostatic interactions between the helix dipoles and the aligned water.

Bechinger: Does the Glx to Lys mutation exhibit pronounced differences in channel characteristics (e.g. life times, open probabilities, etc.)? Can we learn anything from these mutations about the stability of peptide aggregates or their transmembrane residence time?

Woolley: Any of the chemical changes made may affect both permeation through the channel and the stability of the channel structure. I have shown how the lysine modification affects permeation. Somewhat surprisingly, it does not seem to affect channel stability much, at least when the channel is formed from dimers. A channel formed by Lys18 monomers is considerably less stable. The lysine residues are probably not all ionized when the channel is conducting. Also, it is not clear how the presence of water in the channel affects the side chain interactions and thereby, channel stability.

Bechinger: In a previous publication we have used a formalism where charged amino acid side chains are discharged before membrane incorporation (Bechinger 1996). The energy of discharge is a linear function of the difference between the pK value of the amino acid and the pH of the environment. It is therefore possible that channels are observed of only those polypeptides which are in their uncharged state. If this is the case the characteristics of the single channels of the mutations of Glu18 into glutamine or lysine as well as their ion selectivity could all be similar. The probabilities of channel formation, however, might differ and be pH dependent.

Smart: Does changing the pH have any effects?

Woolley: We're looking at that. These experiments were all done at pH 6.8.

Jordan: Have you done any calculations that might suggest what fraction of them are ionized?

Woolley: We have done some crude calculations on the bundle of eight helices without worrying too much about the effects the bilayer, and we find that about half of the lysines are ionized.

Jordan: Have you looked at the bundle of six?

Woolley: No.

Busath: Did you see any evidence of changes in protonation of the lysine analogues during the conductance of a single channel?

Woolley: That's an interesting point. We should go back and look at that.

References

Bechinger B 1996 Towards membrane protein design: pH-sensitive topology of histidine-containing polypeptides. J Mol Biol 263:768–775

Bek S, Jakobsson E 1994 Brownian dynamics study of a multiply-occupied cation channel: application to understanding permeation in potassium channels. Biophys J 66:1028–1038
Bezrukov SM, Vodyanoy I 1993 Probing alamethicin channels with water-soluble polymers. Effect on conductance of channel states. Biophys J 64:16–25
Breed J, Kerr ID, Sankararamakrishnan R, Sansom MSP 1995 Packing interactions of Aib-containing helices: molecular modelling of parallel dimers of simple hydrophobic helices and of alamethicin. Biopolymers 35:639–655
Eisenberg M, Kleinberg ME, Shaper JH 1977 Channels across black lipid membranes. Ann N Y Acad Sci 303:281–294
Gordon LG, Haydon DA 1975 Potential-dependent conductances in lipid membranes containing alamethicin. Philos Trans R Soc Lond B Biol Sci 270:433–447
Gordon LG, Haydon DA 1976 Kinetics and stability of alamethicin conducting channels in lipid bilayers. Biochim Biophys Acta 436:541–556
Hanke W, Boheim G 1980 The lowest conductance state of the alamethicin pore. Biochim Biophys Acta 596:456–462
Hille B 1992 Ionic channels of excitable membranes, 2nd edn. Sinauer Associates, Sunderland, MA
Smart OS, Breed J, Smith GR, Sansom MSP 1997 A novel method for structure-based prediction of ion channel conductance properties. Biophys J 72:1109–1126
Stein WD 1986 Transport and diffusion across cell membranes. Academic Press, Orlando, FL

Lorentzian noise in single gramicidin A channel formamidinium currents

Teresa G. Fairbanks, Chris L. Andrus and David D. Busath[1]

Zoology Department, Brigham Young University, Provo, UT 84602, USA

Abstract. Seoh & Busath (1995) showed that in the presence of formamidinium, single gramicidin A channels were lengthened, had uniformly noisy currents at low voltages and had superlinear current–voltage relationships, all three properties being absent in gramicidin M^- channels in which the interfacial tryptophan residues in gramicidin A are all replaced by phenylalanine. We measured the single channel noise power spectra (PSDs) in small monoolein (GMO) bilayers with formamidinium chloride solutions to help identify the mechanism of noise process. PSDs were Lorentzian with characteristic frequencies of 0.1–1.0 kHz in 0.1 and 0.3 M formamidinium chloride solutions, and from 1–6 kHz in 1 M solution. $S_i(0)$, where measurable, ranged from $\sim$50–200 fA^2/Hz. The time course of the noise process could not be detected in these experiments. The low f_c suggests slow motions or rare states of the blocking 'gates' which, judging from the result with gramicidin M^-, must be equal to or related to the Trp residues.

1999 Gramicidin and related ion channel-forming peptides. Wiley, Chichester (Novartis Foundation Symposium 225) p 74–92

Gramicidin A is an uncharged peptide that forms dimeric cation-selective channels of known three-dimensional structure (Ketchem et al 1997, Arseniev et al 1986) in lipid bilayers. Comprised of 15 amino acids, it forms right-handed $\beta^{6.5}$ helices that join at the N-termini to yield a 25 Å-long cylindrical pore, symmetrical about the centre of the channel, with an inner diameter of 4 Å.

Experiments with peptides in which Trp side chains have been replaced (Becker et al 1991, Daumas et al 1989) or fluorinated (Andersen et al 1998, Busath et al 1998) show that the Trp side chains modulate channel conductance. This is thought to be due to their dipole potential, especially in the centre of the channel (Sancho & Martínez 1991, Hu & Cross 1995, Dorigo et al 1999). Deuterium splitting studies

[1]This chapter was presented at the symposium by David D. Busath, to whom correspondence should be addressed.

constrain the Trp side chain conformational space to four positions, two of which can be ruled out because they have high interaction energies with the peptide backbone and by Raman spectroscopy results (Takeuchi et al 1990, Maruyama & Takeuchi 1997). For the two remaining Trp conformations (G′− and T+), the projection of the dipole moment on the transport path is nearly identical. In both cases, the NH group projects out from the bilayer toward the water phase; they have been labelled o1 and o2, respectively (Dorigo et al 1999). Four other canonical conformations exist for the isolated amino acid. In the context of the right-handed gramicidin A channel, the indole dipole points inward for two of them (i1 and i2) and perpendicular to the channel axis for the other two (p1 and p2). Except for position i1 for Trp9, these other four conformations have all been found to be energetically unfavourable using molecular mechanics (Dorigo et al 1999). Hence, it is not surprising that channels generally demonstrate a stable conductance.

Furthermore, although it is not possible to distinguish from solid-state NMR whether the side chains are in the o1 or o2 conformation, it does appear that they are mostly found in one of the two, because the two conformations would almost undoubtedly have different indole deuteron chemical shifts, but only one set of chemical shifts is observed (Hu et al 1995). When the conformational energy of the helical peptide alone is considered, the o1 position is energetically more stable than o2, especially for Trps 11, 13 and 15, where the difference exceeds 9 kcal/mol (Dorigo et al 1999).

Alkali metal cation current noise in gramicidin A channels has been analysed in considerable detail. Three types of blocks were observed: one incomplete block lasting $\sim$0.5 ms and occurring frequently during channel opening; one lasting an estimated 20 μs and occurring more frequently in decane-inflated bilayers than in hexadecane-inflated bilayers; and one lasting $\sim$0.5 μs detected in the power spectrum and thought to represent occluded conformational states resulting from channel vacancy (Sigworth & Shenkel 1988). The longer, detectable blocks are sparse. Between them, gramicidin A alkali metal conductances are stable, with single channel noise only double the expected shot noise (Sigworth et al 1987). Heinemann & Sigworth (1988) measured the gramicidin A current noise produced by addition of formamide to potassium chloride solution and found it to increase from 1.6 fA^2/Hz for KCl to 4.3 fA^2/Hz for the mixed salt. The noise was white to 20 kHz and kinetic modelling suggested that 1 M formamide blocks the channel with on and off rate constants of $\sim 10^7/s$.

Formamidinium-mediated currents contain considerably more noise, 120 fA^2/Hz in the low frequency range for 1 M formamidinium solution with 50 mV applied potential. When the tryptophans in gramicidin A are substituted with phenylalanines, the formamidinium current noise, current–voltage (I–V) superlinearity, and channel lifetime stabilization effects are eliminated. No single

Trp is responsible for these effects. It is postulated that large Trp dipole motions, stimulated by passage of the large formamidinium cations through the channel, modulate the current to cause the channel current noise. In the process the Trps move into a position where they interact with the lipids causing channel stabilization. Because the current noise and channel stabilization also disappear at high membrane potentials (200 mV), it was proposed that at such voltages, the ions pass too quickly to couple to the characteristic harmonic frequencies of the Trp vibrations (Seoh & Busath 1995).

This chapter presents an analysis of the spectral properties of the formamidinium-induced noise and their dependence upon membrane potential and formamidinium concentration. The noise is well represented by a single Lorentzian with a characteristic frequency in the range of 100–10 000 Hz. These results have been presented in preliminary form (Fairbanks & Busath 1997).

Materials and methods

Formamidine hydrochloride (Aldrich, Milwaukee, WI, USA) solution was prepared fresh daily (to avoid hydrolysis) with purified water (Barnstead NANOpure II, VWR Scientific, San Francisco, CA, USA) and filtered using a 0.2 μm Nalgene filter. The glassware and experimental apparatus were washed with an elutropic series of solvents. Bilayers were formed in the 20 (or 90) μm aperture of a polyethylene pipette using a dispersion (50 mg/ml) of monoolein (NuChek Prep, Elysian, MN, USA) in hexadecane (Aldrich, Milwaukee, WI, USA), following published methods (Busath & Szabo 1981). Membrane potential (V_m) was applied and transmembrane current measured using a single set of Ag–AgCl electrodes immersed >5 mm from the bilayer with asymmetry potential <1 mV. Gramicidin A, purified from gramicidin D (ICN Pharmaceuticals, Costa Mesa, CA, USA) by HPLC, was added from methanol in the usual fashion. Single channel currents were measured with a patch-clamp amplifier (List Electronic EP7, Darmstadt/Eberstadt, Germany). A gain of 10 pA/mV was used to avoid amplifier saturation. The data were filtered at 20 000 Hz and collected at 48 000 points/second using an audio digital tape recorder (CDAT4, Cygnus Technology, Inc., Delaware Water Gap, PA, USA). Currents were measured for 5–7 standard conductance channels in each of three different concentrations of formamidinium solution (0.1, 0.3 and 1 M) for four applied voltages (50, 100, 150 and 200 mV). The channel spectra at 200 mV in all three solutions, and at 150 mV in the 0.3 M solution, did not differ sufficiently from baseline and were not analysed further.

Analysis

Representative channel and baseline traces were identified using a processed trace (i.e. filtered at 100 Hz). The corresponding sections of the original traces were then used to determine the mean, S.D. and single channel noise power spectrum (PSD) of channel and baseline currents with Igor Pro (Wave Metrics, Lake Oswego, OR, USA). PSDs were computed as averages of overlapping subsets of each trace using the Parzen filter and were normalized by multiplying the partially normalized output from IGOR by the number of data points, N, used in the fast Fourier transform (FFT) and the sampling interval used for collecting the data, Δt, to yield the following relationship between the normalized spectral power in the current, $S_i(f)$, and the FFT coefficients, $H(f)$:

$$S_i(f) = \frac{2\Delta t}{N} H^2(f) \tag{1}$$

This normalization yields a PSD with units of A^2/Hz. The PSD method was confirmed by measuring the Johnson current noise of a 9.91 MΩ resistor inserted into the amplifier in place of the membrane. The current noise power was independent of frequency (white) up to 20 000 Hz with $S_i(0) = 1.6\ fA^2/Hz$ as expected from the equation:

$$S_i(f) = \frac{kT}{R}. \tag{2}$$

Overlapping 8192-point PSDs for channels and adjacent baselines were averaged. The average baseline PSD was then subtracted from the average channel PSD. The resulting average single channel PSD was fitted using IGOR to the equation for a single Lorentzian:

$$S_i(f) = \frac{S_i(0)}{1 + (f/f_c)^2} \tag{3}$$

where $S_i(0)$ is the zero-frequency power and f_c the characteristic frequency. To explore whether the 8192-point average PSD yields a robust measure of the $S_i(0)$ and f_c parameters and to check the S.D. of the parameters between channels, we computed 1024-point and 256-point PSDs for individual channels. This approach yielded the mean and S.D. of the parameters for Tables 1 and 2 and Figs 4 and 5. The parameter means were reasonably consistent with those obtained from the 8192-point fit.

Random boxcar channel current noise has a Lorentzian spectrum (DeFelice 1981) with parameters that depend on the two rate constants for gating the channel open (α) and closed (β) and the magnitude of the current change (i) between the two gated states:

$$f_c = \frac{(\alpha + \beta)}{2\pi} \tag{4}$$

$$S_i(0) = \frac{4i^2\alpha\beta}{(\alpha + \beta)^3}. \tag{5}$$

Thus, $2\pi f_c$ can be interpreted directly to represent approximately the rate constant of the fastest process, but interpretation of $S_i(0)$ depends on both $4i^2$ and a function of the gating rate constants that depends on the average duty cycle, $\alpha/(\alpha+\beta)$. In the results reported here, neither i nor the average duty cycle are known, so attention is focused primarily on f_c.

Results

Single channel current transitions can be observed if the current is filtered with a cut-off frequency of $\leqslant$ 3 kHz. An exemplary single channel current trace is shown in Fig. 1 for 1 M formamidinium chloride solution at $V_m = 50$ mV. The noise is similar to that reported previously (Seoh & Busath 1995).

Figure 2 shows the power spectral densities for a representative channel and neighbouring baseline in 1 M formamidinium at $V_m = 100$ mV. Note that for low frequencies the noise in the channel is greater than the noise in the baseline as is expected (with the exception of one aberrant baseline point at ~140 Hz), but as the frequency increases the channel and baseline noise approach each other and merge at around 5000 Hz. This suggests that the characteristic frequency for the channel noise is below 5000 Hz.

The baseline-subtracted power spectrum for the same channel is shown in Fig. 3. Ignoring the spike at ~140 Hz, the noise is Lorentzian with a characteristic frequency of 2600 Hz. Above 10 kHz the noise in the PSD becomes too great to

FIG. 1. Gramicidin A standard single channel currents in 1 M formamidine HCl, at 50 mV. Filtered at 60 Hz. GMO/Hexadecane, 23 ± 1 °C.

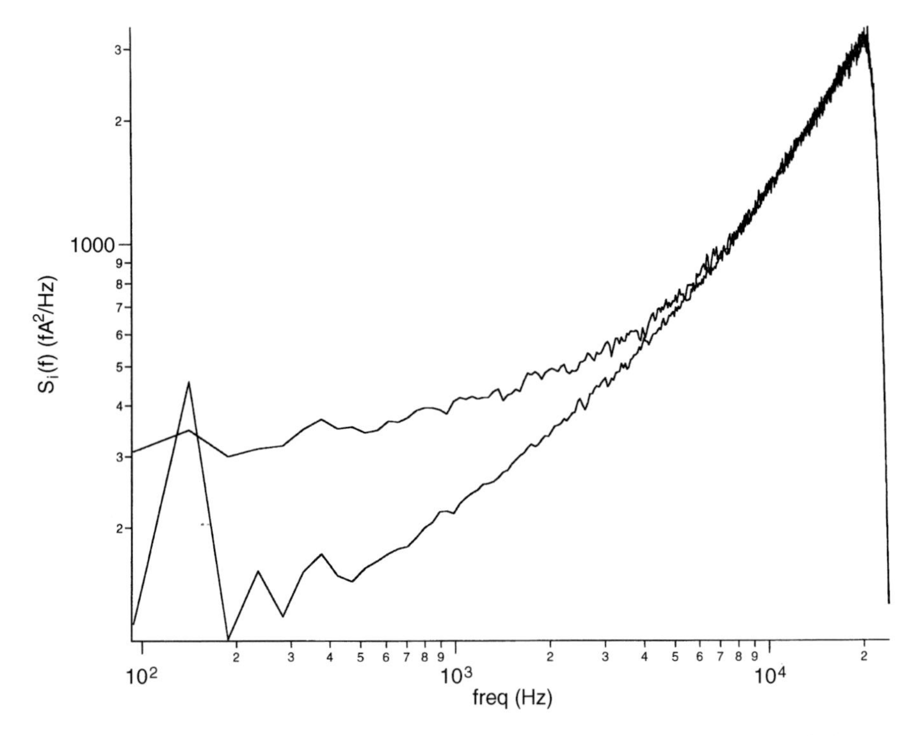

FIG. 2. Power spectral density for gramicidin A standard single channel currents and adjacent baselines in 1 M formamidine HCl at 100 mV applied potential. 1024 point fast Fourier transforms. GMO/Hexadecane, 23 ± 1 °C.

evaluate, but appears to oscillate about $S_i(f) = 0\ \text{fA}^2/\text{Hz}$, necessitating use of the linear vertical scale.

The average value of the limiting power, $S_i(0)$, for standard channels varied with [formamidinium$^+$] and V_m as shown in Fig. 4 and Table 1. Starting at similar levels for the three concentrations tested at $V_m = 50$ mV, it rises at 100 mV and then falls steeply at 150 mV, reaching ~0 fA²/Hz at 200 mV in 0.1 and 1.0 M salt (data not shown). The changes in $S_i(0)$ with [formamidinium$^+$] and V_m were consistent between channels as shown by the error bars.

The characteristic frequency ranged from ~100 Hz to >5 kHz, depending on the [formamidinium$^+$] and V_m as shown in Fig. 5 and Table 2. The value at $V_m = 150$ mV for [formamidinium$^+$] = 0.3 M appeared to be intermediate between 0.1 and 1.0 M, although the low amplitude of the power precludes an accurate estimate with our data. The characteristic frequency rises approximately proportionate to the [formamidinium$^+$], and steeply with V_m.

The average formamidinium current for the channels used is shown in Table 3. The values are similar to those reported for larger populations by Seoh & Busath

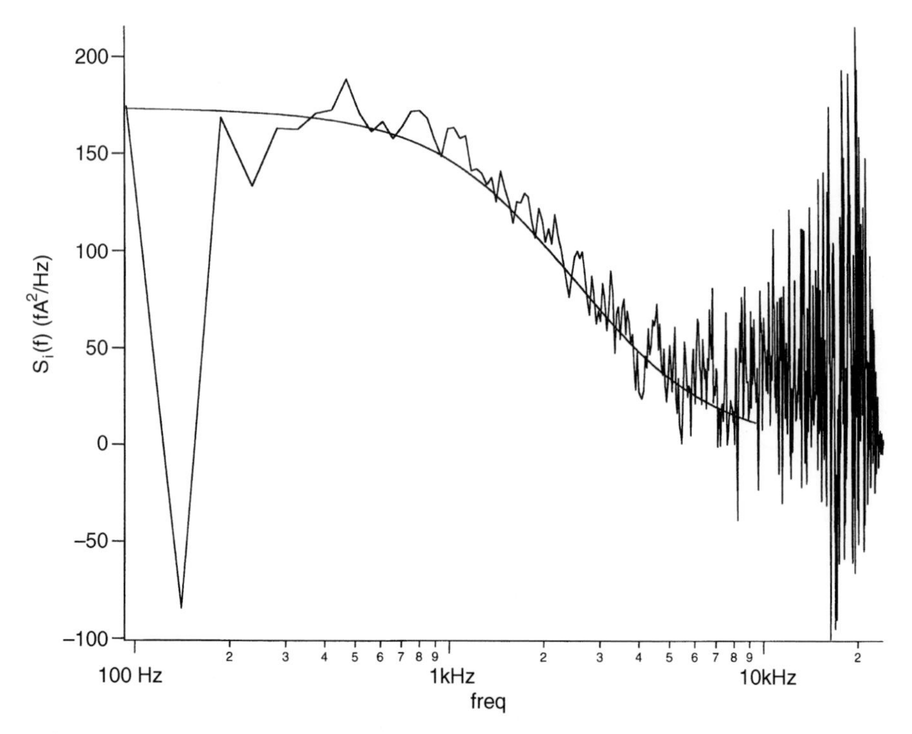

FIG. 3. Single channel power spectrum in 1 M formamidine HCl at $V_m = 100$ mV. GMO/Hexadecane, 23 ± 1 °C. Theoretical curve is Lorentzian with $f_c = 2.6$ kHz, $S_i(0) = 170$ fA^2/Hz.

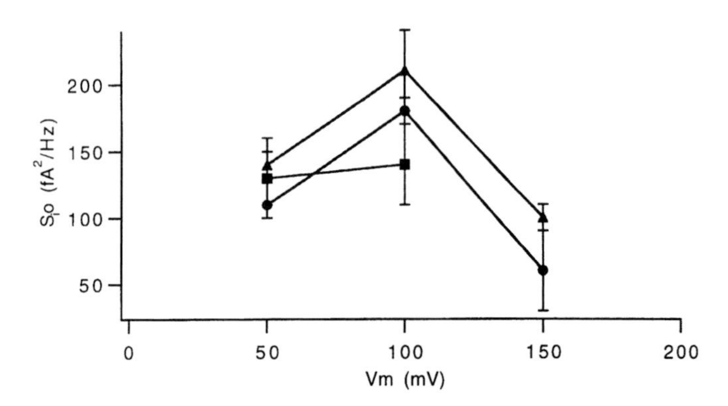

FIG. 4. Limiting current spectral density, $S_i(0) \pm$ S.D., for single channel formamidinium currents as a function of applied membrane potential, V_m, for 0.1 M (circles), 0.3 M (squares) and 1.0 M (triangles) formamidine HCl. GMO/Hexadecane bilayers, 23 ± 1 °C. At $V_m = 200$ mV for all concentrations and $V_m = 150$ mV for 0.3 M, the power was too low to measure reliably in our data sample.

TABLE 1 Average fitted limiting current spectral power, $S_i(0)$ (fA^2/Hz), for formamidinium current noise in single gramicidin A channels

	Concentration of formamidinium		
V_m	*0.1 M*	*0.3 M*	*1.0 M*
50 mV	110[a]	130 ± 30	140 ± 10
100 mV	180 ± 10	140 ± 30	210 ± 30
150 mV	60 ± 30	N.O.[b]	100 ± 10
200 mV	~0	~0	~0

[a]No estimate of S.D. possible due to high noise in single channel noise power spectra.
[b]Estimate not feasible with data set.

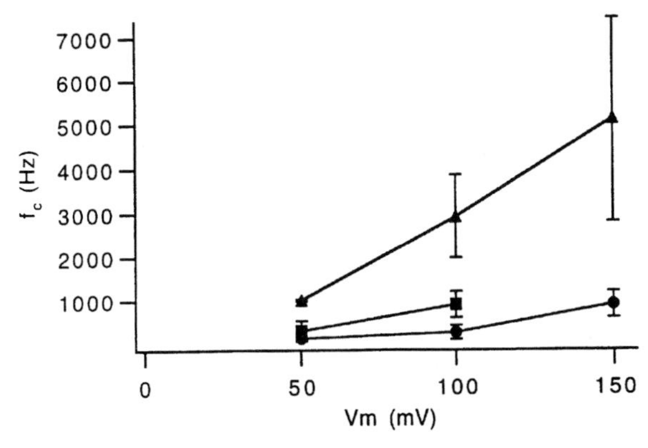

FIG. 5. Lorentzian characteristic frequency, $f_c \pm$ S.D., from single channel formamidinium current power spectra as a function of membrane potential, V_m for 0.1 M (circles), 0.3 M (squares) and 1.0 M (triangles) formamidine HCl. GMO/Hexadecane bilayers, 23 ± 1 °C. See legend for Fig. 4 concerning high voltage data.

(1995). The I–V is superlinear, especially at 1.0 M where the single channel current at 100 mV is 1.46 pA, corresponding to a passage rate of 8.5×10^6/s.

Concluding remarks

The single channel formamidinium current noise spectra have been measured for standard gramicidin A channels under various conditions of membrane potential and formamidinium concentration. The noise is Lorentzian with low (~kHz) characteristic frequency, implying that the underlying gating process occurs on the millisecond time scale. The absence of the noise in gramicidin M channels and presence in each of the four single Trp-to-Phe mutants (Seoh & Busath 1995) implies that each of the Trp side chains is able to participate in the gating.

TABLE 2 Average fitted characteristic frequency, f_c (Hz), ±S.D. for formamidinium current noise in single gramicidin A channels

	Concentration of formamidinium		
V_m	*0.1 M*	*0.3 M*	*1.0 M*
50 mV	130[a]	300 ± 230	1000 ± 20
100 mV	270 ± 160	900 ± 300	2900 ± 940
150 mV	900 ± 300	N.O.[b]	5100 ± 2300
200 mV	N.O.[c]	N.O.[c]	N.O.[c]

[a]No estimate of S.D. possible due to high noise in single channel noise power spectra.
[b]Estimate not possible with data set.
[c]Not observed due to low amplitude of noise.

TABLE 3 Average single standard gramicidin A channel current (pA) ±S.E.M.[a]

	Concentration of formamidinium		
V_m	*0.1 M*	*0.3 M*	*1.0 M*
50 mV	0.17 ± 0.06	0.32 ± 0.05	0.56 ± 0.06
100 mV	0.50 ± 0.05	0.97 ± 0.04	1.46 ± 0.03
150 mV	ND	2.09 ± 0.03	3.40 ± 0.02
200 mV	ND	ND	5.57 ± 0.03

[a]RMS of channel current S.E.M. and baseline current S.E.M. for 3–5 standard channels.
ND, not done.

The change in characteristic frequency with bath [formamidinium$^+$] at a given V_m reflects the effect of the rate of formamidinium passages through the channel on gate motions (as opposed to the effect of passage speed on gate motions). As seen in Fig. 5, f_c rises with bath [formamidinium$^+$], which implies that the gate oscillation increases with increased formamidinium passage rate. However, the proportionality constant is low ($\sim 3.4 \times 10^{-4}$) because the gate oscillation frequency at 100 mV in 1 M solution is 2900 Hz, whereas the mean formamidinium passage rate is 8.5×10^6/s. The rise in characteristic frequency with V_m suggests that the speed of passage of formamidinium ions through the channel also affects the oscillation frequency directly.

The rise in power between $V_m = 50$ mV and $V_m = 100$ mV shown in Fig. 4 could be due to increased current modulations resulting from the gating process. This is expected if the fraction of open channel current block remains approximately constant with V_m because the open channel current increases with membrane potential. Alternatively, the rise may reflect a change in duty cycle ($\alpha/(\alpha+\beta)$, see Equation 5) towards the value of 0.5, at which the $S_i(0)$ reaches a maximum. The decline in $S_i(0)$ at higher membrane potentials indicates either that the percentage of block is reduced or that the duty cycle is approaching an extreme value of either 0 or 1 at high membrane potentials, presumably 1 because a duty cycle of 0 would correspond to a constantly closed channel whereas the I–V relationship is superlinear (Seoh & Busath 1995, Table 3). Temporal resolution of the blocking process is necessary to distinguish between these possibilities. This proved difficult with the current data set for two reasons. If the bath [formamidinium$^+$] is low enough for the noise to be resolvable, the single channel current is too low for temporal resolution. If the [formamidinium$^+$] is sufficiently high, then the gating frequency becomes too high to resolve. Attempts were made to observe a bias in the channel current sample histograms, but they did not reliably differ from the Gaussian distribution (data not shown). We do not deem it worthwhile to fit the data with an assumption about the completeness of block, the duty cycle and the open channel current because the spectral properties depend sensitively on those three properties and none can be ascertained with any degree of confidence from the current data.

The values of $S_i(0)$ reported here were compared to the values of S′ reported by Seoh & Busath (1995) for gramicidin A currents in 1 M formamidinium and found to be consistent (data not shown), confirming that the normalization procedure used here is correct and the experimental result is reproducible.

The noise was noted by Seoh & Busath (1995) to be linked to increased channel lifetime and I–V superlinearity. The formamidinium I–V relationship is extremely superlinear in gramicidin A and the K^+ I–V relationship is linear, whereas in gramicidin M^- both ions yield moderately superlinear I–V relationships. The channel lifetimes are extremely long in formamidinium solution and decrease with V_m, whereas in potassium solution they start low and increase with V_m, reaching the same level as in formamidinium at 200 mV. In gramicidin M^- the lifetimes are the same for the two ions and do not change with V_m. The gramicidin A channel noise is low and constant for potassium, high in formamidinium but drops to the potassium level by 200 mV. The noise is equally low for the two ions in gramicidin M^- channels.

This set of observations was summarized by a model in which the Trp side chains in gramicidin A mediate all three effects, being stimulated to oscillate by the large formamidinium ions passing through the channel. The ion dimensions are larger than the channel diameter (Hemsley & Busath 1991), although flexibility in the channel (Busath et al 1988, Turano et al 1992) renders the transport free energy profile passable according to molecular dynamics computations (Hao et al 1997). The oscillating Trp dipoles could serve as gates that modulate the cation transport inside the channel, could vary in their interactions with the interface to stabilize the channel and increase its lifetime, and the noise could reduce mean channel current at low membrane potentials, creating the appearance of superlinearity in the I–V relationship. The Trp oscillation coupling to formamidinium passages could be eliminated at high membrane potentials as the formamidinium velocity increases, eliminating the effects (Seoh & Busath 1995). This theory has the advantage of explaining all three phenomena simultaneously. It would predict that the frequency of oscillations would increase with increasing formamidinium velocity consistent with an increase in f_c with V_m, but the amplitude would decrease, consistent with our finding that $S_i(0)$ drops and f_c rises as the membrane potential increases.

However, the f_c is remarkably low. As mentioned above, the low f_c implies that there is only one gating event for every 3×10^3 ion passages. It is easily shown that the gating events could not represent chance long formamidinium dwell times sampled from the passage time distribution expected for a constant channel structure. In such a case, the distribution would be exponential with a rate constant equal to the ion passage rate, 18×10^6/s for 1 M formamidinium, $V_m = 150$ mV. The fraction of ions having dwell times long enough to cause blocks, i.e. greater than $1/2\pi f_c$ (assuming that $\alpha >> \beta$), or 31 μs for $f_c = 5100$ Hz, would be $\exp(-18\cdot10^6/\mathrm{s}\times31\cdot10^{-6}\,\mathrm{s})$ or 10^{-242}. Yet, blocks occur every 10^{-3} passages. Therefore, the constant channel model predicts a frequency of random prolonged dwell times which is far too low to be the source of the noise. Furthermore, removal of the Trp side chains eliminates the noise.

We therefore conclude that the noise must result either from modulation of current flow by slow or sparse fluctuations in the Trp side chains which temporarily render the channel impermeable to formamidinium. The blocks would be incomplete if due only to the electric field of the Trp side chain dipoles, but could be complete if they are due to conformational changes in the backbone due to external stresses mediated by the Trp side chains. Under the first hypothesis, blocks would be incomplete akin to the reduction in conductance seen when Trp is replaced by Phe (Becker et al 1991), due to a change in ion side chain interaction (Hu et al 1995, Dorigo et al 1999). The second mechanism would probably produce complete blocks akin to those observed with guanidinium (Hemsley & Busath 1991). We cannot distinguish between these mechanisms with the data presented here, but whatever the mechanism is, the gramicidin M^- and single Trp replacement studies of Seoh & Busath (1995) require that the mechanism involve Trp side chains and rule out the possibility of it depending on any individual Trp.

The low f_c indicates that the formamidinium noise process has a time-scale similar to guanidinium blocks which, at $V_m = 100$ mV, have mean durations of 2 ms (Hemsley & Busath 1991). These authors also reported that gramicidin A channels did not exhibit discrete blocks in 1 M formamidinium, 1 M K^+ solution. When Seoh & Busath (1995) studied gramicidin A currents in pure 1 M formamidinium solution, the currents were obviously noisy, but discrete blocks were not apparent. The reason for this apparent discrepancy may be in part that the characteristic frequency of the noise determined here for 1 M formamidinium at 100 mV, 2900 Hz, is too high to expect to resolve discrete blocks with a high pass filter of 3 kHz or less. The current measurements indicate that the noise process is about one order of magnitude faster than that produced by guanidinium. We do not expect formamidinium noise to be due to intrachannel blockage as is thought to be the case for the other imidium ions (Hemsley & Busath 1991), but we cannot yet rule out the possibility that formamidinium blocks are complete.

The motions responsible for the gating phenomenon, presumably related to Trp motions, are on the millisecond time-scale. Aromatic side chain conformational changes can have a broad range of rate constants. For instance, rate constants for four different Tyr ring flips in bovine pancreatic trypsin inhibitor are calculated to range over 17 orders of magnitude, reflecting differences in freedom of the Tyr due to different packing of the surrounding protein (Brooks et al 1988). Although gramicidin Trp conformations are stable on the NMR (microsecond) time-scale, it is possible that distortions in the channel backbone could induce conformational changes on the millisecond time-scale that have not been apparent heretofore. This would be quite remarkable given the loose packing of the lipid surroundings, which is quite flexible (Bernèche et al 1998).

We have shown that the supposed motions of Trp side chains responsible for current noise, channel stabilization and superlinearity have a low characteristic frequency, which raises difficult questions both about the mechanism of coupling between ion passages and Trp movements and about the nature of the movements themselves. The characteristic frequency rises with bath concentration and membrane potential, as expected for coupling between ion passages and Trp movements, but the ratio of ion passage rate to characteristic gating rate almost seems too high for coupling. The characteristic noise frequency is so low that it suggests a highly activated process, such as normally prohibited side chain conformational changes or disruptions of the backbone hydrogen bonding. These results beg the examination of the temporal course of formamidinium-induced noise. Although the amplitude of the current carried by formamidinium alone is too low to resolve the noise under conditions where the characteristic frequency is within the range to be temporally resolvable, it should be feasible to observe the noise process under conditions where other ions such as K^+, Cs^+ or H^+ are the dominant current carriers and formamidinium functions as an allosteric regulator of the gating. Such experiments should be informative for gating and activation functions in various membrane proteins and enzymes.

Acknowledgements

We are grateful to Fred Sigworth for providing his method of single channel noise power spectrum normalization and to Albert Franco for performing preliminary experiments. This project was supported by National Institutes of Health grant R01 AI23007.

References

Andersen OS, Greathouse DV, Providence LL, Becker MD, Koeppe RE II 1998 Importance of tryptophan dipoles for protein function: 5-fluorination of tryptophans in gramicidin A channels. J Amer Chem Soc 120:5142–5146

Arseniev AS, Lomize AL, Barsukov IL, Bystrov VF 1986 Gramicidin A transmembrane ion-channel. Three-dimensional structural rearrangement based on NMR spectroscopy and energy refinement. Biol Membr (USSR) 3:1077–1104

Becker MD, Greathouse DV, Koeppe II RE, Andersen OS 1991 Amino acid sequence modulation of gramicidin channel function: effects of tryptophan-to-phenylalanine substitutions on the single-channel conductance and duration. Biochemistry 30:8830–8839

Bernèche S, Nina M, Roux B 1998 Molecular dynamics simulation of melittin in a dimyristoylphosphatidylcholine bilayer membrane. Biophys J 75:1603–1618

Brooks CL III, Karplus M, Pettitt BM 1988 Proteins: a theoretical perspective of dynamics, structure, and thermodynamics. Wiley, New York (Adv Chem Phys vol 71)

Busath DD, Szabo G 1981 Gramicidin forms multi-state rectifying channels. Nature 294:371–373

Busath DD, Hemsley G, Bridal T, Pear M, Gaffney K, Karplus M 1988 Guanidinium as a probe of the gramicidin channel interior. In: Pullman A, Jortner J, Pullman B (eds) Ion transport through membranes: carriers, channels and pumps. Kluwer Academic, Dordrecht, The Netherlands, p 187–201

Busath DD, Thulin CD, Hendershot RW et al 1998 Non-contact dipole effects on channel permeation. I. Experiments with (5F-indole)Trp^{13} gramicidin A channels. Biophys J 75:2830–2844

Daumas P, Heitz F, Ranjalahy-Rasoloarijao L, Lazaro R 1989 Gramicidin A analogs: influence of the substitution of the tryptophans by naphthylalanines. Biochimie 71:77–81

DeFelice LJ 1981 Introduction to membrane noise. Plenum Press, New York

Dorigo AE, Anderson DG, Busath DD 1999 Non-contact dipole effects on channel permeation. II. Trp conformations and dipole potentials in gramicidin A. Biophys J 76:1897–1908

Fairbanks T, Busath DD 1997 Lorentzian formamidinium current noise in gramicidin A channels. Biophys J 72:A395

Hao Y, Pear M, Busath DD 1997 Molecular dynamics study of free energy profiles for organic cations in gramicidin A channels. Biophys J 73:1699–1716

Heinemann SH, Sigworth FJ 1988 Open channel noise. IV. Estimation of rapid kinetics of formamide block in gramicidin A channels. Biophys J 54:757–764

Hemsley G, Busath DD 1991 Small iminium ions block gramicidin channels in lipid bilayers. Biophys J 59:901–907

Hu W, Cross TA 1995 Tryptophan hydrogen bonding and electric dipole moments: functional roles in the gramicidin channel and implications for membrane proteins. Biochemistry 34:14147–14155

Hu W, Lazo ND, Cross TA 1995 Tryptophan dynamics and structural refinement in a lipid bilayer environment: solid state NMR of the gramicidin channel. Biochemistry 34:14138–14146

Ketchem RR, Roux B, Cross TA 1997 High-resolution polypeptide structure in a lamellar phase lipid environment from solid state NMR derived orientational constraints. Structure 5:1655–1669

Maruyama T, Takeuchi H 1997 Water accessibility to the tryptophan indole N-H sites of gramicidin A transmembrane channel: detection of positional shifts of tryptophans 11 and 13 along the channel axis upon cation binding. Biochemistry 36:10993–11001

Sancho M, Martínez G 1991 Electrostatic modeling of dipole-ion interactions in gramicidin like channels. Biophys J 60:81–88

Seoh S, Busath DD 1995 Gramicidin tryptophans mediate formamidinium-induced channel stabilization. Biophys J 68:2271–2279

Sigworth FJ, Shenkel S 1988 Rapid gating events and current fluctuations in gramicidin A channels. Curr Top Membr Transp 33:113–130

Sigworth FJ, Urry DW, Prasad KU 1987 Open channel noise. III. High-resolution recordings show rapid current fluctuations in gramicidin A and four chemical analogues. Biophys J 52:1055–1064

Takeuchi H, Nemoto Y, Harada I 1990 Environments and conformations of tryptophan side chains of gramicidin A in phospholipid bilayers studied by Raman spectroscopy. Biochemistry 29:1572–1579

Turano B, Pear M, Busath D 1992 Gramicidin channel selectivity. Molecular mechanics calculations for formamidinium, guanidinium, and acetamidinium. Biophys J 63:152–161

DISCUSSION

Hladky: Have you looked not just at the power spectral density, but also at the actual distribution of the measured conductances? I'm thinking of some of the work that Heinemann & Sigworth (1991) have done on flickers that are too fast to be fully resolved. They looked at the actual shapes of the amplitude distribution curves, and from the bulges on those they were able to obtain information on flickering (see also Yellen 1984). This approach looks ideally suited to your situation.

Busath: We did try that approach, and we didn't see anything other than a Gaussian distribution with our data set. If we had better data we might be able to resolve something. Recently, however, Nathan Bingham, a student in my lab, has measured formamidinium-induced flickers in K^+ currents, so we now know that blocks are discrete, short and complete.

Eisenberg: Which amplifier did you use?

Busath: We used a List Electronic EP7 (Darmstadt/Eberstadt, Germany).

Koeppe: You said that when you applied 200 mV to gramicidin A the formamidinium block disappears, but what happens to the channel lifetime? Is the blocking correlated with the lifetime?

Busath: Yes. The lifetime decreases to a normal level, i.e. the same level as in potassium solution.

Jakobsson: The interaction between Trp residues and the interfacial region of membranes is important for membrane proteins in general. You have an entire class of engineered channels where you have replaced Trp residues with other residues, but do the effects of those substitutions differ in different membranes?

Busath: Fluorination of gramicidin Trp residues doesn't seem to affect lifetimes but we noticed that there were some differences in fluorination effects on conductance. In diphytanoyl phosphatidylcholine (DPhPC) fluorination enhances conductance, but in monoolein it doesn't.

Cross: One of the problems is that we don't know the structure of gramicidin M. The NMR structural studies suggest that gramicidin M is double stranded. Therefore, you may be looking at a non-equilibrium state.

Koeppe: The conducting form of gramicidin M has a single-stranded backbone, and it forms a hybrid channel with gramicidin A. Abad's group has added monomeric, single-stranded gramicidin M to a bilayer and has taken a circular dichroism spectrum as soon as they could before it refolded, and they obtained a spectrum similar to what we attribute to the gramicidin A channel (Salom et al 1998). If they recorded a spectrum of the sample later, they obtained the double-stranded conformation. The conducting events are due to a similar structure to gramicidin A.

Busath: We have to use a 100-fold higher concentration of gramicidin M, and I presume that this is because a low fraction of gramicidin M is in a single helix conformation.

Koeppe: At equilibrium at least 95% is the inert double-stranded conformation.

Busath: This implies that double-stranded channels are simply not visible to us.

Cross: Are there any subtle differences between the gramicidin M-conducting state and the gramicidin A-conducting state? In other words, are there changes within the backbone, rather than the tryptophans, that distinguishes gramicidin A from gramicidin M conductance?

Busath: The channel lifetimes are similar. There are no formamidinium-induced flicker blocks, but there are substate channels in gramicidin M.

Koeppe: The conductance of the gramicidin A/gramicidin M hybrid channel is intermediate between the two parent channels. And there is no energetic cost in forming a hybrid channel between gramicidin A and gramicidin M, as opposed to forming a gramicidin A homodimer (Koeppe et al 1992).

Busath: The answer seems to be that there's nothing that is clearly different.

Koeppe: On the other hand, when the blocking increases the lifetime, I suspect that the six hydrogen bonds which hold the dimer together are affected. There may be a long-range effect of formamidinium, possibly propagated along the backbone.

Busath: Rather than tryptophan interactions with lipids?

Koeppe: The effects of tryptophan interactions with lipids could also be propagated along the backbone.

Woolley: You showed us a series of single Trp mutants that were still noisy, but have you looked at any multiple Trp mutants?

Busath: No. I haven't had a chance to do those experiments.

Arseniev: What happens when the channel closes? NMR spectroscopy in micelles indicates that the weakest point of the channel is six intermolecular hydrogen bonds. Therefore, it is possible that those hydrogen bonds are partially broken.

Busath: The formamidinium may be able to drag the formyl group into the channel behind it, breaking the six hydrogen bonds and temporarily blocking the channel.

Ring: Why are the hydrogen bonds between the heads of the helix more sensitive than any other hydrogen bonds between loops of the channel?

Arseniev: When the temperature is increased, the hydrogen bonds between two monomers break first, in spite of the fact that they are less accessible to the bulk solvent.

Ring: I'm not sure that we know enough about the nature of the correlation between occupancy and lifetime, because there may be long-range effects of channel occupancy on the centre. There is a correlation between ion occupancy and lifetime, because increasing concentrations of potassium and protons increase the lifetime. However, it would be interesting to measure the different long-range effects of potassium and caesium on hydrogen bonding and monomer dimerization.

Koeppe: If you replace Trp9 or Trp11, sometimes you increase the lifetime and sometimes you decrease the lifetime (Becker et al 1991). However, the question is, with occupancy or with tryptophan, is the lifetime a local effect, i.e. exactly where the ion is sitting or where the tryptophan is sitting, or is it propagated to the centre towards the six hydrogen bonds? I originally thought that these were all local effects, but now I've changed my view and I think ultimately the six hydrogen bonds at the dimer junction are affected.

Ring: The voltage dependence is also important, so maybe it's only when you increase the voltage that you observe the differences.

Roux: There are two ways to accelerate any molecular process generated by a rate. You can decrease the barrier height or increase the well depth. Since we don't know what the dimer is doing when it does not conduct ions, we have no idea about the transition state either. Therefore, we don't know whether the presence of protons, caesium, potassium or even the tryptophans stabilizes or destabilizes the well, or whether their presence changes the unknown transition state for channel formation.

Jakobsson: I have a highly speculative theory based on several different calculations, each distantly related to what you're talking about. Another way to close the channel other than by causing a conformational change is to freeze the channel. We have done some computer experiments in which we kept the lumen of the channel at 300 °K and froze the protein (Chiu et al 1991, 1992). This effectively immobilized the channel, and nothing could go through. We also restricted the motion of the gramicidin side chains, with harmonic constraints to their original positions, which we found was as effective as restricting the motion of the entire channel. Therefore, I propose that an explanation for your periodic blocks is that every once in a while the Trp residues fall into a position relative to their surroundings and become stuck. While they're stuck in that position the current is blocked. This is a rare event, and we don't see it in molecular dynamics because it's outside the time-scales. However, what we do know from molecular dynamics is that if the side chains are artificially frozen, one observes what looks like a channel block, even though the mean confirmation is wide open.

Busath: It has to be a tryptophan position or conformational change that results in a structural change in the backbone, which then makes it more difficult for a cation to get through.

Koeppe: It may not be necessary to invoke tryptophan motions to explain these results. All that may be necessary is a tighter binding of formamidinium to gramicidin A, so that the permeation kinetics change, the residence time is then longer and that, in turn, results in a longer channel lifetime. In the case of gramicidin M, the binding is not so tight, so you don't see the noise. For whatever reason, gramicidin A needs to have all the tryptophans present at the local binding site, but that doesn't mean that they have to be dynamic. The blocking site is related to the noise, but it is dependent on the tighter binding of formamidinium.

Jakobsson: But tighter former formamidinium binding would result in a larger formamidinium current in one case than the other. David Busath said that the magnitude of the currents is the same, and it's just the noise that changes.

Busath: Yes, but formamidinium-induced noise must be a spasmodic change. A spasmodic increased affinity causes temporary binding and a temporary block. The passage rate is on average a million per second, and we're seeing one of these blocks every millisecond or so.

Ring: On a related topic, have you thought about the flickers that occur in the absence of a blocking agent?

Busath: Not really. Sigworth & Shenkel (1988) discussed the possibility that the step-up observed when the channel is turned on and the resultant long flickers are due to the dimer junction becoming partially disconnected temporarily or there being five hydrogen bonds instead of six. They observe three types of flickers: the long flickers; intermediate level flickers; and short flickers, which they postulate are due to an empty state of the channel causing the elimination of channel conductance.

Ring: We don't have an explanation for those normally appearing flickers, so the question is, what is the origin of these spontaneous flickers that are seen also in the absence of blocking mechanisms? There could be cross-talk between these two mechanisms.

Busath: That's right, and it would be interesting to see if gramicidin M has those flickers. We have looked for mini-channels in gramicidin M, but we haven't yet looked carefully at flickers.

Eisenberg: The origin of open channel noise is not understood, because no one knows how to calculate the shot noise in a self-consistent way. Until someone does a Brownian motion calculation with electric charge, we can't be sure what the shot noise component is. It may be irrelevant.

Busath: Heinemann & Sigworth (1990) have already modelled it successfully and figured it out with kinetic theory, so are you saying that this is too simplified? They looked at a small amount of noise, but they did it extremely elegantly.

Eisenberg: There are two different issues here. The first is the shot noise present when barriers are not assumed to be large. We published a paper in which we did that model without the assumption that there were large barriers (Barkai et al 1996). It turned out that by using large barriers other workers had left out a group of important terms. The large barriers imply a certain lack of memory, which is extremely unlikely. The point is rather technical but interesting and here is not the place to describe it in detail. What is important is that the high barrier assumption implies things that we did not (as a matter of historical fact) realize. The second issue is that all published models of noise in channels that we know of (including our own) assume a potential profile and do not compute it self-consistently from the charges present. This is potentially a big problem since all the charges present move on the time-scale of open channel noise and their movement can produce open channel noise, at least in principle. What needs to be done is a combination of Brownian motion and a calculation of the potential profile. It is known how to do this, and we are trying to do it, but it hasn't been done yet. Until this is done we don't know what the shot noise should look like, and therefore we can't make any conclusions.

Hladky: We (Wang & Hladky 1994) have also found that there is a long component of the flicker distribution (Sigworth & Shenkel 1988) consisting of partial closures of the channel lasting around 1 ms. These are fully resolved events. They have as much right to be called a channel state as the fully open state. As David Busath noted, Sigworth and Shenkel discuss a much more complicated distribution than that, and the faster ones they report are where I would look for precisely the sort of thing Robert Eisenberg is talking about.

Eisenberg: If you open up the bandwidth, you observe something extremely interesting happening around the shot noise time-scale, i.e. 1–10 μs. In our opinion, it is not possible to interpret this without knowing what the simple rigid structure would do. I suspect that everything that is going on is displaced to the centre and is conducted by the electric field, but we have to prove that.

References

Barkai E, Eisenberg RS, Schuss Z 1996 A bidirectional shot noise in a singly occupied channel. Physiol Rev 54:1161–1175

Becker MD, Greathouse DV, Koeppe RE II, Andersen OS 1991 Amino acid sequence modulation of gramicidin channel function: effects of tryptophan-to-phenylalanine substitutions on the single-channel conductance and duration. Biochemistry 30:8830–8839

Chiu S-W, Jakobsson E, Subramaniam S, McCammon JA 1991 Time-correlation analysis of simulated water motion in flexible and rigid gramicidin channels. Biophys J 60:273–285

Chiu S-W, Gulukota K, Jakobsson E 1992 Computational approaches to understanding the ion channel-lipid system. In: Pullman A, Jortner J, Pullman B (eds) Membrane proteins: structures, interactions, and models. Kluwer Academic, Dordrecht, The Netherlands, p 315–338

Heinemann SH, Sigworth FJ 1990 Open channel noise. V. Fluctuating barriers to ion entry in gramicidin A channels. Biophys J 57:499–514

Heinemann SH, Sigworth FJ 1991 Open channel noise. VI. Analysis of amplitude histograms to determine rapid kinetic parameters. Biophys J 60:577–587

Koeppe RE II, Providence LL, Greathouse DV et al 1992 On the helix sense of gramicidin A single channels. Proteins 12:49–62

Salom D, Pérez-payá E, Pascal J, Abad C 1998 Environment- and sequence-dependent modulation of the double-stranded to single-stranded conformational transition of gramicidin A in membranes. Biochemistry 37:14279–14291

Sigworth F, Shenkel S 1988 Rapid gating events and current fluctuations in gramicidin A channels. Curr Top Membr Transp 33:113–130

Wang KW, Hladky SB 1994 Absence of effects of low-frequency, low-amplitude magnetic fields on the properties of gramicidin A channels. Biophys J 67:1473–1483

Yellen G 1984 Ionic permeation and blockade in Ca^{2+}-activated K^+ channels of bovine chromaffin cells. J Gen Physiol 84:157–186

Can we use rate constants and state models to describe ion transport through gramicidin channels?

S. B. Hladky

Department of Pharmacology, University of Cambridge, Tennis Court Road, Cambridge CB2 1QJ, UK

Abstract. Can we use rate constants and state models to describe ion transport through gramicidin channels? Maybe, but only if rate constants are just proportionality constants between rates and probabilities of observing states of the channel. This approach is natural if the system of channel plus ions (plus water) is almost always in one or another of a small number of identifiable states. Many features of ion transport through gramicidin, including the conductance–concentration relationship, concentration-dependent permeability ratios, anomalous mole fraction effect and to some extent flux ratio exponents, are consistent with a description in which there are four occupation 'states' of the pore: only water; an ion at one end; an ion at the other; and ions at both ends. Current–voltage relationships can (and must) also be fitted, but until there is a theory to predict the potential dependence of the rate constants this success will remain hollow. Other features have resisted interpretation. These include the failures to determine 'binding constants' consistent with all the data; the variation of flux ratio exponents with ion type; and, probably, the variation of the currents with asymmetrical ion concentrations. Nevertheless, state models still have one attractive feature, they allow consideration of the effects that one ion within the pore has on the movements of another.

1999 Gramicidin and related ion channel-forming peptides. Wiley, Chichester (Novartis Foundation Symposium 225) p 93–112

Introduction: low concentrations

A full and precise description of ion transport through a narrow pore still eludes us. The methods of molecular dynamics hold out the promise of a solution in principle, but it is likely to be some time yet before anyone can find a way to make the calculations fast enough to model the hundreds of nanoseconds of real time needed to describe ion transport all the way through a pore. Even when a solution can be obtained, most of us without ready access to supercomputers will still have to make do with approximations for the interpretation of experimental data. These approximate treatments may or may not include the use of rate constants. By rate constant I shall mean a constant of proportionality between a component of a flux and either a concentration or the probability of a state. This is the sense in which rate constants are used in chemical kinetics. For simplicity, in all that follows it is assumed that the transport can be described as movement along a single direction or reaction coordinate.

For an ion to move through a distance of some 2 nm it must undergo a large number of small movements involving many collisions with its neighbours. In the limit of low concentrations this can be described as a sequence of many small jumps (see e.g. Läuger 1973, McGill & Schumaker 1996). Alternatively, it can be written down succinctly in terms of a diffusion constant, $D(x)$, the cross-sectional area of the diffusion path, $A(x)$, and the gradient of the potential energy, $U(x)$, (more precisely a free energy) of the ion as a function of position, x. The diffusion constant describes the local interactions governing how easily the ion can move, while the potential energy gradient describes the longer range influences on the movements, including the consequences of the applied potential difference across the membrane. These are related to the flux (ions per second) and the current by the Nernst–Planck equation:

$$I = zeJ = -zeDA\left[\frac{dc}{dx} + \frac{c}{kT}\frac{dU}{dx}\right] \tag{1}$$

where c is the concentration of the ions. (On a microscopic scale the concentration is defined as the ratio of the probability of finding an ion within a small volume element to the volume of the element.) The Nernst–Planck equation can be integrated all the way from the bulk solution on the left to the bulk solution on the right to yield:

$$I = ze\frac{c_{left} - c_{right}e^{\Delta U/kT}}{\int_{left}^{right}(e^{U/kT}/DA)dx} \tag{2}$$

where the potential energy per ion on the left is set to zero, and the value on the right is ΔU. The current could be predicted if we knew the diffusion constant, the cross-sectional area and the potential energy at all positions through the pore. The basis for these equations and their application to the special case of ion transport through pores have been discussed and reviewed extensively elsewhere (Cooper et al 1985, Levitt 1986, Dani & Levitt 1990, Chen & Eisenberg 1993, Eisenberg 1996, McGill & Schumaker 1996).

In the limit of low ion concentrations it is perfectly permissible to write the transport equation in terms of just two rate constants that are themselves independent of concentration:

$$I = ze(k_{left}c_{left} - k_{right}c_{right}) \tag{3}$$

where

$$k_{left} = \frac{1}{\int_{left}^{right} (e^{U/kT}/DA)dx} \tag{4a}$$

and

$$k_{right} = k_{left}e^{(U_{right}-U_{left})/kT}. \tag{4b}$$

More usefully, a larger number of similar rate constants can be employed as a shorthand whenever it is convenient to consider the transport as occurring in stages or components, e.g. transport across access zones and transport across the pore or membrane proper,

$$\underset{\text{access}}{c_{l0} \underset{k_{l1r}}{\overset{k_{l0f}}{\rightleftharpoons}}} \underset{\text{pore}}{c_{l1} \underset{k_{r1f}}{\overset{k_{l1f}}{\rightleftharpoons}}} \underset{\text{access}}{c_{r1} \underset{k_{r0f}}{\overset{k_{r1r}}{\rightleftharpoons}}} . \tag{5}$$

In the steady state the flux through each stage must be the same, and

$$I = ze(k_{l0f}c_{10} - k_{lr}c_{l1}) = ze(k_{l1f}c_{l1} - k_{rlf}c_{r1}) = ze(k_{r1r}c_{r1} - k_{r0f}c_{r0}) \tag{6}$$

where for instance

$$k_{l1f} = \frac{e^{U_{l1}/kT}}{\int_{l1}^{r1} -e^{\frac{U/kt}{DA}}dx} \tag{7a}$$

and

$$k_{r1f} = k_{l1f}e^{(U_{r1}-U_{l1})/kT}. \tag{7b}$$

Lipid-soluble ions

A clear example of the utility of breaking the transport process into components is provided by the transport of lipid-soluble ions such as dipicrylamine and tetraphenylboron across lipid bilayers (Ketterer et al 1971, Andersen & Fuchs 1975, Hladky 1979, 1992). These ions adsorb strongly to the membrane surfaces. When a potential difference is applied across the membrane the adsorbed ions rapidly redistribute between the surfaces producing a large initial current that decays exponentially with a time constant

$$\tau = 1/(k_{l1f} + k_{r1f}) \tag{8}$$

until it approaches a current, orders of magnitude smaller, supported by movement of the ions through the aqueous phases. The rate constants provide a tidy notation that draws attention to the distinction between the various steps. Consistent sets of rate constants have been derived for the lipid-soluble ions and for the ion carriers trinactin and valinomycin (Benz & Stark 1975, Hladky 1975, 1979, 1992, Knoll & Stark 1975, Hladky et al 1995), but only after it was fully appreciated that the potential dependence of the individual rate constants must be determined from the data, perhaps using simple adjustable energy barrier shapes (Hall et al 1973, Hladky 1974). Use of the assumption that the rate constants vary exponentially with the applied potential seriously compromised early attempts to draw conclusions from the data (see Hladky 1979, 1992).

Studies of the lipid-soluble ions also provide a cautionary tale about the use of smeared charge distributions. For low concentrations of the ions the initial conductance is proportional to the concentration. But for higher concentrations it goes through a maximum and declines. There has been an evolution of theories offered to explain this competition between ions involving: space charge limitation; Gouy–Chapman surface potentials; a finite number of surface-binding sites; the three capacitor model, in which the charge of the adsorbed ions on each side is assumed to be smeared out over an adsorption plane; a hexagonal lattice model; and the mobile discrete charge model. As discussed elsewhere (Andersen et al 1978, Hladky 1979, Tsien & Hladky 1982) while the three capacitor model was a distinct improvement on its predecessors, the smeared charge assumption substantially overstated the repulsion between ions adsorbed to one surface of the membrane. Only by considering the adsorbed ions to be mobile and discrete was it possible to fit the data.

Gramicidin

With lipid-soluble ions and carriers the rate constants for the component processes could be assessed separately from each other by relaxation experiments. For pores, the relaxations are too fast to measure. Thus, less direct and less certain arguments must be used to justify any particular division of the overall transport process into components. When I first obtained data for gramicidin, I rather naïvely hoped that it would be possible to divide the process into potential independent components in the access regions and a component that varied exponentially with potential for transfer through the pore proper. I recorded in my PhD thesis (Hladky 1972) the failure of this approach, but gave it wider circulation only much later when it had become obvious that others were using the exponential approximation in circumstances where it was not appropriate. For gramicidin it is likely that even the access process, which occurs in a restricted region at the mouth of the pore, cannot be described with a rate constant whose potential dependence is a single exponential (Hladky 1987, 1988).

In the absence of ion–ion interactions either the functions that enter into the Nernst–Planck equation or rate constants can be used in the conservative sense described above to describe ion transport. Indeed, this statement can be extended to the case where only one ion can enter the transport process at a time—the so called one-ion pore (Läuger 1973, Levitt 1986, Hladky 1988, Dani & Levitt 1990, McGill & Schumaker 1996). The situation becomes more complex and much more interesting when more than one ion must be considered at a time.

A number of observations suggest that ion–ion interactions are an important feature of transport through gramicidin. These include: a maximum in the conductance–concentration curve; concentration-dependent permeability ratios; the anomalous mole fraction effect with symmetrical solutions; and most conclusively a flux ratio exponent, n, greater than 1 (see Finkelstein & Andersen 1981, Hladky & Haydon 1984, Hladky 1988 for reviews). The flux ratio exponent is most conveniently calculated from the ratio of the conductance for symmetrical solutions, g, and a small applied potential to the unidirectional flux of a tracer at equilibrium, $\underset{\rightarrow}{J^{eq}}$

$$n = \left(\frac{kT}{z^2e^2}\right)\frac{g}{\underset{\rightarrow}{J^{eq}}}. \tag{9}$$

The flux ratio exponent is 1 for any model that satisfies either independence or the one-ion pore assumptions (Levitt 1986, Hladky 1988, McGill & Schumaker 1996).

TABLE 1 Flux ratio experiments

	Flux ratio (0.1 M/1 M) of membrane		
Ion	*Ox brain lipids*[a]	*Diphytanol PC*[b]	*Glyceryl monooleate*[c]
Sodium	1.2/–	1.0/1.0	–/1.0
Rubidium	2.0/1.5	–/–	–/–
Caesium	1.7/–	1.4/1.6	1.2/1.4

[a]Schagina et al (1980).
[b]Procopio & Andersen (1979).
[c]S. B. Hladky, M. Jones, S. Moule & D. Game, unpublished data (1990, 1991, 1993).
PC, phosphatidylcholine.

By contrast, as shown in Table 1, at least for caesium ions there is agreement that values of this flux ratio exponent for gramicidin are greater than 1.

The challenge is to describe transport with ion–ion interactions in a compact form that is sufficiently convenient for routine use. It is also highly desirable that the description reveals as much as possible of the underlying mechanisms and provide a bridge to the more detailed descriptions which it is hoped will result from attempts to predict the function from structure and first principles.

The Nernst–Planck, Poisson, smeared charge approximation

If more than one ion of the same sign attempts to occupy a small space, there will be electrostatic repulsion between them. It is thus natural to enquire whether such repulsion alone can account for the observed ion interactions, and it is worthwhile at least in the spirit of finding a compact description to see if this repulsion can be modelled using classical continuum electrostatics and the Nernst–Planck equation. To do so it is assumed that the interesting features of the variation in the potential energy, U, with position result from variations in the electric potential, V, which in turn varies for three reasons: the applied potential; the distribution of fixed charge on the structure of the pore (which includes charge as part of dipoles etc.); and the presence of the mobile ions. The variations in the potential are calculated from the charge distribution using the Poisson equation, which for present purposes can be written as

$$\nabla^2 U = ze\nabla^2 V = \frac{ze\rho}{\varepsilon\varepsilon_0} \tag{10}$$

where ε is the dielectric constant of the medium (here assumed constant), ε_0 is the permitivity of free space and ρ is the charge density smeared out into a continuum. This equation must be solved over the volume of interest allowing for boundary conditions which will include much of the effects of the charges that are part of the channel and membrane structures. The smeared charge density within the volume of interest is calculated from

$$\rho = \rho_0 + \sum_{\text{all ions}} z_i e c_i \tag{11}$$

where ρ_0 is the density of fixed charge within the volume. Can these equations be used with the Nernst–Planck equation to describe ion interactions within a narrow pore?

As has been noted repeatedly (Cooper et al 1985, Chen & Eisenberg 1993) the Nernst–Planck equation makes a definite prediction for the flux ratio exponent, $(c_{left} = c_{right} = c)$

$$\underset{\rightarrow}{J^{eq}} = \frac{c}{\int_{left}^{right} e^{U/kt}/DA\,dx}$$

$$\begin{aligned} g &= \underset{\Delta V \to 0}{\mathcal{L}} (I/\Delta V) \\ &= \underset{\Delta V \to 0}{\mathcal{L}} \frac{ze}{\Delta V} \frac{c(1 - e^{\Delta U/kT})}{\int_{left}^{right} e^{U/kT}/DA\,dx} \\ &= \frac{\frac{z^2 e^2}{kT} c}{\int_{left}^{right} e^{U/kT}/DA\,dx} \end{aligned} \tag{12}$$

and

$$n = 1. \tag{13}$$

Thus, at least for gramicidin there is at least one important feature of the ion transport process that is not properly taken into account by the Nernst–Planck equation. One factor not taken into account is that ions cannot pass each other in the pore. Another is that their distance of closest approach will be affected by water molecules trapped between them (Levitt 1982).

Even for conditions in which the flux ratio exponent is close to 1, it is hard to believe that the use of the self-consistent smeared charge density to calculate the potentials is an adequate approximation. In the equations as stated the potential energy, $U(x_0)$, at a point, x_0, is calculated using the Poisson equation and a continuous distribution of charge such that within any small interval dx there is a fractional charge:

$$dq = \rho_f(x)A(x)dx = zec(x)A(x)dx \tag{14}$$

where $c(x)A(x)dx$ is the average probability of finding an ion in the interval dx. This average is calculated in a self-consistent manner with an appropriate fractional charge, $zec(x_0)A(x_0)dx$, in the vicinity of x_0. However, when the movement of an ion at position, x_0, is being considered, the potential energy that matters is the potential energy calculated from the distribution of charge when an ion is located at x_0. (As an extra layer of complication beyond that considered here the 'average' distribution needed may vary with time on the same time-scale as the motion of the ion being considered.) Although the potential function needed is still some sort of average, it is not the same as the average when there isn't an ion at x_0, and thus it isn't the same as the overall average calculated from $c(x)$. Thus, while the calculation is self-consistent (and would be correct if the charge were truly present as infinitely many infinitesimal packets), the interactions for real, discrete ions are different (compare Bek & Jakobsson 1994, Jakobsson 1998).

It is difficult to assess the extent to which the approximations used in the Nernst–Planck, Poisson, smeared charge calculations bias the predictions. These equations and the associated functions of position, $U(x)$ and $D(x)A(x)$, have more than enough parameters, whose values can be chosen or fitted, to allow the theory to describe most net flux data (a feature this approach has in common with state models). However, given that the theory permits behaviour which is impossible (e.g. ions passing through each other, Chen & Eisenberg 1993) and makes predictions which are qualitatively wrong, there must remain a real suspicion that a better theory would describe the data with different parameters. However, at least one real strength of this approximation remains—its formalism is closely related to the methods that will be used to derive fluxes from *de novo* predictions of free energies and diffusion constants.

The formalism has been extended to consider discrete charges explicitly by working with the joint probability of simultaneously finding ions at the various positions in the pore (Levitt 1982, 1986, 1987, 1991, Dani & Levitt 1990). A different approach that is being developed is to use Brownian dynamics to simulate the transport (Cooper et al 1985, Jakobsson 1998). However, both of these alternatives entail a substantial increase in the complexity of the calculations.

State models

State models for the transport process deal with ion–ion interaction in a manner closely related to the use of joint probabilities. The two-ion, four-state model for gramicidin simplifies the description by making the assumption that there are only four probabilities that must be considered: the pore occupied only by water; with an ion at one end; with an ion at the other end; or with ions at both ends. The rates of transitions between these states are described by the rate constants for five component processes in the ion transport mechanism: entry of an ion to a pore occupied only by water, AA; transfer of an ion from one end to the other, k; exit from a pore not occupied by another ion, B; entry to a pore already occupied at the far end, DD; and exit from a doubly ion occupied pore, E. This decomposition can be justified only by its success in fitting data or by the application of better theory. This model was never intended to imply that a gramicidin pore possesses just two principal binding sites for an ion (though in fact there may be; Jing et al 1995, Woolf & Roux 1997). It was intended to imply that at most two ions would be found in the pore simultaneously and that ions would occupy the middle of the pore only transiently.

The two-ion, four-state model has been successful in interpreting a wide range of behaviour for gramicidin including (see Hladky & Haydon 1984, Hladky 1988):

the conductance–concentration relationship (which in principle can specify all the free constants for any single species of ion); current–voltage (I–V) relationships (but this is hollow as the predictions are adjustable); concentration dependence of permeability ratios; block of water permeability by ions; anomalous mole fraction effects; and the flux ratio exponent. The strength of this approach is that it provides a convenient, easily comprehended description of the transport which explicitly allows for interaction between discrete ions. At the least it provides a vocabulary with which to describe experimental observations. At best it decomposes a complicated process into components that are amenable to more detailed description.

Shortcomings of the four-state model

Like the Nernst–Planck, Poisson, smeared charge approximation, use of the four-state model is based on assumptions, some of which may be sufficiently in error to distort the interpretation of experimental data. An indication of the defects can be gained by considering aspects of the data that have so far resisted interpretation.

(1) At present gramicidin is modelled as a one-ion pore when considering sodium, but as a two-ion pore when considering potassium, rubidium, caesium and thallium. It is a major shortcoming of both the four-state model and other descriptions of gramicidin (e.g. McGill & Schumaker 1996) that they do not suggest any explanations for this qualitative difference. The rate of entry of ions to the transport process appears to be similar for all ions, to be affected by the viscosity of the external medium and to be only weakly dependent on the applied potential, all of which are consistent with a rate-limiting step outside the pore proper (Andersen 1983, Hladky 1984). It is thus reasonable to propose that for all the ions investigated the rates of entry of first and second ions are similar. For caesium the evidence for second ion entries is clear and the flux ratio exponent is definitely greater than 1. However, for sodium the exponent is not detectably different from 1. Furthermore, for a rate constant of second ion entry close to that for first ion entry, the four-state model cannot simultaneously predict an exponent value close to 1, the shape of the conductance–concentration relation and the concentration dependence of the I–V relationships. If it is accepted on this basis that gramicidin is a one-ion pore when the permeant ion is sodium, then some explanation must be found to explain why attempted second ion entries by sodium ions fail to affect the transport process. If, on the other hand, two sodium ions can enter a gramicidin pore, a view supported by the NMR data (Jing et al 1995, Woolf & Roux 1997), the four-state model fails to explain why the flux ratio exponent is 1.

(2) The theory of a one-ion pore provides an unambiguous prediction that the conductance should vary with concentration (the same on both sides) as:

$$g = \frac{g_{\max} Kc}{1 + Kc} \tag{15}$$

where K is the binding constant for the ions to the pore. Fitting this relation to the conductance–concentration curves for sodium and gramicidin yields $3.2\,M^{-1}$ for diphytanoyl phosphatidylcholine membranes (Finkelstein & Andersen 1981) and $2.7\,M^{-1}$ for monoglyceride membranes (Hladky & Haydon 1984, Wang et al 1995). Binding constants for lithium, sodium, potassium, rubidium, caesium and thallium have been estimated from the ability of these ions to block fluxes of water through the channel (see Fig. 1; Dani & Levitt 1981, Wang et al 1995). On this basis the binding constant for sodium is in the range of $12–20\,M^{-1}$. This stronger binding is consistent with the conductance–concentration curves predicted by the two-ion, four-state model if a second ion can enter the channel,

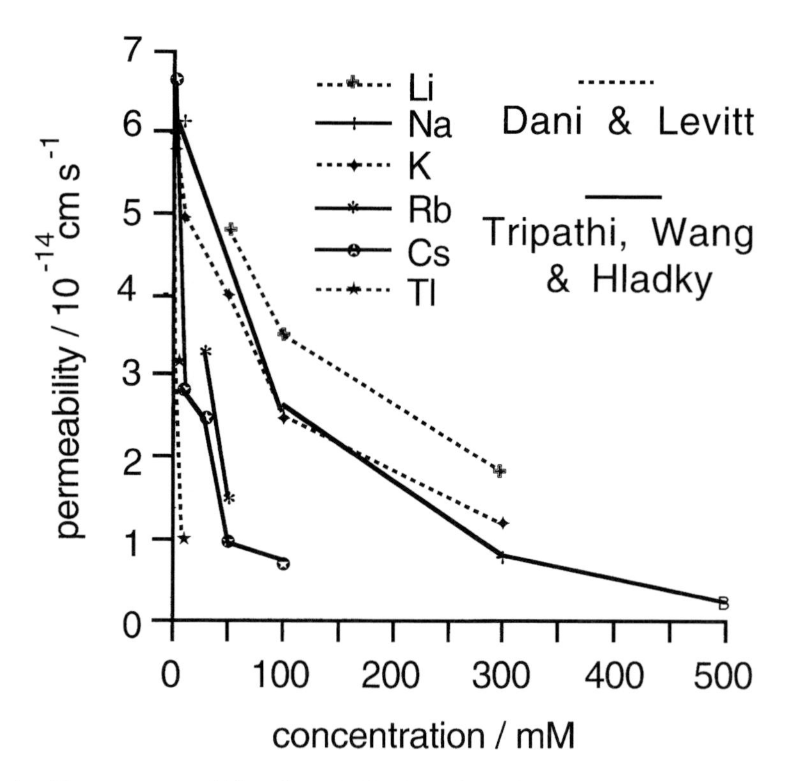

FIG. 1. The water permeability of gramicidin channels as a function of ion concentration. The number of channels was determined from the conductance of the membrane and the single-channel conductance. Water flow was induced by the addition of solute to one side of the membrane. (Drawn from data reported by Dani & Levitt 1981 and Wang et al 1995.)

but it is then not consistent with the flux ratio exponent. How can stronger binding be reconciled with a lack of single filing?

(3) In the form in which it has been used up to now, the four-state model has been based on the assumption that a doubly occupied pore is a blocked pore. That implies that for high concentrations of ions removal of the ions from side 2 of the membrane should allow larger fluxes from side 1 to side 2. Data for 1 molal thallium chloride are reported in Fig. 2. Osmotic flow of water and streaming potentials can seriously affect these measurements. The experiments with glycerol and urea were included as controls to establish that osmotic effects were not important in this instance. Data have also been obtained with sodium, potassium and caesium chloride where the gradients were produced by total replacement with

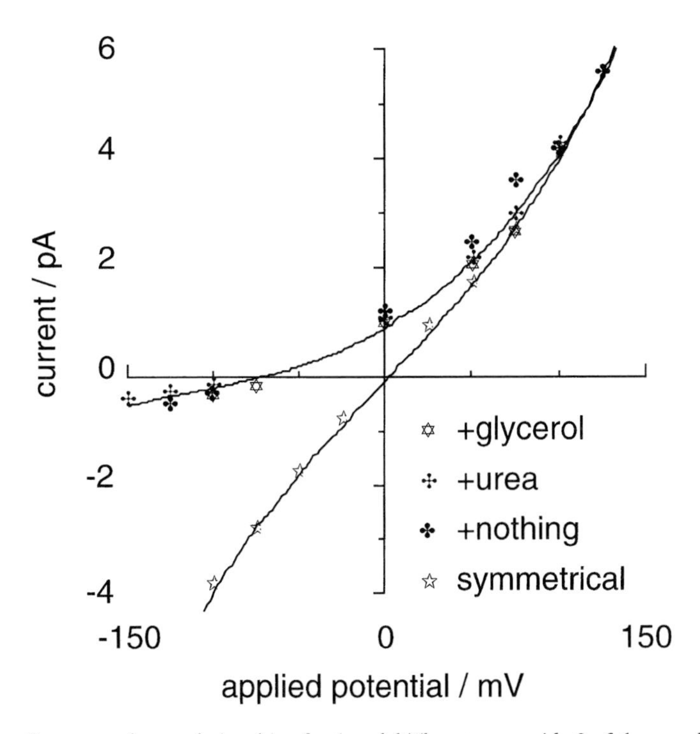

FIG. 2. Current–voltage relationships for 1 molal Tl acetate on side 2 of the membrane. The concentration on side 1 was either 1 molal Tl acetate (symmetrical) or 0.02 molal Tl acetate together with nothing, 2 molal urea or 2 molal glycerol. The electrodes were AgAgCl wires connected to the solutions via two-stage salt bridges with 3 M KCl and 2 M Na acetate to prevent precipitation of TlCl. The expected reversal potential for the current through the pore is 84 mV, consistent with the observed reversals and a difference in electrode tip potentials (calculated using the Henderson equation) of about 12 mV primarily at the 1 molal Tl acetate: 2 M Na acetate junction. Unpublished data of K.-W. Wang & S. B. Hladky (1993).

choline chloride on side 2. The interpretation of these results is not straightforward as a result of electrode tip potentials (3 M KCl salt bridges) and weak block of the pore by choline. However, for all the ions investigated at large positive potentials (side 1 minus side 2) there was no demonstrable increase in current (<5%) when the permeant ions were removed from side 2. This is not surprising since for these potentials the pore would normally be occupied by ions from side 1.

An asymmetrical current index at low potentials can be calculated from g_0, the limiting conductance at zero potential with symmetrical solutions, and I_0, the flux at zero potential when the permeant ions are removed from one side,

$$r = \frac{ze}{kT}\frac{I_0}{g_0}.$$

The four-state model predicts:

$$r = \frac{ze}{kT}\frac{I_0}{g_o} = \frac{\left(B + \frac{D}{2}\right)\left(1 + \frac{2A}{B} + \frac{AD}{BE}\right)\left(1 + \frac{2k}{D+B}\right)}{\left(B + \frac{D}{2}\right)\left(1 + \frac{A}{B}\right) + k\left(2 + \frac{2A}{B} + \frac{AD}{2BE}\right)} \tag{16}$$

where $A = AA \times c$ and $D = DD \times c$. For low concentrations $r = 1$. For a one-ion pore, $DD = 0$,

$$r = \frac{1 + \frac{2A}{B}}{1 + \frac{A}{B}} \tag{17}$$

which varies from 1 to 2. For a two-ion pore second ion entries can reduce r towards 1 for low concentrations, but increase it for high concentrations,

$$r \to \frac{D}{E + k}. \tag{18}$$

The values of r inferred for sodium, potassium and caesium are given in Table 2. These are subject to the caveats noted above. The data for sodium are consistent with transport through a one-ion pore with K between 4 and10 M^{-1}. For thallium, rate constants fitted to the conductances (fit G-a, Hladky & Haydon 1984) predict $r = 2.25$ at 1 molal (activity 0.515). The observed value from Fig. 2 is between 1 and 1.5. For potassium, caesium and thallium these data suggest that the pore is usually occupied by 0.1 M, that second ion entries occur, and that a doubly occupied pore is not completely blocked.

The discrepancies between the predictions of the four-state model and the observations may all arise from two closely related deficiencies in the four-state

TABLE 2 Asymmetrical current index, *r*, calculated from the symmetrical conductance and the asymmetric current at zero applied potential

	Concentration (M)				
Ion	*0.1*	*0.5*	*1.0*	*2.0*	*3.0*
Sodium	1.4	1.7	1.5	1.9	1.9
Potassium	1.3	1.4	1.5	1.7	1.7
Caesium	1.3	1.3	1.4	1.7	1.5

The asymmetry was created by total replacement of the indicated salt with chlorine on one side. Potentials measured were 3 M KCl salt bridges. Caveats about the interpretation of these data are mentioned in the text (W.-C. Leung & S. B. Hladky, unpublished data 1982).

description. The equations used have all assumed that only one ion moves at a time and that ions outside the pore have no effect on those within it. Thus, movements from the left to the right end of the pore are assumed to be independent of attempted entries by additional ions at the left end. Similarly, it has been assumed that an ion on the left can make no progress towards the right until there is no ion at the right end. These assumptions can and probably should be relaxed. It remains to be seen whether simpler or more useful descriptions can be obtained from different starting points.

References

Andersen OS 1983 Ion movement through gramicidin A channels. Studies on the diffusion-controlled association step. Biophys J 41:147–165

Andersen OS, Fuchs M 1975 Potential energy barriers to ion transport within lipid bilayers. Studies with tetraphenylborate. Biophys J 15:795–830

Andersen OS, Feldberg S, Nakadomari H, Levy S, McLaughlin S 1978 Electrostatic interactions among hydrophobic ions in lipid bilayer membranes. Biophys J 21:35–70

Bek S, Jakobsson E 1994 Brownian dynamics study of a multiply-occupied cation channel: application to understanding permeation in potassium channels. Biophys J 66:1028–1038

Benz R, Stark G 1975 Kinetics of macrotetrolide-induced ion transport across lipid bilayer membranes. Biochim Biophys Acta 382:27–40

Chen DP, Eisenberg R 1993 Charges, currents, and potentials in ionic channels of one conformation. Biophys J 64:1405–1421

Cooper K, Jakobsson E, Wolynes P 1985 The theory of ion transport through membrane channels. Prog Biophys Mol Biol 46:51–96

Dani JA, Levitt DG 1981 Binding constants of Li^+, K^+, and Tl^+ in the gramicidin channel determined from water permeability measurements. Biophys J 35:485–499

Dani JA, Levitt DG 1990 Diffusion and kinetic approaches to describe permeation in ionic channels. J Theor Biol 146:289–301

Eisenberg RS 1996 Computing the field in proteins and channels. J Membr Biol 150:1–25

Finkelstein A, Andersen OS 1981 The gramicidin A channel: a review of its permeability characteristics with special reference to the single-file aspect of transport. J Membr Biol 59:155–171

Hall JE, Mead CA, Szabo G 1973 A barrier model for current flow in lipid bilayer membranes. J Membr Biol 11:75–97

Hladky SB 1972 Ion conduction by gramicidin A. PhD thesis, University of Cambridge, Cambridge, UK

Hladky SB 1974 The energy barriers to ion transport by nonactin across thin lipid membranes. Biochim Biophys Acta 352:71–85

Hladky SB 1975 Tests of the carrier model for ion transport by nonactin and trinactin. Biochim Biophys Acta 375:327–349

Hladky SB 1979 The carrier mechanism. Curr Top Membr Transp 12:53–164

Hladky SB 1984 Ion currents through pores. The roles of diffusion and external access steps in determining the currents through narrow pores. Biophys J 46:293–297

Hladky SB 1987 Models for ion transport in gramicidin channels: how many sites? In: Yagi K, Pullman B (eds) Ion transport through membranes. Academic Press, Tokyo, p 213–232

Hladky SB 1988 Gramicidin: conclusions based on the kinetic data. Curr Top Membr Transp 33:15–33

Hladky SB 1992 Kinetic analysis of lipid soluble ions and carriers. Q Rev Biophys 25:459–475

Hladky SB, Haydon DA 1984 Ion movements in gramicidin channels. Curr Top Membr Transp 21:327–372

Hladky SB, Leung JCH, Fitzgerald WJ 1995 The mechanism of ion conduction by valinomycin: analysis of charge pulse responses. Biophys J 69:1758–1772

Jakobsson E 1998 Using theory and simulation to understand permeation and selectivity in ion channels. Methods (Orlando) 14:342–351

Jing N, Prasad KU, Urry DW 1995 The determination of binding constants of micellar-packaged gramicidin A by ^{13}C-and ^{23}Na-NMR. Biochim Biophys Acta 1238:1–11

Ketterer B, Neumcke B, Läuger P 1971 Transport mechanism of hydrophobic ions through lipid bilayer membranes. J Membr Biol 5:225–245

Knoll W, Stark G 1975 An extended kinetic analysis of valinomycin-induced Rb-transport through monoglyceride membranes. J Membr Biol 25:249–270

Läuger P 1973 Ion transport through pores: a rate-theory analysis. Biochim Biophys Acta 311:423–441

Levitt DG 1982 Comparison of Nernst–Planck and reaction rate models for multiply occupied channels. Biophys J 37:575–587

Levitt DG 1986 Interpretation of biological ion channel flux data—reaction-rate versus continuum theory. Annu Rev Biophys Biophys Chem 15:29–57

Levitt DG 1987 Exact continuum solution for a channel that can be occupied by two ions. Biophys J 52:455–466

Levitt DG 1991 General continuum theory for multiion channel. I. Theory. Biophys J 59: 271–277

McGill P, Schumaker MF 1996 Boundary conditions for single-ion diffusion. Biophys J 71:1723–1742

Procopio J, Andersen OS 1979 Ion tracer fluxes through gramicidin A modified lipid bilayers. Biophys J 25:8 (abstr)

Schagina LV, Grinfel'dt AE, Lev AA 1980 Study of interaction between cation fluxes in bilayer phospholipid membranes modified with gramicidin A. Determination of the number of ion-binding sites and their location in the gramicidin channel. Biofizika 25:648–53

Tsien RY, Hladky SB 1982 Ion repulsion within membranes. Biophys J 39:49–56

Wang KW, Tripathi S, Hladky SB 1995 Ion binding constants for gramicidin A obtained from water permeability measurements. J Membr Biol 143:247–257

Woolf TB, Roux B 1997 The binding site of sodium in the gramicidin A channel: comparison of molecular dynamics with solid-state NMR data. Biophys J 72:1930–1945

DISCUSSION

Roux: You said that the entry rate of the second ion is independent of ion type, but the flux data indicate that gramicidin is acting essentially like a one-ion pore in the case of sodium. If the exit rate of the second ion is high, this means that the binding constant of the second ion is low. Does a four-state model that has a low binding constant for the second ion behave like a one-ion pore?

Hladky: No, at least not necessarily. The feature that differentiates the two-ion from the one-ion pore is not the significant fractional occupancy of the second ion, but the number of times the second ion enters successfully. The conductance data are surprisingly insensitive to how long the doubly occupied state persists. The crucial step is whether the second ion succeeds in entering more or less rapidly than the first ion succeeds in leaving. In terms of the empirical constants that I use, AA for entry to an ion-free pore, DD for entry to a pore with an ion at the opposite end, B for exit when only one ion is present and E for exit when there are two; the critical ratio is DD/B not the binding constant, DD/E. This distinction has caused confusion historically. I should also point out that the four-state model, as it has been used in the past, has always assumed that a doubly occupied pore is blocked, i.e. that neither ion could go any further into the pore. However, this may not be the case. Even a small amount of give in the position may be enough to significantly affect rates at high ion concentrations and allow potentials to have an effect on the rates. This was built in to the treatment that Levitt (1982) presented, which is one of the reasons why his treatment might be a better starting point for further progress than the four-state model.

Roux: We compared the binding constant of the first ion in an empty channel to the binding constant of a second ion in a channel that's already occupied, and we found that large second ions are easier to load than smaller ones. Olaf Andersen is in agreement with this (personal communication 1998). Therefore, two sodium ions, relatively speaking, may be thermodynamically less stable than two potassium ions, so perhaps this is why it behaves like a one-ion pore.

Hladky: Unfortunately, most of the interesting effects due to second ion entry are observed when a second ion enters but then one or the other ions leaves quickly. In the model that has been used until now, when two ions are present one has to leave before anything interesting happens, and that is too strong a restriction.

Jakobsson: In contrast to when Cooper et al (1985) first developed the Brownian motion dynamics for channels, computers are now so much faster that it is possible to generate a current–voltage (I–V) curve within a half an hour or so. Our group at the National Center for Supercomputing Applications is starting to think of the web as being a universal operating system, and it is possible to use software from your web browser, whether or not you're running it from your own computer or ours.

I would like to mention some of our results on the simulation of water permeability through gramicidin channels because they relate to some of the points mentioned in Steve Hladky's presentation. The topic of water permeability has tended to be neglected, even though gramicidin is a better water channel than it is an ion channel (for references and critical discussion of this, see Finkelstein 1987). We have done the first quantitatively successful simulation of permeation through a channel starting with molecular dynamics (Chiu et al 1999a,b). We have replicated the studies of Rosenberg & Finkelstein (1978) on water permeability measurements of gramicidin embedded in phospholipid bilayers by analysing the molecular dynamics results in terms of diffusion theory. The analysis breaks down the total permeation resistance into components due to motion in the channel, motion outside up to the channel mouth and the transition in the water molecules' solvation environment. The total permeation resistance (i.e. the inverse of the permeability) is the sum of these. By methods for analysing the molecular dynamics that we have previously published (Chiu et al 1991) we can show that the resistance in the channel is only about 1/6 the total resistance. For diffusion up to the mouth of the channel, the correct expression is:

$$R_{\text{access}} = \int_{r=\infty}^{r=r_c} \frac{dr}{2\pi r^2 [D \cdot \exp(-\Psi)]_{\text{avg}}(r)}$$

Where: r_c = capture radius
D = diffusion coefficient
Ψ = dimensionless free energy (units of kT)
$[\,]_{\text{avg}}$ denotes an averaging over all values of elevation angle ϕ at a given r. Since the diffusing species, water, is not undergoing a net driving force, its concentration can serve as a measure of the free energy via the Boltzmann relationship; i.e. exp $(-\Psi) = C^*$, where C^* = water concentration normalized to bulk = C/C_{bulk}. Thus, the equation for the access diffusive resistance goes to:

$$R_{\text{access}} = \int_{r=\infty}^{r=r_c} \frac{dr}{2\pi r^2 [D \cdot C^*]_{\text{avg}}(r)}.$$

Thus, in the molecular dynamics calculations, we can measure the bulk concentration (C^*), which represents the exp$(-\Psi)$ term, and the mobilities by correlation analysis. We don't know the effective channel radius. There is a physical radius, but this is not the same as the capture radius, or the effective radius, which is something like the physical radius minus the radius of the permeant species, in this case, the water molecule. The point of this is that if we take the numbers that we get from the molecular dynamics for the diffusion coefficient and for C^*, which gives us the free energy profile for permeation up to the channel mouth, and we assume different plausible values of the effective channel capture radius, we obtain a value for the diffusive component of the access resistance that falls far short of the total permeation resistance. Therefore, there is an event at the mouth of the channel, in this case for water but presumably also for ions, that is the transition from being solvated in the extra-channel environment to being solvated in the interior of the channel. This transition event accounts by itself for about 5/6 of the water permeability. I should also say that in the full simulation the way in which we measured the total diffusive resistance was to calculate how many water molecules made the transition from being in the bulk solution to the interior of the channel, during the 1.2 nsec of the simulation.

Note that since the total resistance to channel water permeation is the inverse of permeability, it has units of seconds/cm. The access resistance in the diffusion theory above has units of seconds/cm^3; the ratio between them is the effective cross-sectional area of the channel (for coupling of access resistance in a hemispherical bath to longitudinal diffusion in the channel interior, see Chiu & Jakobsson 1989). In this theory we assume that the radius of the channel cross-section is the same as the capture radius.

Roux: Are you saying that the smaller the capture radius, the slower the water molecules go through?

Jakobsson: The direct effect is, the smaller the capture radius the larger the effective access resistance, and hence the larger total resistance to permeation. That translates into a lower water flux, even if the mobility of the permeant species in the channel is the same. But in this particular case, water flux through gramicidin, the bottom line is that the diffusive description has a major missing component, and this is the discrete transition of water at the mouth of the channel from being bulk water to being internal water.

Ring: There is a dividing line between only using the model as a phenomenological description of the data and believing that it is an approximation of the real situation in the channel. There is an alternative to the multistate models, i.e. the three-barrier, four-site model. What is required of models is that they predict something that we don't already know, i.e. that they are tested by new experimental data, and not that they just fit with the data you already have. I've used the multistate two-barrier models, and they work well, but I do believe that the three-barrier, four-site model is better. Using that model, I tried to predict the occupancy of the channel and test that against the hypothesis that the stability of the channel is dependent on the occupancy of the channel. I was amazed that I could do this with just the simple assumption that the occupancy of the channels stabilizes the channels, and without any fitted parameters. This was true for protons, for caesium, potassium and sodium ions, and for all voltage dependencies without any adjustable parameters at all. Therefore, I'm suggesting that we do not discard the multistate models before we have tested them. One problem is that with more states and more barriers there are more parameters, but it is possible to reasonably fit the I–V curves if you use enough parameters, and you don't need an infinite number of barriers.

Eisenberg: But you cannot fit straight lines with five or six exponentials.

Hladky: If it's perfectly straight you can't fit it, and no one is arguing against that. The key is that it is within a tolerance.

Ring: I would like to ask Steve Hladky why he wants to discard the multi-barrier models.

Hladky: I never said I did. There are several directions one can pursue. One is to invoke a low affinity, virtual binding site at the mouth of the pore. I should point out that in terms of physics this does not contradict the approaches that Bob Eisenberg supports, although he may not like the vocabulary. The Nernst–Planck, Poisson, smeared charge theory is saying that the distribution of ions surrounding the mouth of the pore will effect the transport processes within it, a virtual site is another means of introducing this. The maths, however, are different and people will have different views on how natural these different methods are. With any model that uses a small number of barriers, I wish people would not use exponential potential effects. For individual transition processes between states one should try to use something closer to a solution of the Nernst–Planck equation.

Roux: What do you mean by exponential potential effects?

Hladky: I mean that one should not assume that the rate constants vary exponentially with the applied potential.

Roux: What do you suggest should be done?

Hladky: I have done a curve fit, and used a convenient arbitrary function, the shape of which can be altered—e.g. the prediction of the Nernst–Planck equation with a trapezoid, division of an exponential by a hyperbolic cosine or multiplication of an exponential by an exponential of minus something times the potential squared. It is possible to rationalize the different expressions, but none of them replaces someone trying to solve the rates using the real potentials. I have no objection to people trying to solve the real potentials, but if they do they should not say that the calculations based on fractional charges are correct when we know that the charges present are full charges. It is a difficult problem, and it is also difficult to obtain experimental data that tell us when our theory is good enough and when it isn't.

Roux: Can one replace the voltage dependence of the central barrier by the real integral of the transition rate from one state to the other? This would be more complicated.

Hladky: That's exactly what I'm saying would be good thing to do. I'm advocating a hybrid.

Eisenberg: I can answer Benoît Roux's question with an explicit formula, i.e. how can one do that computation for any shaped potential or any charge distribution? However, the most important point I would like to make concerns the use of rate constants and where the difficulties arise. In my opinion, one ought to use the vocabulary to describe permeation that one uses for condensed basis, i.e. the fundamental processes are diffusion, perhaps convection, perhaps temperature and the electric field. In that formulation the rate constants are output variables, and not input variables. They are simply the flux divided by the concentration. In that formulation one can always define rate constants. The papers that have the most general derivation of rate constants for crossing barriers are the ones I published in 1998 (Eisenberg 1998a,b). The issue is whether the rate constants of the entry step and the exit step are independent. If the rate constant is a function of all the parameters in the problem, then they are likely in many conditions to operate together with the fundamental physical variables. That is, as you change the structure you change more than one rate constant. The point of physical theory is to try to do that explicitly and test it.

Busath: Steve Hladky mentioned that there were several problems with sodium which suggest that it can't doubly occupy a channel, one of which concerns the flux ratio exponent. A high central barrier predicts a flux ratio exponent of one under all conditions.

Hladky: Anderson and Procopio (Andersen & Procopio 1979, Procopio & Andersen 1979) dismissed that possibility. Their reasoning was that at higher sodium concentrations the I–V curves are supralinear, but at sufficiently low sodium concentrations they are markedly sublinear. This implies that the rate-limiting step at low concentrations is not the potential-dependent step which in turn implies that the central barrier is not high. This change in shape of I–V relationships was in fact one of the main reasons that I originally considered double occupancy of the pore and developed the four-state model (Hladky 1972). It is interesting to note that within the model the change in shape requires either that exit of an ion from a singly occupied pore is potential dependent or that a second ion can enter the pore.

References

Andersen OS, Procopio J 1979 The kinetics of Na^+ movement through gramicidin A channels. Biophys J 25:8 (abstr)

Chiu S-W, Jakobsson E 1989 Stochastic theory of singly-occupied ion channels. II. Effects of access resistance and potential gradients extending into the bath. Biophys J 55:147–157

Chiu S-W, Jakobsson E, Subramaniam S, McCammon JA 1991 Time-correlation analysis of simulated water motion in flexible and rigid gramicidin channels. Biophys J 60:273–285

Chiu, S-W, Subramaniam S, Jakobsson E 1999a Simulation study of a gramicidin/lipid bilayer system in excess water and lipid. I. Structure of the molecular complex. Biophys J 76: 1929–1938

Chiu S-W, Subramaniam S, Jakobsson E 1999b Simulation study of a gramicidin/lipid bilayer system in excess water and lipid. II. Rates and mechanisms of water transport. Biophys J 76:1939–1950

Cooper K, Jakobsson E, Wolynes P 1985 The theory of ion transport through membrane channels. Prog Biophys Mol Biol 46:51–96

Eisenberg B 1998a Ionic channels in biological membranes. Electrostatic analysis of a natural nanotube. Contemp Phys 39:447–466

Eisenberg B 1998b Ionic channels in biological membranes: natural nanotubes. Acc Chem Res 31:117–125

Finkelstein A 1987 Water movement through lipid bilayers, pores, and plasma membranes. Theory and reality. Wiley, New York

Hladky SB 1972 Ion conduction by gramicidin A. PhD thesis, University of Cambridge, Cambridge, UK

Levitt DG 1982 Comparison of Nernst–Planck and reaction rate models for multiply occupied channels. Biophys J 37:575–587

Procopio J, Andersen OS 1979 Ion tracer fluxes throughout gramicidin A-modified lipid bilayers. Biophys J 25:8 (abstr)

Rosenberg PA, Finkelstein A 1978 Water permeability of gramicidin A-treated lipid bilayer membranes. J Gen Physiol 72:341–350

The binding site of sodium in the gramicidin A channel

Benoît Roux & Thomas B. Woolf*

*Membrane Transport Research Group (GRTM), Departments of Physics and Chemistry, Université de Montreal, Case Postale 6128, Succursale Centre-Ville, Montreal, Quebec, Canada H3C 3J7, and *Departments of Physiology and of Biophysics and Biophysical Chemistry, Johns Hopkins University School of Medicine, 725 North Wolfe Street, Baltimore, MD 21205, USA*

Abstract. The available information concerning the structure and location of the main binding site for sodium in the gramicidin A channel is reviewed and discussed. Results from molecular dynamics simulations using an atomic model of the channel embedded in a lipid bilayer are compared with experimental observations. The combined information from experiment and simulation suggests that the main binding sites for sodium are near the channel's mouth, approximately 9.2 Å from the centre of the dimer channel, although the motion along the axis could be as large as 1 to 2 Å. In the binding site, the sodium ion is lying off axis, making contact with two carbonyl oxygens and two single-file water molecules. The main channel ligand is provided by the carbonyl group of the Leu10-Trp11 peptide linkage, which exhibits the largest deflection from the ion-free channel structure.

1999 Gramicidin and related ion channel-forming peptides. Wiley, Chichester (Novartis Foundation Symposium 225) p 113–127

Background and perspective

Detailed information about energetically favourable binding sites along the pathway of a permeating ion is essential to understand the function of a transmembrane ion channel at the molecular level. It is possible to further our understanding of the fundamental aspects of the permeation process by studying the interaction of cations with the channel formed by the gramicidin A molecule. One outstanding advantage of the gramicidin A channel is the large amount of structural and functional information presently available. The gramicidin A channel has been the object of numerous experimental as well as theoretical investigations and it is, at this moment, the best characterized molecular pore (for reviews see Hille 1992, Andersen & Koeppe 1992, Roux & Karplus 1994). The goal of this chapter is to review the available information concerning the location and the nature of the sodium-binding sites in the gramicidin A channel.

The first structural evidence of a specific cation-binding site was obtained using nuclear magnetic resonance (NMR) spectroscopy with ^{13}C-labelled gramicidin A incorporated in L-alpha-lysophosphatidylcholine (LPC) lipid vesicles (Urry et al 1982a,b, 1983). The changes in chemical shift induced by the presence of sodium observed for the carbonyl carbon of Trp11 and Trp13 indicate that the binding sites are located near the C-terminus of the monomers. Unfortunately, a simple structural interpretation of the observed isotropic ^{13}C backbone carbonyl chemical shifts of gramicidin A is difficult due to the high sensitivity of the shielding factor on the local electronic structure. In fact, simple structural interpretations of the chemical shifts can be quite misleading. For example, it was concluded initially on the basis of the ^{13}C chemical shifts that the channel was formed by the association of two left-handed β-helical monomers (Urry et al 1982b), whereas it was shown later that the monomers were in a right-handed helical conformation using two-dimensional NMR of gramicidin A incorporated in SDS detergent micelles (Arseniev et al 1985). Further studies of sodium binding to gramicidin A incorporated in LPC vesicles indicated that the largest ^{13}C chemical shift change is observed for the carbonyl group of Trp11 and Trp13 (Jing et al 1995). Tight and weak binding constants, corresponding to singly and doubly sodium occupied channels, were estimated to be, respectively, 67 M^{-1} and 1.7 M^{-1} (Jing et al 1995).

Direct structural information about the location of the cation-binding site was first provided by low angle X-ray scattering on gramicidin A incorporated in oriented dilauroyl phosphatidylcholine (DLPC) bilayers (Olah et al 1991). Analysis of the data shows that there is a binding site for Tl^{+} at 9.6 ± 0.3 Å from the centre of the dimer channel. The experimental conditions were such that singly occupied channels dominated the observations. Although the binding of common ions could not be observed by low angle X-ray scattering due to their poor scattering contrast, it is likely from these measurements that a cationic binding site is located around 9–10 Å.

Solid-state NMR has been used to investigate the binding of sodium to the gramicidin A channel incorporated in oriented dimyristoyl phosphatidylcholine (DMPC) multilayers (Smith et al 1990, Separovic et al 1994, Tian et al 1996). The measurements determine the change, due to the presence of sodium, in the ^{13}C and ^{15}N chemical shift anisotropy (CSA) and ^{15}N-^{2}H dipolar coupling at specific gramicidin A backbone sites. The observations indicate that the CSA of the ^{13}C-labelled carbonyl of Leu10, Leu12 and Leu14 are affected by the presence of sodium, whereas those of Trp11, Trp13, as well as more internal sites, are not (Smith et al 1990, Separovic et al 1994). The CSA of ^{13}C-labelled Leu10 exhibits the largest variation, going from 9 to 0 ppm. This suggests that the sodium cation binds preferentially to the oxygen of this carbonyl group. The chemical shift of ^{15}N-labelled Trp11 exhibits a change of 3–4 ppm (Tian et al 1996), which is consistent with the variations of ^{13}C-labelled Leu10 due to the strong coupling imposed by the peptide linkage. However, the backbone ^{15}N-^{2}H dipolar coupling exhibits a moderate change, going from 3032 kHz to 2840 kHz at a sodium concentration corresponding to double ion occupancy, suggesting that the channel structure is only slightly affected by the presence of sodium in the binding site. The change in dipolar coupling corresponds to a deflection of the Leu10-Trp11 amide plane upon sodium binding on which is in the order of 5 degrees (Tian et al 1996). Based on these observations, it was concluded that channel distortion upon cation binding is almost negligible.

The lack of structural changes upon sodium binding inferred from the solid-state NMR data (Tian et al 1996) contrasts with traditional views of permeation through the gramicidin A channel (Hille 1992). The geometry of the gramicidin A channel is such that only two of those water molecules can remain strongly attached to the ion as it enters the single-file region; the single-file region, in which the ion in making contact with only two water molecules, starts around 10.5 Å away from the channel centre (Roux & Karplus 1993). In bulk solution Na^+ is surrounded by six water molecules, each contributing approximately 24 kcal/mol of interaction energy (Roux & Karplus 1995). The loss of hydration energy is compensated by the backbone carbonyl oxygens which help solvating the permeating ion. Computational studies and molecular dynamics simulations of the gramicidin A channel suggest that the interaction of sodium with the channel backbone is significant. Molecular dynamics simulations have been used to calculate the free energy profile of a single Na^+ along the axis of the gramicidin A channel on the basis of an atomic model of the dimer channel embedded in a simplified membrane (Roux & Karplus 1993). The resulting free energy profile exhibits a local minimum at each extremity of the dimer corresponding to a cation-binding site. The calculated binding site is located at 9.3 Å from the centre of the channel, in agreement with the data from low angle scattering (Olah et al 1991). In the binding site the Na^+ is in close contact with the carbonyl of Leu10 in accord with the solid-state NMR data (Smith et al 1990). According to *ab initio* calculations, the interaction of Na^+ with N-methylacetamide, a realistic model of the channel backbone, is on the order of 37 kcal/mol (Roux & Karplus 1995). Since the channel is intrinsically flexible, spontaneous fluctuations of the peptide plane orientation are on the order of 10 to 20 degrees (Lazo et al 1995), the fact that a cation could induce some local deformation of the backbone structure is reasonable.

Despite the good qualitative agreement between theory and experiments, computational models generally suggest substantial structural distortion upon cation binding. In contrast, rather small structural deformations are inferred on the basis of solid-state NMR data (Tian et al 1996). For a better understanding of the nature of cation–channel interactions, it is important to address the possible causes of the discrepancy.

Results and discussion

In the absence of a high resolution structure from X-ray crystallography, solid-state NMR experiments provide the most direct information on the detailed structure of the sodium-binding site in the gramicidin A channel. However, a direct structural interpretation of the NMR data is not straightforward. The observations result from an averaging process over rapidly fluctuating sodium positions in fast exchange on the time-scale of solid-state NMR. Furthermore, it is important to emphasize that, due to the non-linear functional dependence of the NMR observables on the molecular geometry, such averaging process differs essentially from calculations based on a single average molecular configuration. For a meaningful analysis, it is necessary to consider a dynamical average of the NMR observables over a large number of configurations (Woolf & Roux 1994, 1996). To help interpret the solid-state NMR data, Woolf & Roux (1997) generated a series of 24 dynamical trajectories with a sodium constrained at different position along the channel axis. Details about the simulation systems can be found in the original references. Briefly, the simulation system represents a model for the oriented samples studied in solid-state NMR experiments (Ketchem et al 1993, Smith et al 1989). In the atomic model, this corresponds to one gramicidin A dimer, 16 DMPC molecules and about 700 water molecules, for a total of 4385 atoms. The all-hydrogen force field of CHARMM (Brooks et al 1983), PARAM version 22, for proteins (MacKerell et al 1998) and lipid molecules (Schlenkrich et al 1996) was used for all calculations; the TIP3P water potential was used (Jorgensen et al 1983). The standard CHARMM force field was modified to include both first- and second-order polarization effects induced by the ion on the peptide using a method described previously (Roux 1993, Roux & Karplus 1995, Roux et al 1995).

The average properties of the NMR observables as a function of the sodium ion position along the z-axis were calculated by combining all the information generated from 24 umbrella sampling simulations (Woolf & Roux 1997). The results showed variation in the NMR observables as the ion is moved along the gramicidin A channel axis. These z-dependent averages must be integrated with respect to the z-axis weighted by the probability distribution $P(z)$ to obtain a single number that can be compared directly to the time-averaged value measured experimentally. Under equilibrium conditions, the probability distribution $P(z)$ is related to the free energy profile of sodium $W(z)$ along the channel axis, i.e. $P(z) \propto \exp[-W(z)/kT]$. To calculate the time-average corresponding to the NMR observations, it is necessary to perform a Boltzmann-weighted average using the free energy profile of the ion at various positions along the z-axis. Small deviations in the free energy profile lead to large variations in the probability distribution of the sodium ion in the channel entrance which, in turn, influences the outcome of the weighted average NMR properties. This highlights the difficulties in constructing accurate atomic models and force fields. Due to their sensitivity to the details of the microscopic potential function, properties involving energetic factors such as a free energy profile are difficult to obtain accurately. Nevertheless, in order to interpret the available solid-state NMR data in terms of the location of the sodium-binding site, it is the z-dependence of the solid-state NMR properties rather than an absolute prediction of the energetics of binding that is of interest here.

The backbone dihedral angles surrounding the Leu10-Trp11 peptide linkage are the most affected by the presence of the ion. Respectively, the Ψ10 and Φ11 dihedral angles deviate from the standard values for a β-helix, which are -108 and -111 degrees (Roux & Karplus 1991), to -126 and -94 degrees. The root mean squared (rms) dihedral fluctuations are on the order of 10 degrees, with and without ion. For Leu10 the changes in the angle between the C=O bond and the bilayer normal are most pronounced when the Na is around 9 Å. The largest deflection of the C=O is then around 15 degrees. For Trp11, the largest change in the angle between the N-H bond and the normal to the bilayer occurs when the Na^+ is near 9 Å and is on the order of 10 to 12 degrees. The O-C-N-H trans-peptide dihedral angle ω deviates from planarity in the presence of an ion. It varies from -179 degrees to -174 degrees with rms fluctuations on the order of 8 degrees. The deviation from planarity is consistent with the observed difference of nearly 5 to 10 degrees in the angle distribution of the C=O and N-H bonds relative to the channel axis. The largest change in the CSA of ^{13}C-labelled Leu10 relative to the ion-free channel is observed when the ion is around 9 Å. There is a large change in the ^{15}N chemical shift for Trp11 when the sodium is around 9 Å and smaller change at other locations. A simultaneous variation is expected for the chemical shift of ^{15}N-labelled Trp11 due to the structure of the peptide linkage.

The calculated NMR observables are strongly dependent on the ion position along the channel axis. To find the most plausible location of the main sodium-binding site in accord with the NMR data, a semi-empirical approach combining the information from both the simulations and the ^{13}C experiments was used. The position of the ion-binding site was deduced by empirically matching the magnitude of the observed CSA of ^{13}C-labelled gramicidin A. For simplicity, $P(z)$, the probability distribution of the sodium along the channel axis, was assumed to be a single Gaussian of width 0.1 Å. The position of the Gaussian was then empirically optimized to yield the smallest mean S.D. between the calculated and observed carbonyl ^{13}C solid-state NMR data. The best match was found with the Gaussian centred at 9.2 Å from the centre of the channel, although other functional forms for the probability density could have led to a different result.

In qualitative accord with the solid-state NMR data, the molecular dynamics calculations averaged with a Gaussian centred at 9.2 Å yields a change of 4 ppm for the ^{13}C-labelled Leu10 and 4–5 ppm for the ^{15}N-labelled Trp11 using the best matching Gaussian centred at 9.2 Å. The calculated average deflection of the ^{15}N-^{2}H bond of Trp11 is in the range of 9–10 degrees. However, the experimentally observed maximum distortion estimated from the ^{15}N-^{2}H dipolar coupling at high salt concentration corresponds to 5 degrees (Tian et al 1996). Possible reasons for the discrepancy are discussed below. Nevertheless, the qualitative agreement between the calculations and the experiments is satisfactory.

Limitations and problems with the computational models and the interpretation of the NMR data

The small discrepancy between the data and the molecular dynamics calculations might be due to a combination of subtle factors. It is important to better understand the origin of those factors that could arise from both the calculations and the interpretation of the data. We first outline the limitations of the computational model. We then discuss the possible difficulties in the interpretation of NMR data in terms of a structural model.

(1) The main problem in theoretical studies of ion permeation is the accuracy of the microscopic potential energy function used to describe the atomic systems. The interaction of small metal cations with the carbonyl groups of peptides involve substantial induced polarization effects (Roux & Karplus 1995). The detailed ligand structure around the cation is sensitive to the potential function (Roux 1993). Incorporation or neglect of induced polarization affects the average distance with the oxygen of the nearest carbonyl groups and, in some case, even the number of neighbours (Woolf & Roux 1997).

(2) The present analysis is based on the assumption that the shielding tensor orientation is fixed relative to the local molecular frame and that the component magnitudes are constant. The interpretation of the change in chemical shifts upon cation binding is based entirely on changes in the position of the nuclei. This simplification is necessary in order to determine a three-dimensional structure using solid-state NMR (Ketchem et al 1993). However, the magnitude and the orientation of the components of the chemical shift tensor vary rapidly and are sensitive to the environment (Woolf et al 1995) and the presence of a cation may also induce chemical shifts by perturbing the electronic structure around the nuclei.

(3) The calculated averages depend exponentially on the free energy profile $W(z)$ of the ion along the channel axis (see above). A quantitative agreement

between the calculation and the experimental structural data is difficult to achieve. Small variations in the free energy profile result in a large change in the NMR observables.

(4) The orientational fluctuations of the backbone N-H bond may have been affected because Newton's dynamical equation of motions were integrated with fixed length for all the bonds involving hydrogen atoms using the SHAKE algorithm (Ryckaert et al 1977).

It is also important to explore possible problems with the interpretation of the solid-state NMR data. The present discussion is not complete. It is meant to suggest possible avenues for future investigations of the influence of ions on the gramicidin A channel.

(1) The observed properties results from a time average over many configurations. It is possible that transient distortion of the channel remain undetected on the NMR time-scale if their statistical weight is not dominant. For example, the ion in the binding site could fluctuate rapidly over several positions such that the resulting observed properties reflect small structural distortions. This is illustrated in Fig. 1 where the distribution of the instantaneous values of the CSA of ^{13}C-labelled Leu10 are shown for the ion-free and ion-bound gramicidin A channel. For both states, the average CSA reflects variations over a wide range. In particular, the distribution is not symmetric and the average does not coincide with the most probable value. Thus, one should not expect to interpret the data in terms of a single average structure.

(2) The ^{15}N-^{2}H dipolar coupling of Trp11 does not provide an unambiguous assessment of the orientation of the carbonyl group of Leu10. The calculations indicate that deviations of the peptide linkage from planarity can be on the order of 5–6 degrees with rms fluctuations on the order of 8 degrees. To obtain more information on the orientation of the peptide linkage, a possible, though difficult, experiment would be to observe the ^{13}C-^{19}O dipolar coupling at the Leu10 site using solid-state NMR.

(3) Interpretation of the observed ^{15}N-^{2}H dipolar couplings depends on the magnitude of the intrinsic dipolar interaction, which is assumed to remain constant. The assumed value of the magnitude of the dipolar interaction, based on measurements of dipolar coupling in model compound, is used to interpret the gramicidin A channel data. It is possible that the N-H bond length, and thus the magnitude of the dipolar interaction, is slightly affected in the β-helical hydrogen bonded structure (Trp11 is forming a backbone hydrogen bond with the carbonyl group of Leu4). In addition, the N-H

bond length could also be sensitive to the presence or absence of Na^+ in the binding site.

(4) The observed NMR properties correspond to a superposition of dimer channels with zero, one or two bound sodium ions, with equal contribution from the two identical isotopically labelled monomers. The solid-state NMR experiments were performed with 160 mM (Smith et al 1990) or 150 mM (Tian et al 1996) NaCl. Assuming that the binding constants for singly and doubly occupied channels in the oriented gramicidin A:DMPC samples are similar to those measured with LPC vesicles (Jing et al 1995), there should be, on average, one sodium present per dimer channel at such concentrations. Since the observed NMR spectra represent the superposition of one ion-bound and one ion-free monomer at this concentration, the variations relative to the ion-free channel may be expected to reflect approximately one-half of the total structural perturbation induced in the ion-bound monomer. Similarly, experiments performed with 250 mM NaCl are expected to be dominated by the doubly occupied channels (Tian et al 1996). However, it is possible that, while a first ion binds at ± 9.2 Å, the second binds at ± 10.9 Å, giving rise to a non-symmetric situation (Jing et al 1995). Thus, it is possible that the measurements obtained with 250 mM NaCl do not correspond to the maximum structural distortion of the channel on the ion-bound monomer.

(5) Although the data indicate that the presence of the ion in the binding site has a small influence on the channel structure, larger structural distortions may be possible during the permeation process as the ion is crossing free energy barriers (e.g. between 8.5 and 9.0 Å along the channel axis). Because their statistical weight is necessarily smaller, configurations with the ion at the top of a free energy barrier do not contribute significantly to the time average giving rise to the solid-state NMR properties. The configurations with the ion in the binding site strongly dominate the time average corresponding to the observed NMR properties. Ion movements through the channel involve rare and transient configurations (transition states) that remains undetected on the NMR time-scale.

Comparisons with ion flux data

Traditionally, measurements of the ionic current as a function of an applied transmembrane voltage have been used to gain information about the location of the free energy wells and barriers along the permeation pathway (Hille 1992). It is observed experimentally that the current–voltage relation is supralinear (less than Ohmic), strongly suggesting that the rate-limiting kinetic process opposing the passage of ions through the gramicidin A channel occurs in a region that is not

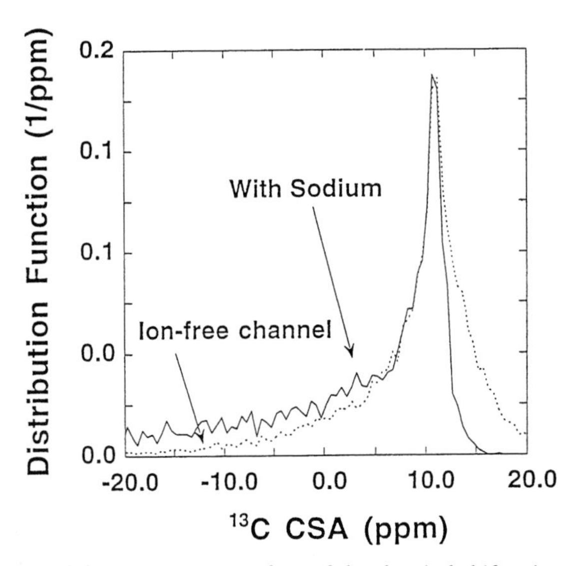

FIG. 1. Distribution of the instantaneous values of the chemical shift anisotropy (CSA) of ^{13}C-labelled Leu10 for the water-filled (dashed line) and singly occupied channel (solid line). The values were calculated on the basis of the configurations generated by molecular dynamics simulations of the gramicidin A channel embedded in a dimyristoyl phosphatidylcholine (DMPC) membrane (Woolf & Roux 1994, 1996, 1997). In the ion-bound case, the sodium was harmonically restrained to remain around $z = 9.2$ Å. The observed value of CSA corresponds to an average over a broad and asymmetric distribution.

sensitive to the applied transmembrane electric field (Eisenman & Horn 1983, O. S. Andersen, personal communication 1995). The interpretation usually relies on an analysis of the data in terms of Eyring rate theory models under the assumption that the transition hopping rates have a simple Arrhenius dependence on the applied voltage (Läuger 1973).

The estimate for the position of the binding site is first obtained in terms of the fraction of the applied transmembrane potential, which is then converted into a position by assuming that the membrane field is constant over the full length of the pore. Based on the flux data several researchers have concluded that the main binding site is located near the entrance of the channel: Andersen & Procopio (1980) found that the binding site for Na^+ is 14% from the channel entry; Eisenman & Sandblom (1983) estimated the distance from the mouth of the channel to the peak of the entry barrier at 6%; Busath & Szabo (1988) reported that the binding site position for K^+ is 15.4% from the channel entry; and Becker et al (1992) found that the binding site position is 14% along the way into the channel (probably the most accurate estimate with this method). Assuming a channel length of 25 Å, a fractional distance of 14% corresponds to a binding site position at 9 Å from the centre of the dimer. This estimate is remarkably consistent with the Tl^+-binding site measured by low angle scattering (Olah et al 1991) and the sodium-binding site deduced from solid-state NMR data (Woolf & Roux 1997).

Conclusions

The analysis of the solid-state NMR data using the configurations generated by molecular dynamics simulations provides good support for the contention that the main binding site for sodium in the gramicidin A channel is around 9.2 Å. This location is in agreement with results from low angle X-ray scattering (Olah et al 1991). It was found to correspond to the lowest free energy minimum in a previous determination of the free energy profile along the channel axis (Roux & Karplus 1993).

In the binding site the sodium is lying off axis, making contacts with the carbonyl oxygen of Leu10, Trp15 and two single-file water molecules. The main channel ligand is provided by the carbonyl group of the Leu10-Trp11 peptide linkage which exhibits the largest deflection from the ion-free channel structure. None the less, there are no large distortions of the channel structure due to the presence of the ion. In particular, the β-helical hydrogen bonding pattern is not strongly perturbed by the presence of the sodium in the binding site. As described previously (Roux & Karplus 1993), the main backbone C=O···H-N hydrogen bonds are C=O(i) to H-N($I+7$) for even i and C=O(i) to H-N($I-5$) for odd i. Thus, Val8-Trp15, Trp13-Val8 and Trp15-Leu10 are forming the backbone C=O···H-N hydrogen bonds near the sodium-binding site. In the β-helical structure, Leu10 is the first C=O group that is pointing toward the bulk solvent with no amide HN partner to for a hydrogen bond. For this reason, it is able to provide a ligand for a cation without losing a backbone hydrogen bond. One might speculate that this is the origin of an energetically favourable site for cation in the entrance of the channel.

Acknowledgement

This work was supported by a grant from the Medical Research Council of Canada.

References

Andersen OS, Koeppe RE II 1992 Molecular determinants of channel function. Physiol Rev 72 (suppl):S89–S158

Andersen OS, Procopio J 1980 Ion movement through gramicidin A channels. On the importance of the aqueous diffusion resistance, and ion–water interactions. Acta Physiol Scand Suppl 481:27–35

Arseniev AS, Barsukov IL, Bystrov VF, Lomize AL, Ovchinnikov 1985 ^{1}H-NMR study of gramicidin-A transmembrane ion channel. Head-to-head right-handed, single stranded helices. FEBS Lett 186:168–174

Becker MD, Koeppe RE II, Andersen OS 1992 Amino acid substitutions and ion channel function. Model-dependent conclusions. Biophys J 62:25–27

Brooks BR, Bruccoleri RE, Olafson BD, States DJ, Swaminathan S, Karplus M 1983 CHARMM. A program for macromolecular energy minimization and dynamics calculation. J Comput Chem 4:187–217

Busath D, Szabo C 1988 Permeation characteristics of gramicidin conformers. Biophys J 53:697–707

Eisenman G, Horn R 1983 Ionic selectivity revisited: the role of kinetic and equilibrium processes in ion permeation through channels. J Membr Biol 76:197–225

Eisenman G, Sandblom JP 1983 Energy barriers in ionic channels. Data for gramicidin A interpreted using a single-file (3B4S″) model having 3 barriers separating 4 sites. In: Spach G (ed) Physical chemistry of transmembrane ion motions. Elsevier Science, Amsterdam, p 329–348

Hille B 1992 Ionic channels of excitable membranes, 2nd edn. Sinauer Associates, Sunderland, MA

Jing N, Prasad KU, Urry DW 1995 The determination of binding constants of micellar-packaged gramicidin A by ^{13}C- and ^{23}Na-NMR. Biochim Biophys Acta 1238:1–11

Jorgensen WL, Impey RW, Chandrasekhar J, Madura JD, Klein ML 1983 Comparison of simple potential functions for simulating liquid water. J Chem Phys 79:926–935

Ketchem RR, Hu W, Cross TA 1993 High-resolution conformation of gramicidin A in lipid bilayer by solid-state NMR. Science 261:1457–1460

Läuger P 1973 Ion transport through pores: a rate-theory analysis. Biochim Biophys Acta 311:423–441

Lazo N, Hu W, Cross TA 1995 Low-temperature solid-state ^{15}N NMR characterization of polypeptide backbone libarations. J Magn Reson B 107:43–50

MacKerell AD Jr, Bashford D, Bellot M et al 1998 All-atom empirical potential for molecular modeling and dynamics studies of proteins. J Phys Chem B 102:3586–3616

Olah GA, Huang HW, Liu WH, Wu YL 1991 Location of ion-binding sites in the gramicidin channel by X-ray diffraction. J Mol Biol 218:847–858

Roux B 1993 Non-additivity in cation–peptide interactions. A molecular dynamics and *ab initio* study of Na^{+} in the gramicidin channel. Chem Phys Lett 212:231–240

Roux B, Karplus M 1991 Ion transport in a gramicidin-like channel. Structure and thermodynamics. Biophys J 59:961–981

Roux B, Karplus M 1993 Ion transport in the gramicidin channel: free energy of the solvated right-handed dimer in a model membrane. J Am Chem Soc 115:3250–3262

Roux B, Karplus M 1994 Molecular dynamics simulations of the gramicidin channel. Annu Rev Biophys Biomol Struct 23:731–761

Roux B, Karplus M 1995 Potential energy function for cations–peptides interaction: an *ab initio* study. J Comp Chem 16:690–704

Roux B, Prod'hom B, Karplus M 1995 Ion transport in the gramicidin channel: molecular dynamics study of single and double occupancy. Biophys J 68:876–892

Ryckaert JP, Ciccotti G, Berendsen HJC 1977 Numerical integration of the cartesian equation of motion of a system with constraints: molecular dynamics of n-alkanes. J Comp Phys 23:327–341

Schlenkrich MJ, Brickman J, MacKerell AD Jr, Karplus M 1996 An empirical potential energy function for phospholipid. Criteria for parameters optimization and applications. In: Merz KM, Roux B (eds) Biological membranes: a molecular perspective from computation and experiment. Birkhauser, Boston, p 31–81

Separovic F, Gehrmann J, Milne T, Cornell BA, Lin SY, Smith R 1994 Sodium ion binding in the gramicidin A channel. Solid-state NMR studies of the tryptophan residues. Biophys J 67:1495–1500
Smith R, Thomas DE, Separovic F, Atkins AR, Cornell BA 1989 Determination of the structure of a membrane-incorporated ion channel. Solid-state nuclear magnetic resonance studies of gramicidin A. Biophys J 56:307–314
Smith R, Thomas DE, Atkins AR, Separovic F, Cornell BA 1990 Solid-state ^{13}C-NMR studies of the effects of sodium ions on the gramicidin A ion channel. Biochim Biophys Acta 1026:161–166
Tian F, Lee K-C, Hu W, Cross TA 1996 Monovalent cation transport: lack of structural deformation upon cation binding. Biochemistry 35:11959–11966
Urry DW, Prasad KU, Trapane TL 1982a Location of monovalent cation binding sites in the gramicidin channel. Proc Natl Acad Sci USA 79:390–394
Urry DW, Walker JT, Trapane TL 1982b Ion interactions in (1–13C)D-Val8 and D-Leu10 analogs of gramicidin A, the helix sense of the channel and location of ion binding sites. J Membr Biol 69:225–231
Urry DW, Trapane TL, Prasad KU 1983 Is the gramicidin A transmembrane channel single-stranded or double stranded helix? A simple unequivocal determination. Science 221:1064–1067
Woolf TB, Roux B 1994 Molecular dynamics simulation of the gramicidin channel in a phospholipid bilayer. Proc Natl Acad Sci USA 91:11631–11635
Woolf TB, Roux B 1996 Structure, energetics, and dynamics of lipid–protein interaction: a molecular dynamics study of the gramicidin A channel in a DMPC bilayer. Proteins 24:92–114
Woolf TB, Roux B 1997 The binding site of sodium in the gramicidin-A channel. Comparison of molecular dynamics with solid-state NMR. Biophys J 72:1930–1945
Woolf TB, Malkin VG, Malkin OL, Salahub DR, Roux B 1995 The backbone ^{15}N chemical shift tensor of the gramicidin channel. A molecular dynamics and density functional study. Chem Phys Lett 239:186–194

DISCUSSION

Jakobsson: I would like to mention that doing Brownian dynamics is relatively easy. Steve Hladky coupled the Levitt theory (Levitt 1982) and Brownian dynamics together, and showed how to evaluate the integrals in the Levitt theory. However, if you solve the fluxes with the Levitt theory, the computation time increases geometrically with the number of ions in the channel, whereas for Brownian dynamics it increases linearly. Therefore, when the channels are multiply occupied, it is more computationally efficient to solve the fluxes by Brownian dynamics.

Roux: Performing stochastic Brownian dynamics simulations is equivalent to solving the multi-dimensional diffusion equation, and not the only way. In both cases the physics is the same.

Eisenberg: We have published two papers on this (Barcilon et al 1993, Eisenberg et al 1995). In the first, we did an enormous simulation with four different methods. This was a general simulation, and we did not make any assumptions about there being only one ion. In the second, we did the probability theory without any simulation. They gave precisely the same answer.

Roux: For multiple ion systems, it would be more effective to carry out stochastic simulations than trying to attempt to solve this highly coupled system of different equations. If one wants to look at the effects of just a single ion, then the differential equation is not so difficult to solve. But a one-ion pore is already an approximation, and one may not want to do it.

Smart: What was the dielectric representation in your modified Poisson–Boltzmann calculations?

Roux: In the pore region the dielectric constant is one because all the atoms are present, i.e. the protein, the water and the ions.

Smart: Are they present in the sample average?

Roux: For the outside, you assume a certain screening length and dielectric constant. It's a valid approximation because it is far away from the interior of the channel. But the theory does not treat anything as a continuum in the pore region.

Koeppe: The binding of potassium or caesium is more favourable than for sodium, and the permeation of those ions is more favourable than for sodium. The size of those ions more closely matches the gramicidin channel dimension, so you would expect potassium or caesium to be closer to the centre so that the ion may follow a straighter trajectory through the channel.

Roux: It is true that small ions would tend to follow a trajectory that deviates from a straight line, but the rate of transport across the channel is dominated by the free energy landscape and not by the geometry of the path. We tend to think in terms of there being one barrier and two sites, or perhaps four sites, but for sodium there are probably many smaller barriers along the channel axis. This is probably not the case for potassium and larger cations.

Koeppe: I would like to ask Tim Cross if he thinks the helical path is inhibitory for potassium as well as for sodium.

Cross: Yes for potassium, but perhaps not for caesium.

Jakobsson: The lower mobility of the smaller ions is related to the fact that the charge centres between the ion and the carbonyl oxygen are closer together, which is essentially the same reason why small ions have a lower mobility in bulk water.

I have another comment along these lines. Benoît Roux showed all the potential wells for sodium permeation, and each of those is at a distance of one L–D pair (Roux & Karplus 1991). A few years ago we calculated the nature of the crossing of those small barriers (Chiu et al 1993). Benoît mentioned the two extremes: the diffusion theory and the rate theory. It is possible to look at the balance of these two extremes by comparing the effective mean free path of the transported species with the width of the barrier. If the mean free path is large compared to the barrier width, the rate theory fits; whereas if the mean free path is small compared to the barrier width, the diffusion theory fits. However, what we found for barriers about the width that pertains in gramicidin, which would be typical for the binding site of any channel because it's about the size of a side chain, is that you have an intermediate case. The crossing of those elementary barriers in the L–D pair is not at the rate theory limit or the diffusive limit, but is intermediate between the two. Therefore, it is necessary to compute by detailed simulation the coefficients for getting across the channel.

Cross: I would like to ask a question about the binding site. You said that it doesn't interact with Val8 because the carbonyl of Val8 is hydrogen bonded to Trp15. However, you said that it does interact with Trp15, but the carbonyl of Trp15 is hydrogen bonded to the NH in the second turn of the helix.

Roux: I'm not saying that a cation cannot interact with a carbonyl oxygen that is hydrogen bonded. I'm just saying that there is a small price to pay to interact with carbonyls that are hydrogen bonded.

Cross: But then there's no evidence that the cation interacts with the Trp15 carbonyl.

Roux: These are the results of trying to optimize the position of the sodium to make a least square fit. There is a small change in chemical shift anisotropy (CSA) at that position, and this is not a prediction. The conclusion comes out of the least square fitting procedure.

Cross: A small change in CSA means that there's little interaction.

Roux: No, in that calculation it just means that it's not deflected.

Cross: In your calculation that's true. However, we have observed that there's a change in the CSA due to the binding of cations, and so there's no evidence that that cation is interacting significantly with the Trp15 carbonyl.

Davis: When you determined the location of that binding site, and demonstrated that it was in agreement with the NMR spectroscopy data, did you calculate changes in the tensor elements or did you simply assume that they undergo a change in orientation?

Roux: The only way one can compute changes in the tensor elements is do to an electronic structure calculation. We assumed that the components of the tensor are constant.

Davis: Was the orientation of the Leu10 an average or a snapshot of a simulation?

Roux: I think it was a snapshot.

Davis: The carbonyl of the Leu10 seemed to be deflected substantially.

Cross: Does it interact with the cation when it's not deflected?

Roux: Sometimes it interacts with the cation and sometimes it doesn't. I agree that the deflections in this picture are substantial, but I just showed it as an example of the interaction. I could show you pictures in which it is not significantly deflected.

Cross: The reason why the Val8 carbonyl is highly deflected is because it's so far into the channel that it's interacting with Trp15, and it is below the level of the

Leu10 carbonyl oxygen. Because you only have two waters interacting with this cation, you have eliminated the potential barrier at the bottom of the binding site, i.e. the elimination of the third water molecule from the primary hydration sphere of the cation.

Roux: You would need to move the ion by at least 1 to 2 Å to begin to observe hydration by a third water molecule. We find two wells, one around 9.3 Å and one around 11.0 Å (Roux & Karplus 1993). In our calculations their relative energy differs by about 0.25 kcal/mol.

Cross: But this is not compatible with the NMR spectroscopy data.

Roux: This is the result of the least square fitting of all computed NMR data, assuming that the components of the tensor are fixed, which is the way you interpret the geometry.

References

Barcilon V, Chen D, Eisenberg RS, Ratner M 1993 Barrier crossing with concentration boundary conditions in biological channels and chemical reactions. J Chem Phys 98:1193–1212

Chiu SW, Novotny JA, Jakobsson E 1993 The nature of ion and water barrier crossings in a simulated ion channel. Biophys J 64:98–109

Eisenberg RS, Klosek MM, Schuss Z 1995 Diffusion as a chemical reaction: stochastic trajectories between fixed concentrations. J Chem Phys 102:1767–1780

Levitt DG 1982 Comparison of Nernst–Planck and reaction-rate models for multiply occupied channels. Biophys J 37:575–587

Roux B, Karplus M 1991 Ion transport in a gramicidin-like channel: dynamics and mobility. J Phys Chem 95:4856–4868

Roux B, Karplus M 1993 Ion transport in the gramicidin channel: free energy of the solvated right-handed dimer in a model membrane. J Am Chem Soc 115:3250–3262

The mechanism of channel formation by alamethicin as viewed by molecular dynamics simulations

Mark S. P. Sansom, D. Peter Tieleman* and Herman J. C. Berendsen*

*Laboratory of Molecular Biophysics, The Rex Richards Building, Department of Biochemistry, University of Oxford, South Parks Road, Oxford OX1 3QU, UK, and *BIOSON Research Institute and Department of Biophysical Chemistry, University of Groningen, Nijenborgh 4, 9747 A G Groningen, The Netherlands*

Abstract. Alamethicin is a 20-residue channel-forming peptide that forms a stable amphipathic α-helix in membrane and membrane-mimetic environments. This helix contains a kink induced by a central Gly-X-X-Pro sequence motif. Alamethicin channels are activated by a *cis* positive transbilayer voltage. Channel activation is suggested to correspond to voltage-induced insertion of alamethicin helices in the bilayer. Alamethicin forms multi-conductance channels in lipid bilayers. These channels are formed by parallel bundles of transmembrane helices surrounding a central pore. A change in the number of helices per bundle switches the single channel conductance level. Molecular dynamics simulations of alamethicin in a number of different environments have been used to explore its channel-forming properties. These simulations include: (i) alamethicin in solution in water and in methanol; (ii) a single alamethicin helix at the surface of a phosphatidylcholine bilayer; (iii) single alamethicin helices spanning a phosphatidylcholine bilayer; and (iv) channels formed by bundles of 5, 6, 7 or 8 alamethicin helices spanning a phosphatidylcholine bilayer. The total simulation time is *c.* 30 ns. Thus, these simulations provide a set of dynamic snapshots of a possible mechanism of channel formation by this peptide.

1999 Gramicidin and related ion channel-forming peptides. Wiley, Chichester (Novartis Foundation Symposium 225) p 128–145

Ion channels are formed in lipid bilayers by integral membrane proteins. Channels enable selected ions to move rapidly (*c.* 10^7 ions/s per channel) and passively (i.e. down their electrochemical gradients) across membranes. Ion channels are important in numerous cellular processes (Hille 1992). To understand the physical events underlying the biological properties of channels, one must characterize their dynamic behaviour. This is far from easy, as a crystallographic structure is known for only one ion channel protein, a bacterial K^+ channel (Doyle et al 1998). Thus, peptide models such as gramicidin A may provide insights into the structural basis of channel function (Woolley & Wallace 1992). However, the structural idiosyncrasies of gramicidin redirect attention to other peptide models that may more closely mimic ion channel proteins. Many channel proteins contain a central pore lined by a bundle of approximately parallel α-helices (Montal 1995). Alamethicin, a largely hydrophobic 20-residue peptide, whose structural properties have been studied in considerable detail (Cafiso 1994, Sansom 1993, Woolley & Wallace 1992), provides a good model for channels based on this structural motif.

The crystal and solution structures of alamethicin in a non-aqueous environment are strikingly similar (Esposito et al 1987, Fox & Richards 1982). The largely α-helical conformation of alamethicin is stabilized by the presence of a number of α-amino isobutyric acid (i.e. α-methyl alanine, Aib) residues in its sequence:

Ac-Aib-Pro-Aib-Ala-Aib-Ala-Gln7-Aib-Val-Aib-Gly-Leu-Aib-Pro14-Val-Aib-Aib-Glu18-Gln-Phol

NMR amide exchange data demonstrate that the largely α-helical conformation of alamethicin when dissolved in methanol is retained when it interacts with lipid bilayers (Dempsey & Handcock 1996). Amide exchange data also suggest that alamethicin in methanol undergoes hinge-bending motion about the central kink induced by the Gly-X-X-Pro14 motif (Gibbs et al 1997).

Early studies of the interactions of alamethicin with membranes using circular dichroism spectroscopy on oriented multibilayers (Vogel 1987) suggested that the orientation of the helix relative to the bilayer normal was sensitive to the hydration state and phase of the lipid. Subsequent work (Huang & Wu 1991) stressed the dependence of helix orientation on the peptide-to-lipid ratio, an increase in peptide favouring an inserted orientation over a surface associated orientation. Other investigators (North et al 1995) have concluded that alamethicin helices insert into lipid bilayers. Overall, these data suggest that alamethicin exists in a dynamic equilibrium between a surface-associated and a bilayer-inserted form.

Channel formation by alamethicin is voltage dependent. The resultant channels switch rapidly (on a *c*. 10 ms time-scale) between multiple conductance levels. The multi-conductance behaviour of alamethicin channels is generally explained in terms of the barrel stave (i.e. helix bundle) model (Baumann & Mueller 1974) in which several (*c*. five to 10) alamethicin helices form a bundle surrounding a central pore. Different conductance levels correspond to different numbers (N) of helices per bundle. Surface-bound alamethicin is believed to be the first step towards channel formation. Subsequent insertion of alamethicin helices is aided by the application of a transmembrane voltage difference (Biggin et al 1997). Inserted helices then self-assemble into bundles that form pores through which ions may flow (Breed et al 1996, 1997).

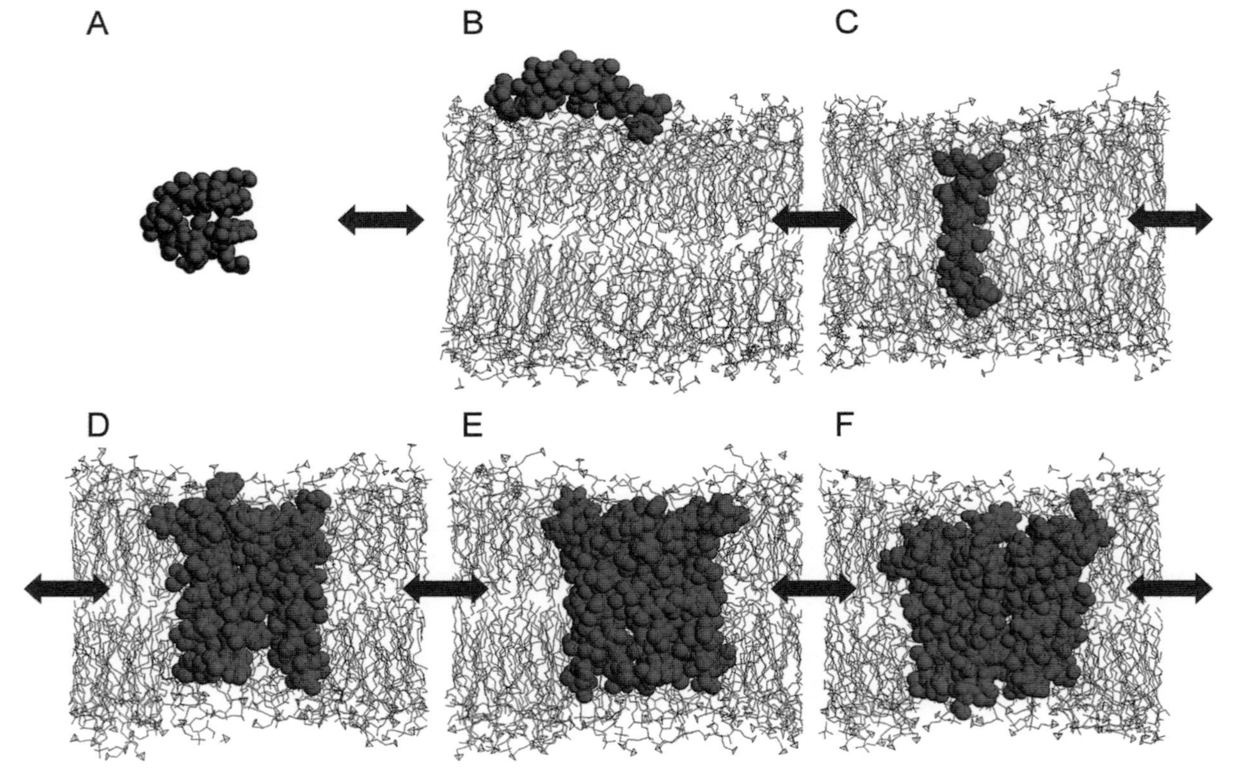

FIG. 1. Summary of the alamethicin molecular dynamics simulations: (A) single alamethicin molecules in water (alamethicin/water); (B) at the surface of a palmitoyloleoyl phosphatidylcholine (POPC) bilayer (alamethicin/surface); (C) spanning a POPC bilayer (alamethicin/transmembrane); and alamethicin helix bundles in a POPC bilayer with (D) $N=5$; (E) $N=6$; and (F) $N=7$ helices bundle. In each case alamethicin is shown in space-filling format, the POPC molecules are shown in bonds format, and the water molecules are omitted for clarity.

TABLE 1 Summary of alamethicin simulations

Simulation	*Components*	*Number of atoms*	*Duration (ns)*
Alamethicin/water	Alamethicin, 3467 waters, 1 Na^+	10569	2
Alamethicin/MeOH	Alamethicin, 1682 methanols, 1 Na^+	5157	1
Alamethicin/surface	Alamethicin[a], 128 POPC, 3552 waters, 1 Na^+	17480	2
Alamethicin/transmembrane	Alamethicin[b], 127 POPC, 3822 waters, 1 Na^+	18238	2
Alamethicin/N5	Alamethicin N=5 bundle[b], 103 POPC, 3511 waters, 1 Na^+	16729	3.7
Alamethicin/N6	Alamethicin N=6 bundle[b], 102 POPC, 3527 waters, 1 Na^+	16893	4
Alamethicin/N7	Alamethicin N=7 bundle[b], 96 POPC, 3524 waters, 1 Na^+	16740	3.7
Alamethicin/N8	Alamethicin N=8 bundle[b], 95 POPC, 3548 waters, 1 Na^+	16928	4

[a]The initial structure of alamethicin for these simulations was generated by restrained molecular dynamics simulations *in vacuo* (Biggin et al 1997). The alamethicin model thus generated is close in its backbone conformation to the X-ray structure (Fox & Richards 1982).
[b]Initial helix bundle models were generated by *in vacuo* restrained molecular dynamics simulations as described by (Breed et al 1997) with minor modifications.

To fully understand the mechanism of pore formation by alamethicin, one needs atomic resolution simulations of: alamethicin at the bilayer surface (Tieleman et al 1999a); single alamethicin helices spanning a bilayer (Tieleman et al 1999b); and bundles of alamethicin helices (Tieleman et al 1999c,d; Fig. 1). In this chapter we survey the use of molecular dynamics (Tieleman et al 1997) simulations of alamethicin, as an isolated molecule and in helix bundles (see Table 1), to help understand the mechanism of channel formation. These studies exploit the methods used by Tieleman & Berendsen (1998) in simulations of porin in a lipid bilayer. The methods are described in Tieleman et al (1999b, c, d). Simulations were run using Gromacs (http://rugmd0.chem.rug.nl/~gmx/gmx.html).

Single alamethicin molecules

The first two simulations, alamethicin/water and alamethicin/MeOH, were chosen to examine the stability of an alamethicin helix in aqueous solution and in non-aqueous (but isotropic) solution. Alamethicin/water simulates alamethicin in aqueous solution before binding to and inserting into bilayers. Alamethicin/MeOH is of interest as methanol is commonly used as a membrane-mimetic solvent in solution studies of membrane-active peptides and peptide fragments of membrane proteins.

Major changes in the conformation of alamethicin occur in water, whereas the alamethicin X-ray structure is largely maintained in methanol (Fig. 2A,B). During the alamethicin/water simulation the N-terminal segment (residues 1 to 10) retains

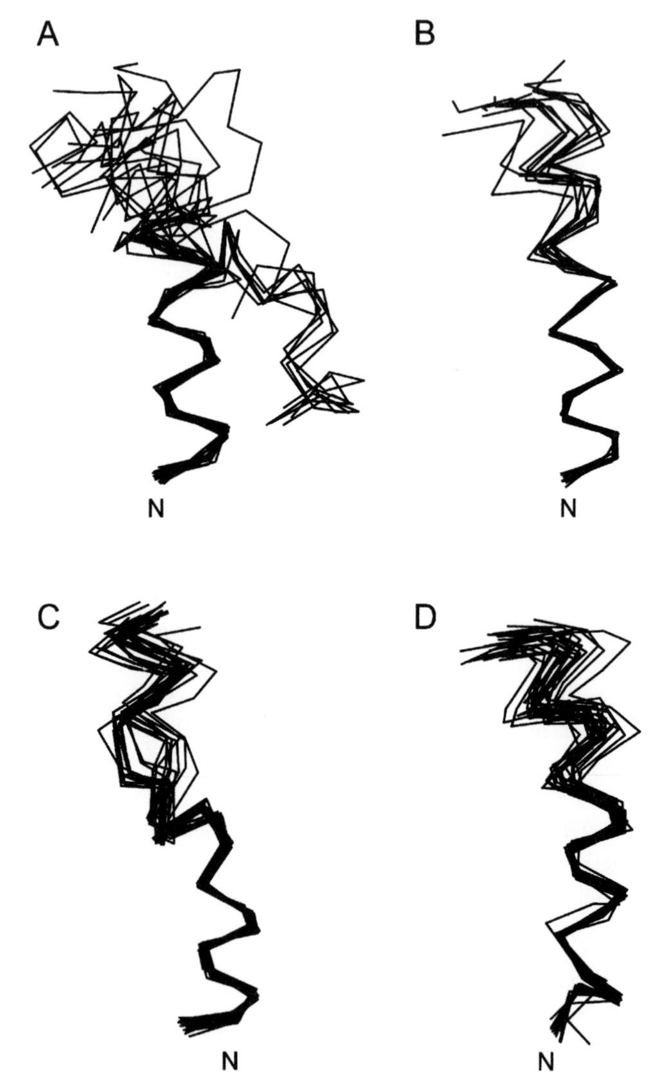

FIG. 2. $C\alpha$ traces, corresponding to structures saved every 200 ps, for: (A) alamethicin/water; (B) alamethicin/MeOH; (C) alamethicin/surface; and (D) alamethicin/transmembrane. In each case the N-terminal helices (residues 1 to 10) were superimposed.

its initial α-helical conformation but there are marked conformational changes about the Gly-X-X-Pro hinge and in the C-terminal segment. In particular, in the second half of the simulation the alamethicin molecule folds back on itself. Analysis of the time-dependent secondary structures reveals that in alamethicin/MeOH the largely α-helical conformation of the peptide is maintained throughout the simulation. Thus, in isotropic solution the alamethicin helix seems to require a non-aqueous solvent for stability, at least on a nanosecond time-scale.

Molecular dynamics simulations were also used to compare alamethicin/water with alamethicin bound to the surface of a palmitoyloleoyl phosphatidylcholine (POPC) bilayer (alamethicin/surface). The surface simulation corresponds to a loosely bound alamethicin molecule that interacts with lipid headgroups but does not penetrate the hydrophobic core of the bilayer. In alamethicin/surface, loss of helicity was restricted to the C-terminal third of the molecule and the rod-shaped structure of the peptide was retained (Fig. 2C). About 10% of the peptide/water H-bonds present in alamethicin/water were replaced by peptide/lipid H-bonds in alamethicin/surface. It seems that some degree of stabilization of the alamethicin α-helix occurs at a bilayer surface even without interactions between hydrophobic side chains and the acyl chain core of the bilayer. Interestingly, similar results have been seen in simulations of dermaseptin (an antimicrobial peptide from frog skin) (P. La Rocca & M. S. P. Sansom, unpublished results 1998) and of melittin (a membrane-active peptide from bee venom; A. Baumgärtner & J. Lin, personal communication 1998) when loosely bound to the surface of POPC bilayers.

Molecular dynamics simulations of a single transmembrane alamethicin helix (alamethicin/transmembrane) reveal similar conformational dynamics to that of an alamethicin helix in methanol. There was little change from the initial helical conformation of the peptide, but the molecule underwent hinge-bending motion about its central Gly-X-X-Pro sequence motif. Analysis of H-bonding interactions revealed that the polar C-terminal side chains of alamethicin provided an 'anchor' to the bilayer/water interface via formation of multiple H-bonds which persisted throughout the simulation. This may explain why the preferred mode of helix insertion into the bilayer is N-terminal, which is believed to underlie the asymmetry of voltage activation of alamethicin channels.

Molecular dynamics simulations in the presence of a simple mean field approximation, in which a bilayer is represented by a hydrophobicity potential and which included a transbilayer voltage difference, have been used to explore possible mechanisms of helix insertion (Biggin et al 1997). Using this admittedly somewhat simplistic representation of a bilayer, an alamethicin helix inserted spontaneously in the absence of a transbilayer voltage. Application of a *cis* positive voltage decreased the time to insertion. Insertion of the helix resulted in a decrease in the mean kink angle, thus helping the alamethicin molecule to span the bilayer.

In summary, molecular dynamics simulations of isolated alamethicin helices suggest that: (i) the alamethicin helix is stable in transmembrane and membrane-mimetic environments; (ii) the alamethicin helix is not stable in water; (iii) some stabilization of an alamethicin helix occurs at the bilayer surface; and (iv) the C-terminus of the helix acts as an anchor to the bilayer surface, thus favouring N-terminal insertion in response to a *cis* positive voltage. Once inserted, the helices are believed to assemble to form a helix bundle. As yet this (presumably slow) process has not been investigated by simulations. However, molecular dynamics simulations have been used to explore the structure and dynamics of the bundles thus formed.

Alamethicin helix bundles

Channels formed by approximately parallel bundles of alamethicin helices containing between $N=4$ and 8 helices per bundle have been modelled using brief *in vacuo* molecular dynamics simulations (which included inter- and intra-helix distance restraints; Breed et al 1997). Channel conductances were predicted on the basis of pore radius profiles (Smart et al 1997), and suggested that the $N=4$ bundle did not form a continuous pore, whereas bundles with $N>5$ helices corresponded to open channels.

The *in vacuo*-generated models of alamethicin helix bundles included a ring of Glu18 side chains at the C-terminal mouth of the channel. In this respect, alamethicin is similar to the helices of the pore-lining M2 bundle of the nicotinic acetylcholine receptor (nAChR) channel (Adcock et al 1998). Studies of the ionization state of the ring of glutamate residues at the C-terminal mouth of the nAChR suggest that these residues are not fully ionized because their pK_as are shifted due to their location at the C-terminus of the α-helix dipole and their proximity to one another. Prediction of pK_as of the ring of Glu18 side chains at the C-terminal mouth of the alamethicin pore suggests that at neutral pH all but one of these side chains will remain protonated (Tieleman et al 1999d).

Molecular dynamics simulations of $N=5$, 6 (Fig. 3), 7 and 8 bundles of alamethicin helices in a POPC bilayer have been run (Table 1). In each case, a single Glu18 side chain in the helix bundle was assumed to be ionized. Earlier 2 ns simulations of an $N=6$ bundle revealed that if all of the Glu18 side chains were ionized, the bundle was unstable; if none of the Glu18 side chains was ionized the bundle was stable (Tieleman et al 1999c). In each of the simulations with a single ionized Glu18 the bundle remained stable throughout 4 ns. Analysis of fluctuations from the average structure during the molecular dynamics trajectories indicated that the C-terminal half of the alamethicin molecule within a bundle underwent greater fluctuations than the N-terminal half. These fluctuations are less marked than those for alamethicin/water but greater than for alamethicin/transmembrane or alamethicin/MeOH. A similar pattern was seen in the secondary structure of the helices. This presumably reflects the anisotropic environment of an alamethicin molecule within a bundle, with its more apolar face interacting with the lipid but its more polar face forming H-bonds with water molecules inside the pore.

Pore radius profiles (Fig. 4A) were determined every 50 ps for the alamethicin bundle simulations. The $N=5$ bundle is constricted (minimum pore radius *c.* 0.15 nm) at its C-terminal mouth, i.e. in the vicinity of the Glu18 side chain ring. There is a lesser degree of constriction in the vicinity of the ring of Gln7 side chains (at z *c.* 2.8 nm). This pattern is inverted in the $N=6$, 7 and 8 bundles, with the narrower constriction in the region of the Gln7 ring, and a lesser degree of constriction at the C-terminal mouth. Thus, the smallest helix bundle, thought to correspond to the lowest conductance level of the channel, differs somewhat in its pore geometry from the others.

There is a continuous column of water within the lumen of the alamethicin helix bundles. The longitudinal (i.e. along the pore, z, axis) diffusion coefficients of waters within the pore are markedly reduced relative to those of waters in the bulk region (Fig. 4B). In the narrowest regions of all four pores the water diffusion coefficients fall to less than a 10th of the bulk water value. Thus, the reduction in the translational motion of water within narrow pores seen in simple (no bilayer) simulations (Breed et al 1996) is reproduced in the current simulations. By comparison with other systems (Smith & Sansom 1998), this reduction in water diffusion coefficients suggests that diffusion of ions will also be slower through alamethicin pores than in bulk solution.

The dipoles of the water molecules within all four pores are oriented by the surrounding parallel helix dipoles. Thus, within the pore the mean z component of the water dipoles is nearly 1.8 Debye (Fig. 4C), which should be compared with a dipole moment of 2.3 Debye for a single SPC water. There is some dependence of the degree of water dipole orientation on the number of helices per bundle.

The observed degree of orientation of the water dipoles may be employed to calculate the local field experienced by water molecules within the pore (Sansom et al 1997, Tieleman et al 1999c,d). Thus, the values of μ_z averaged along the pore when compared with μ_0 can be used to estimate the field strengths within each pore due to the aligned helix dipoles. Using the Langevin equation:

$$\mu_z = \mu_0\left[\coth\left(\frac{\mu_0 E_z}{k_B T}\right) - \left(\frac{k_B T}{\mu_0 E_z}\right)\right]$$

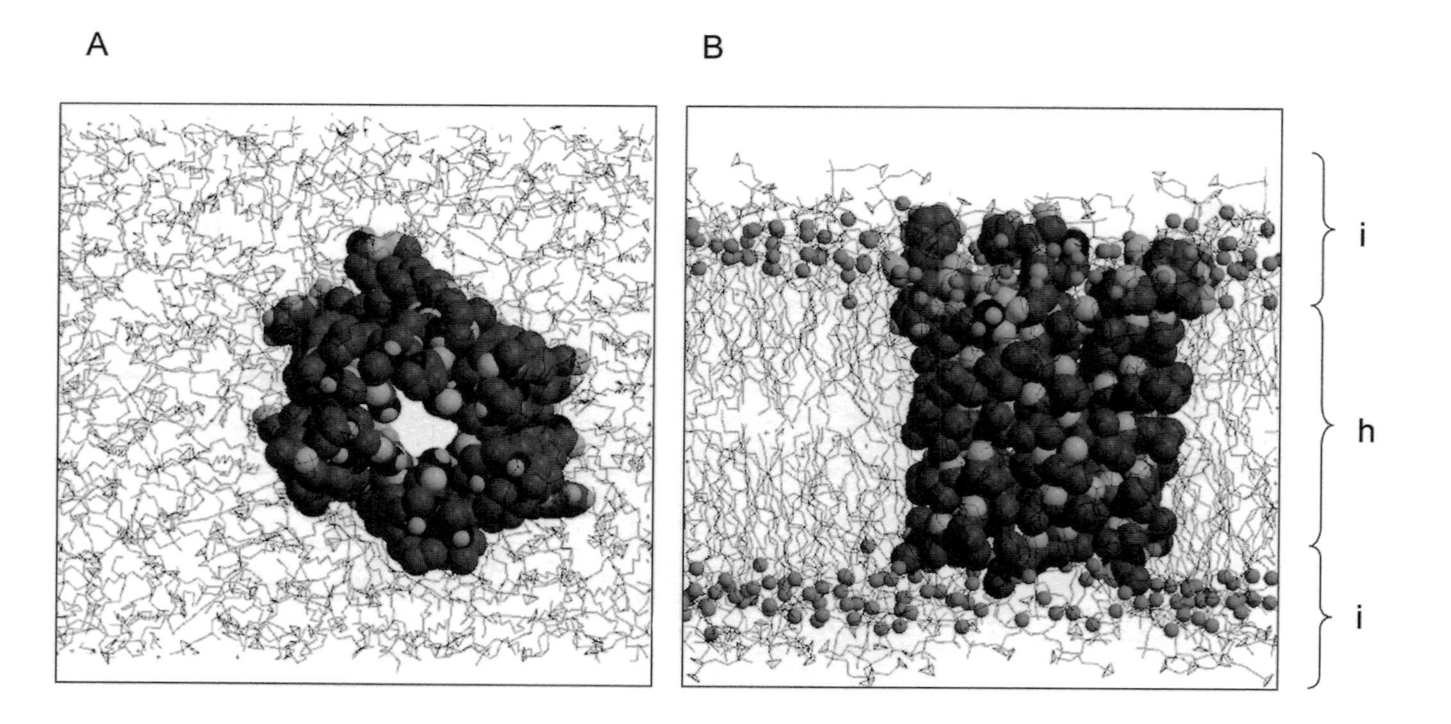

FIG. 3. Images of the alamethicin/N6 simulation at $t = 4$ ns. Peptide atoms are shown in space-filling format, and the carbonyl oxygens of the phospholipid molecules as small mid-grey spheres. (A) View down the pore (z) axis, with the N-terminal mouth of the pore towards the reader; (B) view perpendicular to the pore axis, with the C-terminal mouth of the pore uppermost. The hydrophobic core and interfacial regions of the bilayer are represented by h and i, respectively.

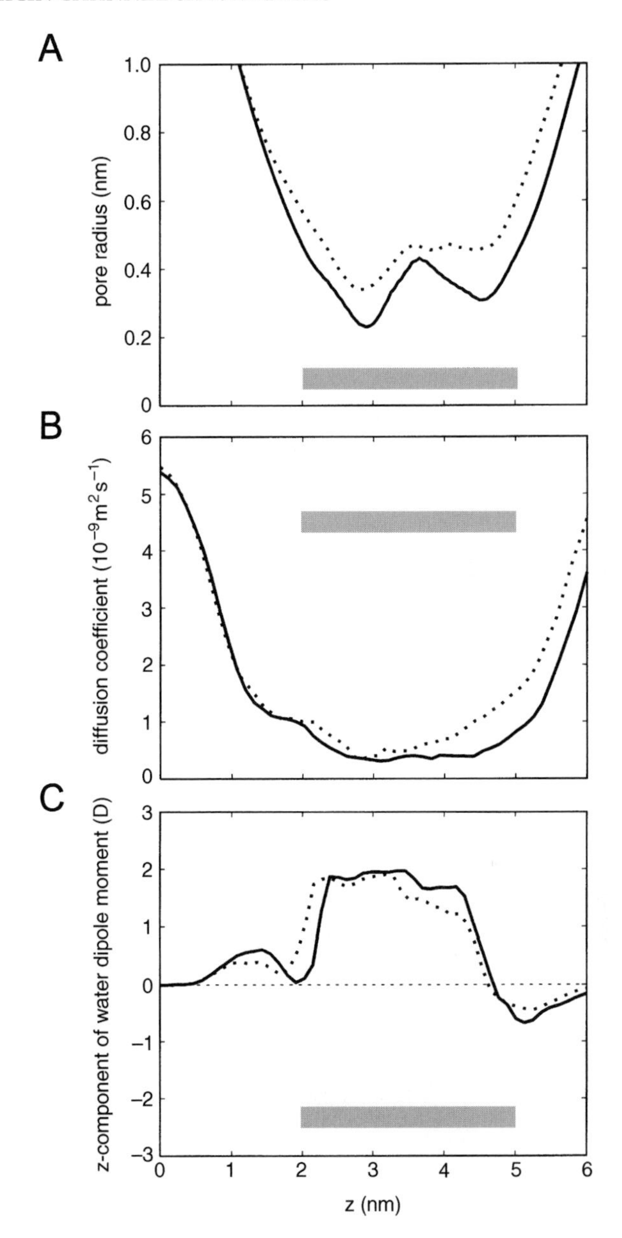

FIG. 4. (A) Pore radius profiles. The profiles were averaged from structures saved every 50 ps. The N-termini of the helices are at z *c.* +2 nm and the C-termini at z *c.* +5 nm (as indicated by the horizontal grey bar). (B) Water diffusion coefficients (D_z) as a function of position along the pore (z) axes. (C) Projection of water dipole moments onto the pore axes. For each graph, the following convention is used: solid line for alamethicin/N6; broken line for alamethicin/N8.

where E_z is the z component of the electrostatic field due to the helix dipoles, k_B and T are the Boltzmann constant and temperature, respectively, and where μ_0 is the dipole moment of water and μ_z is its projection along the z (pore) axis. This yields fields of $E_z = 2.4, 2.3, 1.8$ and 1.8×10^9 V/m, respectively, for the $N = 5, 6, 7$ and 8 pores. Such strong interactions between water dipoles and aligned helix dipoles will contribute to the stability of the helix bundles (Fig. 5).

Other channels

These studies of alamethicin show that molecular dynamics simulations in a fully hydrated bilayer environment may be used to characterize the structural and dynamic properties of ion channels formed by bundles of α-helices. This work on alamethicin may be extended to other channels formed by peptides (e.g. the LeuSer peptides of DeGrado and colleagues [Lear et al 1988, Mitton & Sansom 1996]) and to 'simple' ion channels encoded by viruses, such as the M2 protein from influenza A (Forrest et al 1998). Such simulations may be used to refine initial models of these channels developed using restrained *in vacuo* simulations.

Alamethicin helix bundles may be considered as a simple paradigm for more complex ion channel proteins. Therefore, it is likely that fluctuations similar to those seen in the helices of an alamethicin bundle within a bilayer will occur in more complex ion channels. Conformational changes in pore-lining helices have

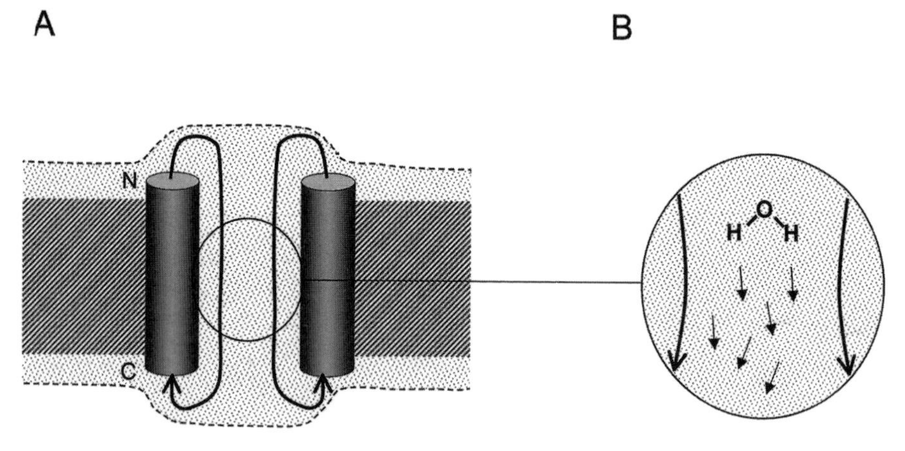

FIG. 5. (A) Schematic diagram of a pore formed by a bundle of parallel α-helices (grey cylinders) embedded in a lipid bilayer (diagonal shading). The boundary between bulk water and 'immobilized' water (stippled) is indicated by a broken line. The two arrows indicate the electrostatic field created by the helix bundle. (B) The inset shows the arrangement of water dipoles (small arrows) in response to the electrostatic field (large arrows) created by the helix bundle.

been implicated in gating of, for example, nAChR (Unwin 1995) and of K^+ channels (Perozo et al 1998). Our studies suggest that molecular dynamics simulations in a lipid bilayer help to explain such changes.

Future directions

What remains to be done to obtain a complete picture of alamethicin channel formation and properties? The current simulation times are relatively short. For example, the mean dwell time of an ion passing through a channel is *c.* 100 ns and the mean time for a lipid molecule to exchange with its lateral nearest neighbour is *c.* 10 ns. Encouragingly, simulations of small proteins of up to 1 μs are now possible (Duan & Kollman 1998). However, longer simulations alone will not be sufficient to understand, for example, ion permeation. To understand better the energetics of pore/ion interactions, free energy profiles (Roux 1996) for the ion as it moves along the pore should be calculated.

Current simulations do not provide any real description of the forces driving alamethicin helix insertion into the bilayer and aggregation within the bilayer. To understand these driving forces, we will require an extension of current work to a larger time and length scale, possibly using more approximate methods. For example, one might ask: to what extent are the processes of insertion aggregation of alamethicin helices coupled? To understand this, we will require simulations with multiple alamethicin helices. Furthermore, peptides such as alamethicin alter their position, orientation and/or conformation at a bilayer/water interface when a transbilayer voltage difference of *c.* +100 to +200 mV is applied. Although there have been preliminary studies of how to include a transbilayer voltage in simulations (Biggin et al 1997, Roux 1997) further methodological work is required before such methods may be applied to simulations of peptide/water/bilayer systems. Once all these problems have been solved, we may be able to fully describe alamethicin channel formation and properties at atomic resolution.

Acknowledgements

Work in MSPS's laboratory is supported by The Wellcome Trust. DPT was supported in part by the European Union under contract CT94-0124.

References

Adcock C, Smith GR, Sansom MSP 1998 Electrostatics and the selectivity of ligand-gated ion channels. Biophys J 75:1211–1222

Baumann G, Mueller P 1974 A molecular model of membrane excitability. J Supramol Struct 2: 538–557

Biggin P, Breed J, Son HS, Sansom MSP 1997 Simulation studies of alamethicin-bilayer interactions. Biophys J 72:627–636

Breed J, Sankararamakrishnan R, Kerr ID, Sansom MSP 1996 Molecular dynamics simulations of water within models of ion channels. Biophys J 70:1643–1661

Breed J, Biggin PC, Kerr ID, Smart OS, Sansom MSP 1997 Alamethicin channels—modelling via restrained molecular dynamics simulations. Biochim Biophys Acta 1325:235–249

Cafiso DS 1994 Alamethicin: a peptide model for voltage gating and protein–membrane interactions. Annu Rev Biophys Biomol Struct 23:141–165

Dempsey CE, Handcock LJ 1996 Hydrogen-bond stabilities in membrane-reconstituted alamethicin from amide-resolved hydrogen-exchange measurements. Biophys J 70:1777–1788

Doyle DA, Cabral JM, Pfuetzner RA et al 1998 The structure of the potassium channel: molecular basis of K^+ conduction and selectivity. Science 280:69–77

Duan Y, Kollman PA 1998 Pathway to a folding intermediate observed in a microsecond simulation in aqueous solution. Science 282:740–744

Esposito G, Carver JA, Boyd J, Campbell ID 1987 High resolution ^{1}H NMR study of the solution structure of alamethicin. Biochemistry 26:1043–1050

Forrest LR, DeGrado WF, Dieckmann GR, Sansom MSP 1998 Two models of the influenza A M2 channel domain: verification by comparison. Folding & Design 3:443–448

Fox RO Jr, Richards FM 1982 A voltage-gated ion channel model inferred from the crystal structure of alamethicin at 1.5 Å resolution. Nature 300:325–330

Gibbs N, Sessions RB, Williams PB, Dempsey CE 1997 Helix bending in alamethicin: molecular dynamics simulations and amide hydrogen exchange in methanol. Biophys J 72:2490–2495

Hille B 1992 Ionic channels of excitable membranes, 2nd edn. Sinauer Associates, Sunderland, MA

Huang HW, Wu Y 1991 Lipid–alamethicin interactions influence alamethicin orientation. Biophys J 60:1079–1087

Lear JD, Wasserman ZR, DeGrado WF 1988 Synthetic amphiphilic peptide models for protein ion channels. Science 240:1177–1181 (erratum: 1989 Science 245:1437)

Mitton P, Sansom MSP 1996 Molecular dynamics simulations of ion channels formed by bundles of amphipathic α-helical peptides. Eur Biophys J 25:139–150

Montal M 1995 Design of molecular function: channels of communication. Annu Rev Biophys Biomol Struct 24:31–57

North CL, Barranger-Mathys M, Cafiso DS 1995 Membrane orientation of the N-terminal segment of alamethicin determined by solid-state ^{15}N NMR. Biophys J 69:2392–2397

Perozo E, Cortes DM, Cuello LG 1998 Three-dimensional architecture and gating mechanism of a K^+ channel studied by EPR spectroscopy. Nat Struct Biol 5:459–469

Roux B 1996 Valence selectivity of the gramicidin channel: a molecular dynamics free energy perturbation study. Biophys J 71:3177–3185

Roux B 1997 Influence of the membrane potential on the free energy of an intrinsic protein. Biophys J 73:2980–2989

Sansom MSP 1993 Structure and function of channel-forming peptaibols. Q Rev Biophys 26:365–421

Sansom MSP, Smith GR, Adcock C, Biggin PC 1997 The dielectric properties of water within model transbilayer pores. Biophys J 73:2404–2415

Smart OS, Breed J, Smith GR, Sansom MSP 1997 A novel method for structure-based prediction of ion channel conductance properties. Biophys J 72:1109–1126

Smith GR, Sansom MSP 1998 Dynamic properties of Na^+ ions in models of ion channels: a molecular dynamics study. Biophys J 75:2767–2782

Tieleman DP, Berendsen HJC 1998 A molecular dynamics study of the pores formed by *Escherichia coli* OmpF porin in a fully hydrated POPE bilayer. Biophys J 74:2786–2801

Tieleman DP, Marrink SJ, Berendsen HJC 1997 A computer perspective of membranes: molecular dynamics studies of lipid bilayer systems. Biochim Biophys Acta 1331:235–270

Tieleman DP, Berendsen HJC, Sansom MSP 1999a Surface binding of alamethicin stabilises its helical structure: molecular dynamics simulations. Biophys J 76:3186–3191
Tieleman DP, Sansom MSP, Berendsen HJC 1999b Alamethicin helices in a bilayer and in solution: molecular dynamics simulations. Biophys J 76:40–49
Tieleman DP, Berendsen HJC, Sansom MSP 1999c An alamethicin channel in a lipid bilayer: molecular dynamics simulations. Biophys J 76:1757–1769
Tieleman DP, Breed J, Berendsen HJC, Sansom MSP 1999d Alamethicin channels in a membrane: molecular dynamics simulations. Faraday Disc 111:209–223
Unwin N 1995 Acetylcholine receptor channel imaged in the open state. Nature 373:37–43
Vogel H 1987 Comparison of the conformation and orientation of alamethicin and melittin in lipid membranes. Biochemistry 26:4562–4572
Woolley GA, Wallace BA 1992 Model ion channels: gramicidin and alamethicin. J Membr Biol 129:109–136

DISCUSSION

Woolley: Although a static model of the tetramer does not appear to contain much of a pore, I wonder if during a simulation of the tetramer in a bilayer you might see transient openings, so that the tetramer could be the lowest conducting state in single-channel recordings?

Sansom: We have run $N=4$ bundle simulations in a bilayer and the channel does not open, even after *c.* 2 ns.

Jordan: You said that the protonation state of alamethicin is determined more by the interaction with the peptide dipole than by the direct interactions between the glutamates. If you substituted glutamate with lysine, would you expect the lysine to be more protonated than the glutamate?

Sansom: Glutamate is easy to analyse, because all the electrostatic effects altering the pK_a are pushing in the same direction. We have only done some preliminary analysis of Drew Woolley's derivative in which the glutamates are substituted by lysines, but the situation does seem to be more complicated. The helix dipoles are pushing for the lysine to be ionized, but the interactions of the slightly longer lysine side chains with one another across pore are tending to suppress ionization. The calculations to sort this out are rather fiddly. Briefly, if you do the intrinsic pK_a measurements, where you just calculate solvation terms, you find that the lysines want to be ionized, but as soon as you introduce the other terms corresponding to Lys-Lys interactions, you find that they don't. On average, about 50% of them are ionized. Drew, would it be fair to say that it is sensitive to the conformation?

Woolley: We see about the same fraction ionized for a number of different conformations. This also fits with the level of charge selectivity we see.

Jakobsson: I have two questions about the glutamates. The argument you made about the fully charged glutamates having a role in binding rather than in permeation would certainly pertain if only one ion were present, but if there was a knock on/knock off method, in which the second ion could knock off the first, then presumably there could be a high rate of permeation consistent with deep wells. Could you respond to that point? And my second question is, is there a relationship between the charge of one of the glutamates and the developing asymmetry of the channel?

Sansom: Your second point is neat, and I hadn't thought about it. It is testable, but if it proved to be the case, one would wonder whether it is really happening, because even if we say that on average only one glutamate is ionized, the protons could hop between glutamates on a relatively short time-scale.

In terms of your first question, I was almost reluctant to show the graph of the potential energy because it is so crude. My feeling as to why the conductance data indicate that we are correct in assuming that not all glutamates are ionized is that the experimental differences between the Glu18 alamethicin and Gln18 alamethicin conductances are not huge, so one can assume that not all the glutamates are ionized. My guess is that if we assume a ring of charge -6 as opposed to zero, we should expect to observe differences in single channel conductance between the Glu18 and Gln18 alamethicins, although the differences may not be large.

Koeppe: I have a couple of questions about symmetry. When you make a model, do you start with a perfectly symmetrical model in which all the α-helices are the same? And do you do special docking procedures to fit the subunits together?

Sansom: The models are built from a restrained molecular dynamics technique, rather like that used for building structures on the basis of NOE-derived constraints in NMR experiments. We start off with a symmetrical α-carbon template, and we then build in the side chains such that they will automatically select conformations compatible to the secondary structure of the monomers. We then do an *in vacuo* simulation, in which we allow the α-carbon positions to move, so there's either no symmetry restraints imposed or at the best a weak α-carbon symmetry restraint. Having developed those models, which are not exactly symmetrical but have a 'history' of symmetry, we do the bilayer simulations, and in these simulations there are no restraints at all. We are just watching the drift from initial structures that are approximately symmetrical.

Koeppe: If there is a partial charge on the glutamates, do you put an equal partial charge on each of them?

Sansom: We worried about that, and in the end we thought which ever way we do it it's wrong, but given that the force field doesn't allow protons to come on and come off, we put it on one particular glutamate.

Koeppe: And finally, do you have any general conclusions about what the side chains are doing during your simulations?

Sansom: We've looked at the aromatics, but the simulations aren't long enough to say much. Peter Tieleman carried out a more detailed analysis of the tryptophans, the phenylalanines and the tyrosines in his porin simulations. We have also looked at tryptophans in the influenza M2 simulations, and the phenylalaninols in the alamethicin simulations. The conclusion from all these studies (see Tieleman et al 1998) is that they can adopt many different orientations. On balance, we haven't run the simulations for long enough to sample side chain conformations properly, but certainly there doesn't seem to be any clear preferential orientation.

Busath: Your helices appear to go straight through with the axis perpendicular to the bilayer, and the channels are a little bit shorter than the thickness of the alamethicin peptide, so there wasn't any imposed tilting of the helices with respect to each other. Do you expect there to be none, because of the discrepancy between the length and thickness of the bilayer? Why did you not impose tilted axes?

Sansom: We didn't impose tilted axes because no one has any strong evidence that they are present. We initially thought that we might see a pronounced tilt in the initial model-building exercise, but the presence of the second methyl group in the α-aminoisobutyric acid (Aib) residues makes the helices more slippery and less likely to pack in a ridges-and-grooves fashion. One thing that we haven't done yet, because 2 ns is not long enough to do it, is to compare the internal mobility of the helix bundle, i.e. the extent of the fluctuations in helix–helix parameters, with the fluctuations of the helix bundle in the bilayer as a whole. If we run the simulations for about 4 ns, we do see that the helix bundle starts to move as a rigid body, such that the principle axis of the pore is no longer perpendicular to the bilayer hemisphere, i.e. it is tilted by 10–15°. However, we need to do longer simulations on all systems to be certain about that. It does suggest, however, that these simulations give us the chance to find out what aromatic side chains are doing when they are trying to lock a membrane protein into the bilayer.

Separovic: In your experiment where you showed the pore alignment in different environments, was the standard structure of the monomer based on X-ray data?

Sansom: We wondered about starting with the X-ray structure, but there are three different monomer structures in the asymmetric unit, so we went back to some earlier work (Biggin et al 1997). In these studies we used a simulated annealing procedure starting with a linear α-helix template. This gave us an alamethicin monomer model, and the deviation of this from any of the three crystal monomers was no bigger than that of any of the three crystallographic monomers from the other two. Therefore, it's a model structure, but I suspect within the range of ensemble that the crystallographic structure is also sampled from.

Separovic: How many lipids are there per monomer?

Sansom: It depends how big a hole you cut to put the bundle in, but about 100.

Separovic: What I found interesting is that the long axis of melittin, for example, in methanol is relatively straight, with a bend of about 15°, and that this angle was also about 15° in the transbilayer form.

Stein: I was interested in the suggestion that the water molecules seemed to be oriented. Can you comment on how this is likely to effect the water permeability, and have you done any calculations on the possible water permeabilities of the alamethicin channels? How do these compare with experimental numbers, if they are available?

Sansom: I'm not sure whether anyone has measured the water permeabilities of alamethicin.

Jakobsson: I'm sure that although no one has measured the water permeability, it is possible to predict it from a fluctuation analysis of the data you already have.

Bechinger: The highly ordered water face observed when several monomers are present suggests that there is a loss of entropic energy. How does this compare to the gains from favourable interactions?

Eisenberg: Mark Sansom changes the temperature, so your question is unanswerable.

Huang: At low alamethicin concentrations, alamethicin stays on the membrane surface. It forms a channel only transiently, although it is possible to increase the possibility of channel formation by increasing the potential. At high alamethicin concentrations, however, it forms stable pores. Therefore, when studying the stability of the alamethicin channel, one has to be careful because it's not a stable channel under all conditions.

Bechinger: But once you know how much alamethicin is present, you can calculate the possibility of channel formation.

Hladky: It is hard to see how to make an experimental measurement of the water permeability of an alamethicin pore because there is no way to tell how many pores are present, so we can't divide the measured permeability by the number of pores.

Stein: At any concentration it is possible to do electrical measurements, and interpret these in terms of the relative probability of channel formation.

Hladky: But the problem is that when you have enough channels to be able to produce a water permeability you can measure, you are beyond the point at which you can do noise analysis. Therefore, we have no way of obtaining the single channel conductance value that we require to normalize the data. The reason you can do this with gramicidin is that you can resolve the single channel conductance and you have a reasonable chance that this is close to the channel conductance under the conditions of the water permeability measurement.

Woolley: I would like to comment on the voltage dependence of activation (i.e. channel opening) for alamethicin. The voltage dependence decreases when alamethicin molecules are chemically linked together, so it may be possible to create a situation where the alamethicin channel is open in the absence of an applied field and it is still possible to resolve single-channel events.

Roux: Have you considered the influence of voltage on the channel?

Sansom: We have run some simulations in the past with a simple helix bundle plus a water 'droplet', but with no bilayer, and we looked at the water orientation to try to get feeling for the dielectric (Sansom et al 1997). However, in terms of to what extent voltage stabilizes the alamethicin bundle, which would address your question, we admit that we are doing a simulation on an entity that is at best transiently stable in the absence of voltage. Peter Tieleman has run a simulation on a single alamethicin helix interacting with a bilayer in which he applied a linear voltage, but nothing much happens (P. Tieleman, personal communication 1998). We are hoping to look at a helix bundle with or without a voltage and see whether there are any obvious differences.

Jakobsson: Peter Jordan (1987) has done electrostatic calculations with idealized but relatively realistic channels in dielectric media. He showed that the potential profile across the channels was relatively linear.

References

Biggin P, Breed J, Son HS, Sansom MSP 1997 Simulation studies of alamethicin–bilayer interactions. Biophys J 72:627–636

Jordan PC 1987 How pore mouth charge distributions alter the permeability of transmembrane ionic channels. Biophys J 51:297–311

Sansom MSP, Smith GR, Adcock C, Biggin PC 1997 The dielectric properties of water within model transbilayer pores. Biophys J 73:2404–2415

Tieleman DP, Forrest LR, Sansom MSP, Berendsen HJC 1998 Lipid properties and the orientation of aromatic residues in OmpF, influenza M2 and alamethicin systems: molecular dynamics simulations. Biochemistry 37:17554–17561

General discussion II

The Poisson–Nernst–Planck model

Eisenberg: I would like to suggest a simple working hypothesis that works for a variety of channels (Eisenberg 1998a,b). It extends the work of Benoît Roux in that it addresses non-equilibrium conditions. This is necessary since few people measure current–voltage (I–V) relationships with potentials smaller than 25 mV and few people use concentration gradients that are negligible. The working hypothesis is that the only variables important in determining permeation are the distribution of fixed charge along the wall of a channel and a diffusion coefficient of an ion. The advantage of this hypothesis is that it is specific, and by mathematics alone one can predict a current point, so there is no vagueness about predicting I–V relationships at any concentrations. The hypothesis can be tested in a variety of channels by saying, can I find a single function, $P(x)$, that predicts all the measured I–V relationships? We have now done this in about seven types of channels that have different qualitative properties and different values for $P(X)$. In the calcium release channel, one needs a spatial uniform charge to predict a I–V relationship from 20 mM to 2 M lithium, sodium, potassium and rubidium and caesium at −200 mV to +200. We were able to do a similar calculation for porin. Three or four parameters were required to describe the charge profile. We can do mutations where we know what the change in charge is. We can analyse the two experiments independently, and we can recover the correct charge within about 8%. The physics is extremely simple. This working hypothesis is equivalent to saying that if you do Poisson–Boltzman calculations with flux consistently by mathematics, you are then able to account for a wide range of channel data. We have just started to work on gramicidin because the data have just been made available, and early results look good. The model we have used is one-dimensional.

Jordan: I would like to comment on your point about equilibrium versus non-equilibrium conditions. In statistical mechanics, if you perturb the conditions around equilibrium and use fluctuation analysis, you get a linear domain for transport coefficients. Therefore, as a practical tool the equilibrium phenomena analyses and fluctuation analyses give you information about transport.

Eisenberg: Although what you say is correct, I disagree with you in the context of the channels. The qualitative properties of channels, e.g. selectivity, occur far from equilibrium. You can study a field effect transistor with zero voltage, but you won't see any amplification and you won't see its qualitative properties. In my opinion, to see the qualitative property of selectivity, which is what most of us are interested in, you have to have current flow.

Roux: I disagree. Selectivity doesn't have to reflect non-equilibrium conditions; selectivity may be seen as the propensity of an ion to occupy a region, and it can be defined using the reversible Nernst potential. There can be a large concentration gradient and a large potential at equilibrium.

Eisenberg: Every experimental measurement of selectivity has been done when there are large current flows, because almost none of the channels are perfectly selective.

Hladky: There are aspects near equilibrium that can be discussed, and the binding experiments that the NMR methods are addressing should be in this category. There are other aspects that cannot, and the actual current selectivity of gramicidin is far from equilibrium in all the measurements we can actually do. Partly, this is because it is access limited, and the selectivity of conductance that we can measure has little to do with the selectivity of first-ion binding. The binding shows greater selectivity.

Jordan: Isn't it possible to get around these problems by dealing with fluctuations much as one does in bulk systems? Clearly channel behaviour depends on ionic concentrations in the bathing solutions. The reference state for the fluctuations isn't just the channel, but the solutions as well. Thus, as solution properties change, the reference state changes. As long the system is a true equilibrium (for example, gramicidin with symmetric electrolytes at zero transmembrane voltage), fluctuation analyses using the proper ionic concentrations in the bulk solutions should be adequate for describing transport. For bi-ionic conditions things aren't as simple. Onsager reciprocity is only applicable for fluctuations about equilibrium. This isn't the case for permeability measurements with bi-ionic solutions.

Hladky: You're probably correct. If you know both the fluctuations and the equilibrium, you will be able to predict the low voltage conductance.

Eisenberg: There is no doubt that in a system which is qualitatively like a machine, in that flows are necessary for its function, it is not possible to use equilibrium analysis.

Jakobsson: I have wrestled with this for a while, and I'm inclined towards Peter Jordan's view. The movement of individual ions and water molecules inside the channel is not affected by the bulk concentrations. This is a boundary condition, and it tells you the likelihood of the channel being occupied. There is a difference in how we should treat the chemical potential, as opposed to the electrical potential. The electrical potential changes the physics of what's happening inside the channel to some extent and has to be taken into account when considering the microscopic physics. The chemical potential, however, is a boundary condition, and it tells you the likelihood of the ion being inside the channel in one state or another.

Roux: Most non-equilibrium theories concerning transport processes in liquids begin by defining the equilibrium structure, i.e. a free energy profile or potential mean force. It is then assumed that the probability distribution deviates from that of equilibrium. A non-equilibrium state is conceptualized as the relaxation back to the proper distribution. Often it is assumed that the energy surface is constant during this process, or it is assumed that it relaxes through a linear response process. Perhaps, what I'm saying is a technical point, but in all non-equilibrium processes it is always useful to define the free energy profile at equilibrium.

Sansom: Part of the problem may depend on where you place the boundaries. In a single file channel, like gramicidin or the selectivity filter of the potassium channel, the boundaries are distant and the only non-equilibrium part is the electrostatic field across the channel. However, in porin or alamethicin, for example, it is not clear where the boundary between the outside and the inside lies. In the latter case, there is a region where the chemical potential does need to be considered because ions are able to arrive and leave, and this region is more difficult to treat with an equilibrium theory.

Ring: We know how to interpret the data with multistate models. We may assume that there is single filing and perhaps a knock on/knock off mechanism. When an ion is moved, other ions may also move, and the association step may even be associated with the translocation step, but how useful is this when the ion concentrations of different ions are changing?

Eisenberg: In the calcium release channel, which is the only channel in which we have selectivity data, if you say that the channel has a charge of 1.02, and you assign one diffusion coefficient for each ion, you can predict all the I–V relations of lithium, sodium, potassium, rubidium and caesium, and all mixtures, from 20 mM to 2 M. This cannot be done with any barrier model.

Sansom: I accept that this works, but what worries me is that although you have a large number of points in your I–V data, they are all highly correlated. So, I'm not convinced that there may not be other sets of parameters that fit the data equally well.

Roux: When a model ignores a part of the reality that we know is there, it is always a concern. For instance, Mark Sansom showed that waters are re-oriented in the pore, and this affects the electric distance and the way in which the motion of ions in the pore couples to the applied voltage. A continuous model cannot deal with this because it does not take into account the intrinsic polarization of the solvent. Therefore, you may need explicit water molecules in the pore region.

Eisenberg: I agree that molecular motions underlie this, but I would like to point out that Henderson et al (1979) have explicitly addressed the issue that you have just raised. They showed that those terms in other geometries are present but are small. Therefore, in this kind of model, we are saying that these are the dominant terms, and we're not saying that the other terms are not present.

The biological activities of gramicidin D

Smart: I would like to bring up some work by Henry Paulus and co-workers undertaken some time ago and now generally forgotten (Sarkar & Paulus 1972, Sarkar et al 1977, 1979, Paulus et al 1979, Mandl & Paulus 1985). Gramicidin D is a mixture of similar peptides synthesized by *Bacillus brevis* during sporulation. As well being able to form ion channels, gramicidin D was also found to be an inhibitor of transcription by RNA polymerase (Sarkar & Paulus 1972, Sarkar et al 1977). To investigate its mode of action, Paulus produced gramicidin-negative mutants of *B. brevis* that could not sporulate normally (Paulus et al 1979). It was found that the addition of gramicidin to these bacteria could restore normal activity. Gramicidin derivatives were then tested, with the result that the ability to induce normal sporulation and the channel activity of the derivative were not correlated. Instead, a relation was found between the potency of *in vitro* transcription inhibition and the ability to induce normal sporulation. Furthermore, it appears that the antibiotic activity of the peptide is probably also unrelated to channel formation (Banerjee & Sengupta 1981, Smart & Wallace 1999). Gramicidin is a competitive inhibitor of transcription, which means it probably binds to either DNA or to the enzyme complex in the binding site (Sarkar et al 1979). I have some preliminary evidence that it does not bind directly to DNA, and this is in agreement with the early work done by Ristow et al (1975), but which they later put into question (Bohg & Ristow 1986). However, we have shown that it binds the DNA intercalator actinomycin D (Smart & Wallace 1999). The importance of this is that gramicidin D can reverse the inhibition of transcription by actinomycin D, despite the fact that that they are both RNA polymerase inhibitors. Paulus suggested that this was not due to a direct interaction (Sarkar et al 1979). However, both u.v. visible absorbance spectroscopy and filter binding show that gramicidin in an aqueous suspension in a concentration range of 1–100 μM binds to actinomycin D (Smart & Wallace 1999). The actinomycin is taken out of solution and forms part of the suspension. There are one to two gramicidin-binding sites per actinomycin, suggesting a specific interaction (Smart & Wallace 1999). The biological relevance of the interaction maybe that gramicidin may control the activity of another DNA intercalator tyrocidine. Tyrocidine is a complex of three closely related cyclic decapeptides, produced with gramicidin D by *B. brevis* during sporulation. There is some indirect evidence that gramicidin and tyrocidine interact not only in an aqueous environment (Bohg & Ristow 1986) but also in lipid (Aranda & de Kruijff 1988). We are currently undertaking further studies with the aim of elucidating the mode of the varied actions of gramicidin and tyrocidine.

Roux: Does it never act as a channel in nature? Because each time one modifies gramicidin, its ability to conduct decreases by only a small amount, so it seems as though nature has optimized something like a channel function.

Smart: Yes, Paulus found a sporulation-inducing role (Mandl & Paulus 1985). In the normal bacterial population, it is best if all bacteria sporulate at the same time, but a subpopulation does not. Gramicidin is able to induce sporulation in that subpopulation, but non-channel-forming derivatives cannot do this. So, the channel activity may be related to gramicidin acting as a bacterial messenger (Mandl & Paulus 1985). Paulus et al (1979) showed that the induction of sporulation in the gramicidin-negative mutants could be achieved by many non-channel derivatives down to nine residues in length, so presumably the channel-forming ability performs some role for *B. brevis*.

Busath: Is there a connection between the ability of gramicidin to inhibit RNA polymerase and its ability to induce sporulation?

Smart: Yes, there is a general correlation but, unfortunately, these experiments were done before modern molecular biology techniques were available. This means that the exact mode of action has not been determined. Not all transcription can be shut down because presumably the bacterium has to synthesize some proteins to form spores, but there must be a control mechanism. It is interesting that tyrocidine also inhibits transcription, although they presumably affect different promoters.

Cross: If I remember those papers correctly, they stated that gramicidin binds to the sigma subunit of RNA polymerase and causes the vegetative form of the sigma subunit to fall off the polymerase, thus making way for the sporulation form of the sigma subunit to proceed.

Smart: This point is still unclear. Paulus showed that transcription inhibition was not sigma dependent (Sarkar et al 1979), but soon after this was contradicted by Fisher & Blumenthal (1982).

Hladky: I recollect that there was a Russian paper on a related subject in the early 1980s (Ivanov & Sychev 1982). They suggested that its main activity was not to form pores.

Jakobsson: The antibiotic effect may not be related to its ability to form pores, but are the toxic side-effects related to its pore-forming ability?

Smart: There have been attempts (Banerjee & Sengupta 1981) to make derivatives that had antibiotic but not haemolytic activity. They found that it is possible to change the relative activities, but not eliminate one or the other entirely. This may depend on the assay conditions, because it may act as a detergent at high concentrations.

Cross: One of the early arguments was that because of its tight binding to RNA polymerase, cells wouldn't need to produce the vast amount of gramicidin that they do, so it was suggested that there is another function, hence the channel function.

Ring: I think that cyclic gramicidin also inhibits the Na^+/K^+ ATPase, but I don't remember the reference for this.

Smart: I do not know whether gramicidin has any effect on this system. However, gramicidin is known to be a potent uncoupler of oxidative phosphorylation through a specific interaction with the H^+ ATPase (Rottenberg 1990).

Hladky: Chappell's group in Bristol studied the effects of a number of ion-transporting antibiotics on oxidative phosphorylation and the swelling of mitochondria, and compared this with their ability to make erythrocytes and lipid vesicles permeable to cations like sodium and potassium (refs in Henderson et al 1969). They concluded that the effects of gramicidin on mitochondrial swelling were accounted for by the increased permeability to small cations, and proposed that this action also accounted for the inhibition of oxidative phosphorylation. The effects on mitochondria indicate that gramicidin has an important biological activity separate from inhibition of RNA polymerase. Other polycyclic peptides can also be isolated from *B. brevis* including the tyrocidines and the chemically similar compound gramicidin S. These are also membrane active, making membranes leaky to ions, but I don't know of any evidence that these leaks occur by any specific ion transport mechanism.

Smart: Subsequent work by Rottenberg has shown that the effect on oxidative phosphorylation is separate from channel-forming activity (Rottenberg 1990). In particular, desformyl gramicidin has the same activity as normal gramicidin but is unable to form channels (Rottenberg 1990). It is noteworthy that desformyl gramicidin has been shown to have similar antibiotic potency to normal gramicidin (Ishii & Witkop 1964).

Hladky: That makes three important actions for gramicidin!

References

Aranda FJ, de Kruijff B 1988 Interrelationships between tyrocidine and gramicidin A′ in their interaction with phospholipids in model membranes. Biochim Biophys Acta 937:195–203

Banerjee PC, Sengupta S 1981 Chemical modification of some peptide antibiotics in relation to their biological activities. J Sci Ind Res 40:246–266

Bohg A, Ristow H 1986 DNA-supercoiling is affected *in vitro* by the peptide antibiotics tyrocidine and gramicidin. Eur J Biochem 160:587–591

Eisenberg B 1998a Ionic channels in biological membranes. Electrostatic analysis of a natural nanotube. Contemp Phys 39:447–466

Eisenberg B 1998b Ionic channels in biological membranes: natural nanotubes. Acc Chem Res 31:117–125

Fisher R, Blumenthal T 1982 An interaction between gramicidin and the σ subunit of RNA polymerase. Proc Natl Acad Sci USA 79:1045–1048

Henderson D, Blum L, Lebowitz JL 1979 An exact formula for the contact value of the density profile of a system of charged hard spheres near a charged wall. J Elec Chem 102:315–319

Henderson PJ, McGivan JD, Chappell JB 1969 The action of certain antibiotics on mitochondrial, erythrocyte and artificial phospholipid membranes. The role of induced proton permeability. Biochem J 111:521–535

Ishii S-I, Witkop B 1964 Gramicidin A. II. Preparation and properties of 'seco-gramicidin A'. J Am Chem Soc 86:1488–1853

Ivanov VT, Sychev SV 1982 The gramicidin A story. In: Snatzke G, Bartmann W (eds) Biopolymer complexes. Wiley, New York, p 107–125

Mandl J, Paulus H 1985 Effect of linear gramicidin on sporulation and intracellular ATP pools of *Bacillus brevis*. Arch Microbiol 143:248–252

Paulus H, Sarkar N, Mukherjee PK et al 1979 Comparison of the effect of linear gramicidin analogues on bacterial sporulation, membrane permeability, and ribonucleic acid polymerase. Biochemistry 18:4532–4536

Ristow H, Schazschneider B, Vater J, Kleinkauf H 1975 Some characteristics of the DNA-tyrocidine complex and a possible mechanism of the gramicidin action. Biochim Biophys Acta 414:1–8

Rottenberg H 1990 Decoupling of oxidative-phosphorylation and photophosphorylation. Biochim Biophys Acta 1018:1–17

Sarkar N, Paulus H 1972 Function of peptide antibiotics in sporulation. Nat New Biol 239:228–230

Sarkar N, Langley D, Paulus H 1977 Biological function of gramicidin: selective inhibition of RNA polymerase. Proc Natl Acad Sci USA 74:1478–1482

Sarkar N, Langley D, Paulus H 1979 Studies on the mechanism and specificity of inhibition of ribonucleic acid polymerase by linear gramicidin. Biochemistry 18:4536–4541

Smart OS, Wallace BA 1999 Linear gramicidin binds actinomycin D in an aqueous environment, in prep

Ionic interactions in multiply occupied channels

Vladimir L. Dorman, Stefano Garofoli and Peter C. Jordan[1]

Department of Chemistry, MS-015, Brandeis University, PO Box 9110, Waltham, MA 02454-9110, USA

Abstract. A significant number of physiologically important ion channels function via multi-ion mechanisms where repulsion between ions at slightly separated locations is believed to be critical for permeation. We apply the semi-microscopic Monte Carlo approach and analyse how multiple occupancy affects permeation energetics and ion–water–peptide correlations. We consider double occupancy in idealized models of two systems: gramicidin A and the KcsA K^+ channel. We focus on the excess repulsion energy due to ion–water and ion–peptide correlations (repulsion energy adjusted for direct ion–ion interaction). Gramicidin, where multiple occupancy is marginally important functionally, is ideal for correlating structure and ion interactions. Pair occupancy is stabilized by interaction with bulk solvent, destabilized by interaction with both the channel water and, as binding sites are far apart, the peptide backbone. In the KcsA K^+ channel, double occupancy is promoted by the uneven spacing and the large ion–water separations in the selectivity filter. The carbonyls forming the binding cavities are equally important for pair stabilization. Due to the binding pocket's design, net ionic repulsion is ~25–30% of what it would be in a gramicidin-like structure with the same interionic spacing.

1999 Gramicidin and related ion channel-forming peptides. Wiley, Chichester (Novartis Foundation Symposium 225) p 153–169

For many ion-selective channels multiple occupancy or multi-ion mechanistics appears intrinsic to their functionality, an especially noteworthy property of voltage-activated potassium, sodium and calcium channels (Hodgkin & Keynes 1955, Hille & Schwartz 1978, Hess & Tsien 1984, Almers & McCleskey 1984, Pongs 1993), although some of the interpretative arguments used for deducing such behaviour from conductance measurements have recently been questioned (Nonner et al 1998). In other systems, such as the valence-selective channel gramicidin, multiple occupancy occurs but is not critical functionally (Hladky & Haydon 1972, Neher et al 1978, Urban et al 1980). A conventional electrostatic

[1] This chapter was presented at the symposium by Peter C. Jordan, to whom correspondence should be addressed.

picture would imply that multiple occupancy combined with high selectivity requires a long, narrow constricted region with well-separated binding sites to reduce direct coulombic repulsion in low permittivity surroundings (Hodgkin & Keynes 1955, Hille & Schwartz 1978, Jordan 1986). This describes gramicidin, where the binding sites are located at the opposite mouths of the channel, some 20 Å apart (Andersen & Procopio 1980, Becker et al 1992, Woolf & Roux 1997).

However, in the voltage-activated calcium, sodium and potassium channel families structural inference and channel sequence homologies indicate that the constriction must be short (MacKinnon & Yellen 1990, Yool & Schwarz 1991, Yellen et al 1991, Pongs 1993). Depending on how the channel is modelled, the narrow pore would be estimated in the range of 5–10 Å (Lipkind & Fozzard 1994, Guy & Durrell 1995). In a recent *tour de force*, MacKinnon and his co-workers (Doyle et al 1998) have solved the structure of a bacterial K channel. Its constriction is ~12 Å long and contains two ions that are separated by as little as ~7.5 Å. The following question then arises. How can two ions so close together be stable in relatively non-permittive surroundings where the direct coulombic repulsion would be ~90 kJ/mol? The MacKinnon structure provides a number of clues. The pore is sandwiched between the highly permittive solvent domain and a large intrachannel water pool. The structure of each binding pocket is beautifully crafted, formed by eight carbonyl oxygens creating regions of high local negative charge density. None the less, there remain unfavourable features; the ions are close together, with a single intermediate water apparently unable to effectively co-ordinate either one.

This chapter studies the energetic ramifications of K^+ channel architecture, contrasting them with those in gramicidin. What features of narrow channels promote multiple occupancy? Which exact energy penalties? How can two ions that are close together be stabilized (but not too effectively)? To do this we study highly simplified models of the channels that, we believe, describe critical features of the constrictions of each. We first sketch the computational approach and our channel abstractions. We then consider a doubly occupied gramicidin channel, identifying features that encourage multiple occupancy in long channels. We finally focus on the KcsA K^+ channel, determine how it differs from gramicidin and identify various properties lowering the energetic cost of double occupancy.

Models and computational details

We apply the semi-microscopic Monte Carlo technique to computational models of both gramicidin and the KcsA K^+ channel. The general computational method has been fully described previously (Sections 2 and 3 of Dorman et al [1996]); only an outline is given here.

Semi-microscopic modelling analyses solvent rotational relaxation and its effect on dielectric stabilization; electronic degrees of freedom (the basis for high frequency permittivity) are treated by embedding the molecular model in a background dielectric with $\varepsilon = 2$. The interfacial Helmholtz layer accounts for the finite size of bulk water molecules and their orientational ordering (reduced permittivity) at the membrane–bulk water boundary. Channel description is limited to the polar species presumed most important for permeation energetics: the ions and waters in the single-filing region and selected moieties of the channel former itself. Electrical interaction between these groups and bulk water (solvent) is fully accounted for by image computations.

For gramicidin, the molecular model, illustrated in Fig. 1, is exceedingly simple: nine equally spaced intra-channel sources (zero to two ions and, correspondingly,

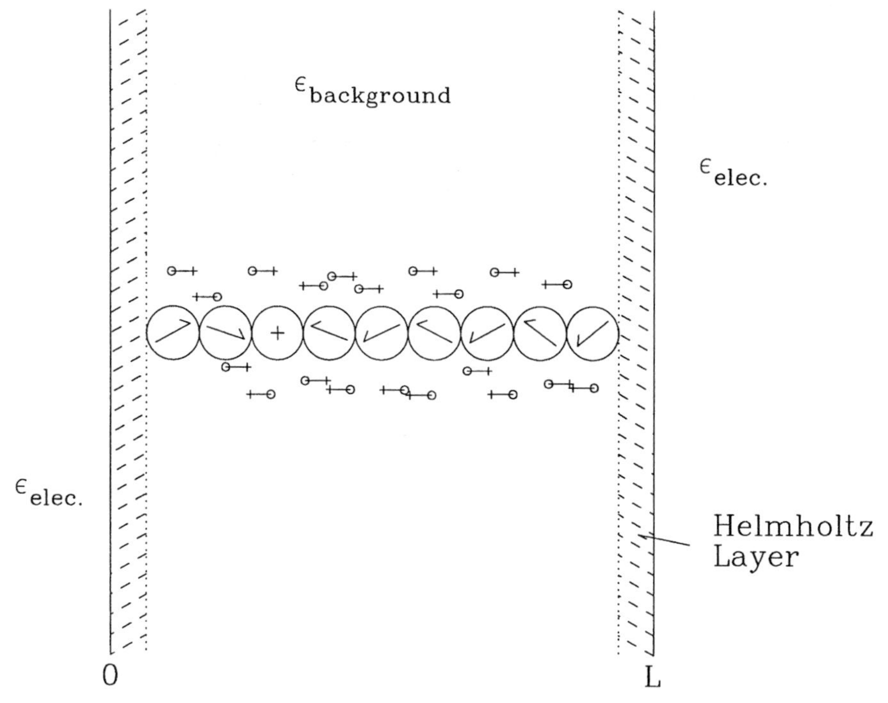

FIG. 1. Semi-microscopic model of the ion–water–gramicidin–lipid–electrolyte ensemble. The Helmholtz layer accounts for orientationally ordered bulk water. The region between 0 and L forms the background dielectric ($\varepsilon = 2$). External regions contain electrolyte in which $\varepsilon_{elec} \equiv \infty$. The water dipoles are aligned by the ion, in this diagram located at site 3. Surrounding the ion–water chain is a planar projection of part of gramicidin's native carbonyl distribution (carbonyl group reorientation due to interaction with the ion is suppressed); only 20 of the 32 carbonyl groups and none of the amide groups are depicted. Carbon atoms are denoted (+) and oxygen atoms (O). (Reprinted from Dorman et al (1996) with the permission of the Biophysical Society.)

nine to seven waters) and the 32 carbonyl and amide groups of the surrounding channel backbone. As implemented to date, waters are treated as dipoles ($\mu = 1.86$ D) in spherical cavities (R = 1.5 Å). All intrachannel groups, while fixed axially, are free to rotate. The idealized reference backbone structure was determined by Koeppe & Kimura (1984), and is inverted to form a right-handed β-helix. Backbone C and N atoms are fixed; carbonyl ($q_C = 0.5e_o$, $k_{bend} = 7.5 \times 10^{-20}$ J) and amide ($q_N = -0.25e_o$, $k_{bend} = 7.1 \times 10^{-20}$ J) bending is opposed by an harmonic restoring force. Charges are taken from AMBER and k_{bend} is representative of low energy peptide torsions (Weiner et al 1984).

The simplified KcsA K^+ channel model is based on the structure determined by Doyle et al (1998) for the ion-occupied channel. The critical crystallographic features of the selectivity filter are the single file, two cations entrapping a water, and four sets of four backbone carbonyl groups (from residues 75–78) forming cavities that stabilize the ions. To determine if there is more single file water, we used InsightII (Biosym/MSI, AMBER force field [Weiner et al 1984]) to minimize the ion-occupied channel with additional water (restraining all backbone C_αs); after minimization two extra waters were sited nearly axially, each in van der Waals contact with its neighbouring ion. These are included in the five-site molecular model of Fig. 2. To establish a reference structure for the unoccupied channel, we minimized the MacKinnon structure void of water and ions, all backbone C_αs held fixed. As the four strands were no longer identical, the carbonyl conformation of equivalent residues on each strand was symmetrized. Backbone amide groups are sufficiently far away from the pore to be energetically of little consequence; they are not included in the model. The gramicidin structural parameters (qs, μ and k_{bend}) were used. The experimental structure has an elongated pool of water, ~10 Å in diameter containing ~60–100 waters and a third ion, internal to the second ion. Ideally this would be treated as a finite domain, dielectrically much like bulk water. As the pool is large its ion is quite distant from the selectivity filter; thus its influence on filter properties is fairly small and, to a first approximation, can be ignored. We approximate the water pool in two ways: (1) siting the Helmholtz layer just internal to the single file (overestimating the pool's stabilizing influence on the selectivity domain); and (2) adding four single file waters to the system (probably underestimating the pool's influence, although the single file imposes some additional correlation; Partenskii & Jordan 1992).

In the semi-microscopic method (Dorman et al 1996), permeation energetics is decomposed into transfer and stabilization terms, ΔG_{trans} and ΔG_{stab}. ΔG_{trans}, the free energy of exchanging an aqueous ion for a water molecule in the background dielectric continuum ($\varepsilon = 2$), is the sum of four terms: (1) free energies of ionic hydration and (2) of aqueous vaporization and (3) Born energies of transfer of both ion and (4) water from vacuum ($\varepsilon = 1$) to the dielectric background. The

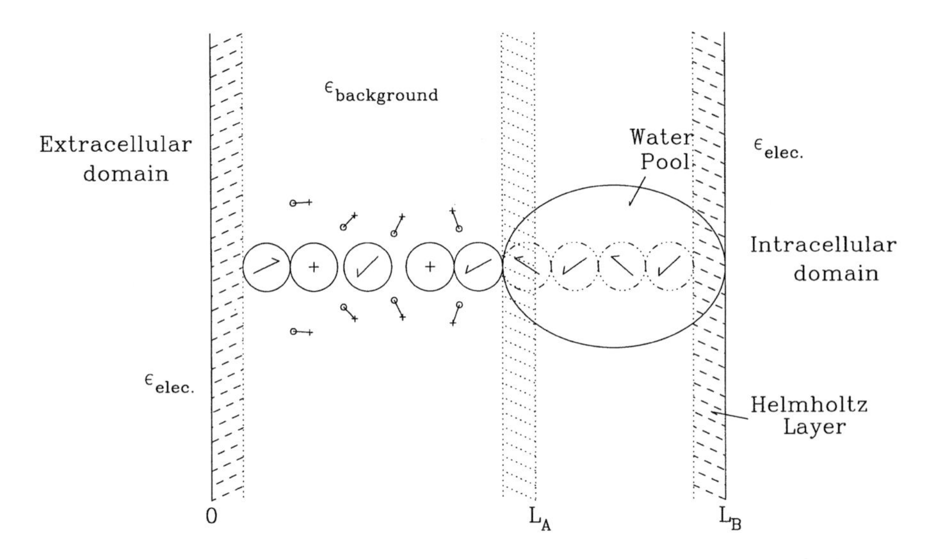

FIG. 2. Semi-microscopic model of selected features of the ion–water–KcsA K^+ channel–lipid–electrolyte system. The conventions of Fig. 1 apply. The ions are 7.3 Å apart (the second ion is situated at the closer of the two possible site 2 positions), with the water 3.4 Å from the outer ion. Surrounding the ion–water chain is a planar projection of the carbonyl groups forming the binding cavities (as shown, corresponding to minimized locations in the empty channel); only eight of the 16 carbonyl groups are depicted. Two approximate treatments of the water pool are illustrated. In case A, the region to the right of L_A is treated as electrolyte containing no explicit water (the boundaries and the associated Helmholtz layer are denoted with dotted lines). In case B, the low ε domain extends to L_B with four additional explicit waters (denoted by - - - -) included to roughly approximate the stabilizing influence of the water pool.

Born term is an interaction energy between charges in a solvation cavity and the reaction field induced in the surrounding dielectric. Only if the charge is centred in the cavity and the cavity is spherical (Beveridge & Schnuelle 1975, Roux et al 1990, Nina et al 1997, P. C. Jordan, unpublished work 1998) is this energy determined solely by a 'Born radius'; both conditions can be violated in the asymmetrical solvation environment of ion channels. ΔG_{stab} computes the energy of transmuting a channel water into an ion while simultaneously transmuting the ion in the low dielectric continuum into a water molecule. For doubly occupied channels, two ions and two waters participate in the transmutation process, etc.

Double occupancy in gramicidin

Our focus is the repulsion free energy in doubly occupied channels, $\delta G_k(m,n) \equiv G_k(m,n) - G_k(m) - G_k(n)$; $G_k(m,n)$ and $G_k(m)$ are stabilization free

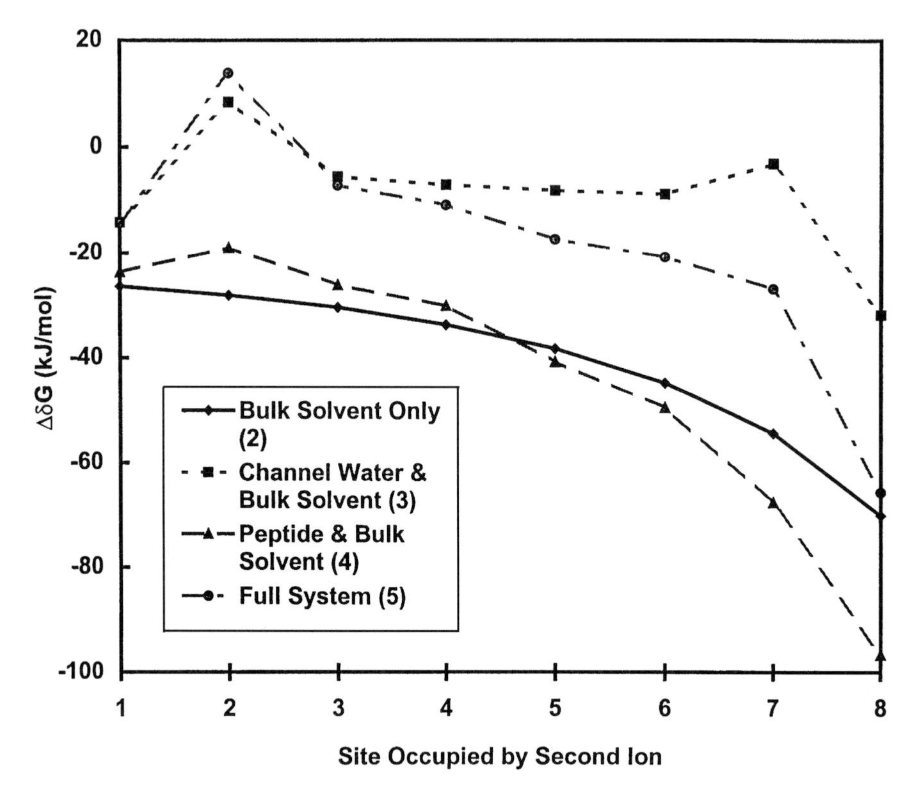

FIG. 3. Reduction of repulsion free energy in gramicidin-like models due to interaction stabilization. $\Delta\delta G(n)$ is the pair repulsion energy with ions at sites n and 9 with the direct coulomb contribution for an infinite membrane, $e^2/\varepsilon R_{n9}$, subtracted out. The effect of the various interactive components is presented separately: bulk solvent (2); channel water and bulk solvent (3); peptide and bulk solvent (4); full system (5).

energies for channels with ions at sites m and n (transfer contributions cancel in computing $\delta G_k[m,n]$). Each k value describes a distinct contribution to the repulsion free energy: (1) direct repulsion (infinite continuum dielectric where $\delta G_k(m,n) = e^2/\varepsilon R_{nm}$); (2) stabilization by bulk solvent (finite continuum dielectric); (3) destabilization by channel water; (4) stabilization by peptide; (5) co-operative channel water–peptide influences. Terms 2–5 include interactive components. To highlight these, Fig. 3 presents interaction contributions to the repulsion free energy, $\Delta\delta G_k(n) \equiv \delta G_k(n,9) - \delta G_1(n,9)$ (the direct repulsive term has been subtracted); one ion is fixed at mouth site (9) and the second is located variably ($n = 1$ to 8).

Interaction with bulk solvent, $\Delta\delta G_2(n)$, is always stabilizing. With one ion at site 9, the absolute effect intensifies as the second ion approaches because ion–other image forces increase; stabilization of two neighbouring ions is roughly that for one doubly charged ion located at their midpoint. It is relatively greater the closer ions are to the solvent. With both ions at mouth sites the nearby solvent reduces net repulsion by $>90\%$; interactions with images of the ion at the opposite mouth nearly cancel the direct repulsion. For gramicidin, with a ~ 31 Å thick low ε region, interaction with bulk solvent is always significant, lowering net repulsion by 15% for neighbouring ions in mid-membrane.

Interaction with channel water destabilizes double occupancy; $\Delta\delta G_3(n)$ is always noticeably larger than $\Delta\delta G_2(n)$. The second ion substantially disrupts dipolar correlation in the water chain. The physics is self evident. In single occupancy, distant water dipoles are aligned with their oxygen atoms directed toward the ion; with a second ion present, orientation of these waters is controlled by the second ion, thus forcing reorientation and increasing the net repulsion energy. The energetic cost of this disruption can be quite large, sometimes exceeding the stabilization component due to ionic interaction with bulk solvent ($\Delta\delta G > 0$).

$\Delta\delta G_4(n)$ illustrates the peptide's influence on repulsion. It is dependent on the interionic distance and is only large when ions are close together. Comparison of $\Delta\delta G_4(n)$ and $\Delta\delta G_2(n)$ shows that, for well separated ions, there is added peptide-mediated repulsion; for nearby ions, net repulsion is greatly reduced. Why? Consider a carbonyl group near ion 2 with its oxygen oriented toward ion 1 (see Fig. 4). For ions far apart, carbonyl deflection is governed by interaction with the nearer ion 2. Defining the plane normal to the channel axis and passing through carbon 2 as $z = 0$, the oxygen's position is $(R - d_{CO}\sin\theta,\ 0,\ d_{CO}\cos\theta)$, with R the radial position of the carbonyl carbons (3.5 Å), d_{CO} the carbonyl bond length and θ the deflection angle. For an ion at z_I, ion–oxygen distance is:

$$D(z_I,\theta) = [(z_I - d_{CO}\cos\theta)^2 + (R - d_{CO}\sin\theta)^2]^{1/2} \quad (1)$$

In the absence of the second ion, $\theta \sim 0^\circ$. Depending on ionic position, deflection can increase or decrease ion–oxygen distance; the distance change is $\Delta D(z_I,\theta) \equiv D(z_I,\theta) - D(z_I,0)$. With z_I small enough, deflection diminishes separation ($\Delta D[z_I,\theta] < 0$) increasing attraction between ion 1 and carbonyls near ion 2. The switch occurs when $z_I = R\cot(\theta/2)$. For typical deflection angles of 30–40°, this happens when $z_I \sim (2.7-3.7)R$, i.e. $z_I \sim 9.5-13$ Å, ions separated by 2 or 3 waters, just as observed. As ionic separation decreases each ion interacts favourably with more carbonyl oxygens, reducing net interionic repulsion. For two ions in mid-channel, reduction is maximal, $\sim 25\%$.

The combined effects of channel water, peptide and bulk solvent, $\Delta\delta G_5(n)$, are non-additive. Adding channel water to the ion–peptide system always increases net repulsion; but $\Delta\delta G_5(n) - \Delta\delta G_4(n)$ is less than $\Delta\delta G_3(n) - \Delta\delta G_2(n)$, differential repulsion without the peptide. The effect of adding peptide to the ion–water

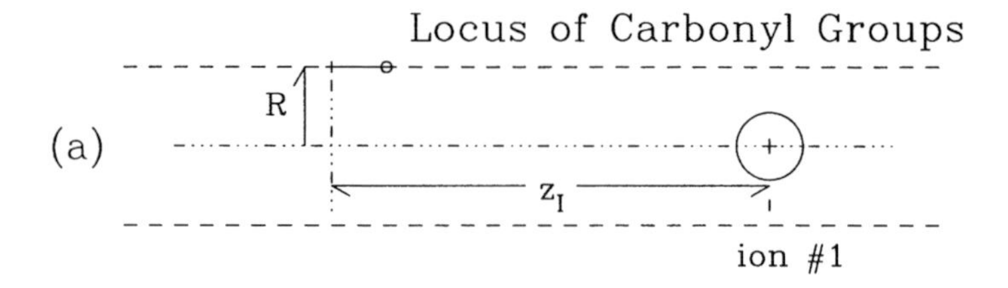

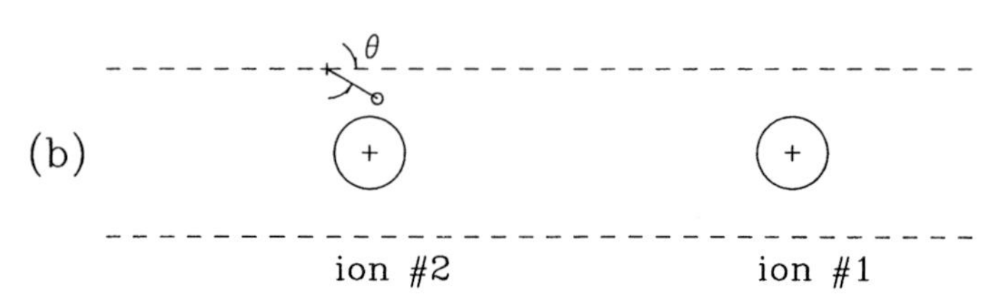

FIG. 4. The influence of a second ion on interaction between the first ion and a distant carbonyl in gramicidin. The channel radius is R; z_I is the axial distance between ion 1 and the carbonyl carbon. Only a single carbonyl group is illustrated. In (a) the carbonyl group is not deflected. In (b) the distance between oxygen and ion 1 has increased due to deflection by θ, thus reducing ion–oxygen attraction. Were the interionic separation smaller, deflection would decrease this distance and promote stabilization.

system depends on ionic separation. With ions far apart, $\Delta\delta G_5(n) - \Delta\delta G_3(n)$ is again positive; net repulsion increases. At smaller ionic distances, the difference is negative but switching occurs at larger ionic separations, with ions five or six waters apart.

Overall, pair interactions in gramicidin are greatly stabilized by interaction with the continuum electrolyte, significantly destabilized due to disruption of ion–channel water correlation, and, for ions sufficiently close together, restablized by extra favourable interaction with peptide. Pair energetics do not simply separate into independent ion–water and ion–peptide terms. There is an additional stabilization due to interaction of re-oriented water with backbone carbonyls.

For long channels there are other effects due to interaction with water. From Fig. 3 differential repulsion is especially large with ions located at sites 2 and 9. In fact, as seen in Fig. 5, the total stabilization free energy of the system, $G_k(m,n)$, can behave counterintuitively as interionic separation decreases. With an ion fixed at interfacial site 9, both G_3 and G_5 decrease as ion 2 moves from site 2 to site 3; comparable, but less pronounced effects are noted with the first ion fixed at site 8. This is a consequence of the long single file. There are three competing

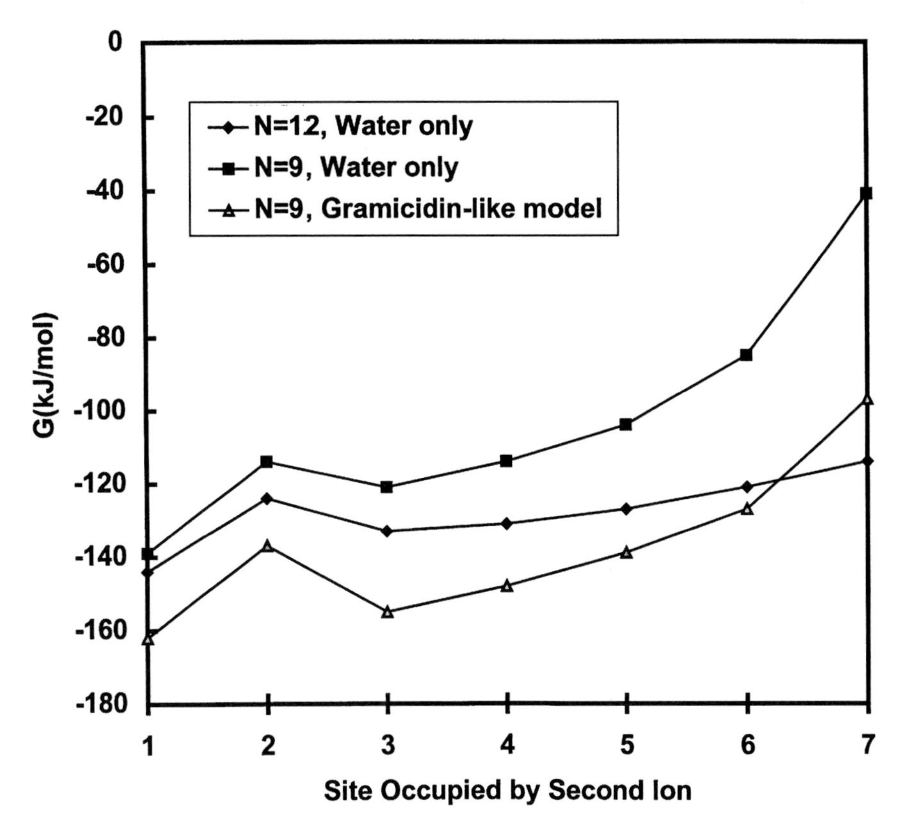

FIG. 5. Stabilization free energy, $G_{stab}(j,N)$, for doubly occupied single-file narrow channels with gramicidin-like geometries. It increases as channel length increases; however, differences between the 12-site and 9-site stabilization free energies are only substantial when there are three or fewer interionic waters (the second ion is at site 5–8) in the 9-site system. Peptide stabilization increases as the interionic separation drops.

contributions as ion 2 moves inward: ion–bulk solvent attraction decreases; ion channel–water attraction increases (ion 2 aligns waters between it and the near interface); ion–ion repulsion increases. For long channels, the latter is insignificant; only the first two terms matter. Decreased ion–solvent attraction is more than compensated for by increased ion channel–water attraction. Comparison of data for nine- and 12-site ion–water single files (Fig. 5) is illustrative. As the ion moves inward from site to site, all energy increases are smaller for the longer single file; in going from site 2 to site 3 the energy decrease is greater. Destabilization due to distant electrical features (the other ion and unfavourably aligned waters) have been attenuated. Differences in stabilization energy only grow rapidly with three or fewer waters between the ions. In our

TABLE 1 Differential repulsion (adjusted to account for solvent contribution, see text) free energy in gramicidin-like channel at selected sites, $\delta\delta G_k(m,n)$, in kJ/mol

Ion-occupied sites	*Peptide only* $\delta\delta G_4(m,n)$	*Channel water only* $\delta\delta G_3(m,n)$	*Full system* $\delta\delta G_5(m,n)$
6 & 9	−5	36	24
7 & 9	−13	51	28
8 &9	−27	38	4
3 & 6	−2	35	24
4 & 6	−13	49	28
5 & 6	−48	34	−17

treatment a single ion aligns distant waters, but the energy penalty for realignment is fairly small for waters far away from the ion. These relatively weak (energetically) long range correlations are consistent with observations that long range water alignment may be model dependent (Chiu et al 1991, Roux & Karplus 1994, Duca & Jordan 1998).

There is unusually large net repulsion when ions are separated by a single intervening water (e.g. site 7 of Fig. 3), clearest in the differential repulsion free energy with solvent contributions eliminated, $\delta\delta G_k(m,n) \equiv \delta G_k(m,n) - \delta G_2(m,n)$, presented in Table 1 for surface and interior sites. With only peptide present ($k = 4$), the extra ion–peptide attraction steadily increases net stabilization (*vide supra*). However, with channel water present ($k = 3$ or 5) the interaction component peaks with one interionic water. The sandwiched water cannot orient preferentially toward either ion; each loses one of its two strongest ion–dipole stabilization components. Thus, there is a large decrease in the pair stabilization free energy and a corresponding increase in net repulsion.

Finally, to show that our results are quantitatively reasonable, we compare repulsive interaction for ions at the mouth sites (1 and 9) with experimental data (Urban et al 1980, Becker et al 1992, Jing et al 1995). Studies of doubly occupied gramicidin provide estimates of K_1 and K_2, equilibrium constants for ion binding to empty and singly occupied channels, respectively. Since $\delta G_5(1,9) = kT\ln(K_1/K_2)$, the experiments (based on quite different methods and somewhat different interpretative schemes) yield values ranging from 6–19 kJ/mol, all compatible with our result, ~ 15 kJ/mol. It is worth noting here that the computed repulsive energy between other symmetric pairs of sites (2 and 8, 3 and 7, etc.) exceeds 50 kJ/mol, i.e. only for ions at the end of the single file is pair repulsion low enough to be consistent with experiment.

Double occupancy in the KcsA K^+ channel

This channel contrasts sharply with gramicidin. The single file is short. The two binding sites are close to one another with two stable inner site positions (Doyle et al 1998); we focus on the configuration with the smaller interionic spacing (see Fig. 2). The carbonyl groups which define them form a fairly rigid assembly (Doyle et al 1998) and the intermediate, asymmetrically placed water is relatively far from both of the single file ions. A potassium channel is highly discriminating: comparably permeable to K^+ and Rb^+, permitting passage of Cs^+ and discriminating almost absolutely against Na^+ and Li^+ (Hille 1992). Its multiple ion architecture is presumed crucial to its function (Hodgkin & Keynes 1955, Hille & Schwartz 1978). Our focus is on how channel structure affects its two ion properties, in particular via interaction with the channel water and the binding cavity carbonyls. From Fig. 3 it is clear that in gramicidin, ion–water interactions greatly destabilize pair free energies. Interaction with peptide does not compensate. Does the same hold for the K channel? And if not, what structural features are the bases for any differences?

We consider four questions. How important is the short single file? How important energetically is its placement near the extracellular surface? How important are the (fairly) long ion–water separations and the intermediate water's asymmetrical placement in the selectivity filter? How important is the peptide for pair stabilization?

Table 2 addresses the first three questions. The water pool and the subsequent interior channel is described in two ways—as dielectrically equivalent to bulk solvent (case A) or as an extended single file region (case B or C)—approximations roughly bounding the (ion-free) pool's stabilizing influence. The structure of the selectivity filter is probed by contrasting behaviour with equal (3 Å apart) and realistic (K channel, as in Fig. 2) spacing of the three crystallographic sites. Sites 1 and 2 are the exterior and interior binding sites respectively with stabilization free energies G_1 and G_2. Differences attributable to the length of the single file or its placement in the ensemble (compare equal spacing cases A, B and C or K channel spacing cases A and B) are insignificant. With site spacing invariant, there are only small changes in ion–water interaction energies. Treating the (ion-free) water pool as bulk solvent or an extended single file has little effect on filter energetics. The filter would be equally effective deeper in the membrane.

However, differences between properties of the unequally spaced K channel structure and the equally spaced model structure are substantial. Both ionic spacing and ion–water separation appear crucial. Regardless of how we describe the water pool, the interior ion is less well stabilized but the doubly occupied single file is stabilized. Binding site separation promotes double occupancy. Net repulsion is greatly reduced, by $\sim 50\,\mathrm{kJ/mol}$ (comparing K channel and equal

TABLE 2 **Stabilization and repulsion free energies (G and δG, in kJ/mol) for single or double occupancy of water-filled single-files with K channel reminiscent geometries**

	Equally spaced sites			*K channel site spacing*	
	Case A[a]	*Case B*[b]	*Case C*[c]	*Case A*[a]	*Case B*[b]
G_1	−77	−78	−76	−70	−70
G_2	−77	−76	−76	−49	−64
G_{12}	−21	−30	−19	−42	−52
$\delta G \equiv G_{12} - G_1 - G_2$	+133	+124	+133	+77	+82

[a]Water pool treated as continuum bulk solvent.
[b]Water pool treated as four molecule single file.
[c]Selectivity domain symmetrically sited.

TABLE 3 **Stabilization and repulsion free energies (G and δG, in kJ/mol) for single or double occupancy of water-filled single-files with K channel reminiscent geometry (upper data set, freely rotating COs; lower data set, restrained CO rotation)**

	Case A[a]			*Case B*[b]		
	Water only	*Peptide only*	*Full system*	*Water only*	*Peptide only*	*Full System*
G_1	−70	−131	−143	−70	−128	−141
G_2	−49	−124	−115	−64	−112	−118
G_{12}	−42	−215	−232	−52	−191	−227
$\delta G \equiv G_{12} - G_1 - G_2$	+77	+39	+26	+82	+49	+32
G_1	−70	−120[c]	−133[c]	−70	−114[c]	−130[c]
G_2	−49	−54[c]	−50[c]	−64	−32[c]	−51[c]
G_{12}	−42	−142[c]	−159[c]	−52	−108[c]	−153[c]
$\delta G \equiv G_{12} - G_1 - G_2$	+77	+32[c]	+24[c]	+82	+38[c]	+29[c]

[a]Water pool treated as continuum bulk solvent.
[b]Water pool treated as four molecule single file.
[c]Restrained CO rotation (see text).

spacing results). By separately investigating infinite continuum and solvent only problems, the observed changes appear roughly equally reflective of: less ion–ion repulsion; reduced disruption of ion–solvent interaction; and reduced disruption of interaction with the intermediate single file water.

Table 3 documents the influence of the carbonyl groups forming the binding pockets. Since molecular mechanics calculations (S. Garofoli & P. C. Jordan, unpublished work 1998) suggest the structure is fairly flexible, we contrast two extreme possibilities for carbonyl group relaxation, one freely rotating and the other constrained to the initial locations described in the models section. Consistent with the protein-free results, there are no substantive differences between cases A and B; treating the water pool as bulk dielectric or as a single file has no discernible influence; naturally with water absent (peptide only) both G_2 and G_{12} increase as the width of the low ε region increases. The peptide greatly stabilizes the ion at site 1, $\sim$70 kJ/mol and, depending upon the orientational freedom attributed to cavity 2, may be comparably influential for ionic stabilization at site 2. Overall, the contrast with gramicidin, where the additional peptide contribution never exceeds $\sim$10 kJ/mol (Dorman et al 1996), is striking, reflecting binding site structure design in the K channel. The carbonyl groups are, by and large, initially well aligned to accept ions; relatively little distortion is required to permit binding. The four separate peptide strands permit much more extensive ion–oxygen co-ordination than in gramicidin where the cation coordinates no more than four oxygens (Roux & Karplus 1993, Duca & Jordan 1997). Regardless of the binding cavities' deformability, channel architecture has a major influence on pair energetics. Net repulsion energy is tiny ($\sim$20–30 kJ/mol) which should not be affected by the pool ion as that ion is too far from the filter to influence ion–water or ion–carbonyl correlations; the repulsion free energy between comparably separated sites in a gramicidin-like structure may be estimated to be in the range of 75–90 kJ/mol, three- to fourfold greater. Repulsion energy is reduced by $\sim$50 kJ/mol due to the asymmetry of the selectivity domain; it is reduced another 50 kJ/mol because carbonyl co-ordination is optimized.

Summary

Pair occupancy in gramicidin is greatly stabilized by ionic interaction with bulk solvent. The roughly equal spacing of the binding sites leads to maximum destabilization due to ionic interaction with the channel waters. Ion–peptide interaction would only promote double occupancy for closely separated ions, in which case co-operative peptide–water interactions would provide additional stabilization. In the KcsA K^+ channel, double occupancy of the closely spaced binding sites is promoted by the uneven spacing and the large ion–water separations in the selectivity filter. The peptide is equally important for promoting pair stabilization. Due to the unusual design of the K channel binding pocket, net ionic repulsion is only 25–30% of what it would be for a gramicidin-like structure with the same interionic spacing. The selectivity filter's proximity to bulk solvent is of relatively minor importance. The detailed structure of the water pool has no great effect on properties in the constriction; what matters is the presence of interior waters.

Acknowledgements

This work has been supported by the National Institutes of Health grant GM-28643. We thank M. B. Partenskii and C. Miller for their suggestions.

References

Almers W, McCleskey EW 1984 Non-selective conductance in calcium channels of frog muscle: calcium selectivity in a single-file pore. J Physiol (Lond) 353:585–608

Andersen OS, Procopio J 1980 Ion movement through gramicidin A channels. On the importance of aqueous diffusion resistance, and ion–water interactions. Acta Physiol Scand Suppl 481:27–35

Becker MD, Koeppe RE II, Andersen OS 1992 Amino acid substitutions and ion channel function. Model-dependent conclusions. Biophys J 62:25–27

Beveridge DL, Schnuelle GW 1975 Free energy of a charge distribution in concentric dielectric continua. J Phys Chem 79:2562–2566

Chiu SW, Jakobsson E, Subramaniam S, McCammon JA 1991 Time-correlation analysis of simulated water motion in flexible and rigid gramicidin channels. Biophys J 60:273–285

Dorman VL, Partenskii MB, Jordan PC 1996 A semi-microscopic Monte Carlo study of permeation energetics in a gramicidin-like channel: the origin of cation selectivity. Biophys J 70:121–134

Doyle DA, Cabral JM, Pfuetzner RA et al 1998 The structure of the potassium channel: molecular basis of K^+ conduction and selectivity. Science 280:69–77

Duca KA, Jordan PC 1997 Ion–water and water–water interactions in a gramicidin-like channel: effects due to group polarizability and backbone flexibility. Biophys Chem 65:123–141

Duca KA, Jordan PC 1998 Comparison of selectively polarizable force fields for ion–water–peptide interactions: ion translocation in a gramicidin-like channel. J Phys Chem B 102:9127–9138

Guy HR, Durrell SR 1995 Structural models of Na^+, Ca^{2+} and K^+ channels. In: Dawson DC, Frizzell RA (eds) Ion channels and genetic diseases. Rockefeller University Press, New York (Society of General Physiologists Symposium Vol 50) p 1–16

Hess P, Tsien RW 1984 Mechanism of ion permeation through calcium channels. Nature 309:453–456

Hille B 1992 Ionic channels of excitable membranes, 2nd edn. Sinauer Associates, Sunderland, MA

Hille B, Schwartz W 1978 Potassium channels as multi-ion single file pores. J Gen Physiol 72:409–442

Hladky SB, Haydon DA 1972 Ion transfer across lipid membranes in the presence of gramicidin A. I. Studies of the unit conductance channel. Biochim Biophys Acta 274:294–312

Hodgkin AL, Keynes RD 1955 The potassium permeability of a giant nerve fibre. J Physiol (Lond) 128:61–88

Jing N, Prasad KU, Urry DW 1995 The determination of binding constants of micellar-packaged gramicidin A by ^{13}C- and ^{23}Na-NMR. Biochim Biophys Acta 1238:1–11

Jordan PC 1986 Ion channel electrostatics and the shape of channel proteins. In: Miller C (ed) Ion channel reconstitution. Plenum Press, New York, p 37–55

Koeppe RE II, Kimura M 1984 Computer modeling of β-helical polypeptide models. Biopolymers 23:23–28
Lipkind GM, Fozzard HA 1994 A structural model of the tetrodotoxin and saxitoxin binding site of the Na^+ channel. Biophys J 66:1–13
MacKinnon R, Yellen G 1990 Mutations affecting TEA blockade and ion permeation in voltage-activated K^+ channels. Science 250:276–279
Neher E, Sandblom J, Eisenman G 1978 Ion selectivity, saturation and block in gramicidin A channels. II. Saturation behavior of single channel conductances and evidence for the existence of multiple binding sites in the channel. J Membr Biol 40:97–116
Nina M, Beglov D, Roux B 1997 Atomic radii for continuum electrostatics calculations based on molecular dynamics free energy simulations. J Phys Chem B 101:5239–5248
Nonner W, Chen DP, Eisenberg B 1998 Anomalous mole fraction effect, electrostatics, and binding in ionic channels. Biophys J 74:2327–2334
Partenskii MB, Jordan PC 1992 Theoretical perspectives on ion-channel electrostatics: continuum and microscopic approaches. Quart Rev Biophys 25:477–510
Pongs O 1993 Structure–function studies on the pore of potassium channels. J Membr Biol 136:1–8
Roux B, Karplus M 1993 Ion transport in the gramicidin channel: free energy of the solvated right-handed dimer in a model membrane. J Am Chem Soc 115:3250–3260
Roux B, Karplus M 1994 Molecular dynamics simulations of the gramicidin channel. Annu Rev Biophys Biomol Struct 23:731–761
Roux B, Yu HA, Karplus M 1990 Molecular basis for the Born model of ion solvation. J Phys Chem 94:4683–4688
Urban BW, Hladky SB, Haydon DA 1980 Ion movements in gramicidin pores. An example of single-file transport. Biochim Biophys Acta 602:331–354
Weiner SJ, Kollman PA, Case DA et al 1984 A new force field for molecular mechanical simulation of nucleic acids and proteins. J Am Chem Soc 106:756–784
Woolf TB, Roux B 1997 The binding site of sodium in the gramicidin A channel: comparison of molecular dynamics with solid-state NMR data. Biophys J 72:1930–1945
Yellen G, Jurman M, Abramson T, MacKinnon R 1991 Mutations affecting internal TEA blockade identify the probable pore-forming region of a K^+ channel. Science 251:939–942
Yool AJ, Schwarz TL 1991 Alteration of ionic selectivity of a K^+ channel by mutation of the H5 region. Nature 349:700–704

DISCUSSION

Jakobsson: Does the size of the ion influence these results? And could you account for the size of the ion by adjusting the space the ion occupies? Because this may be a fairly simple extension of your work, and it may provide an intuitive physical understanding of why larger ions are more likely to admit a second ion into the gramicidin channel than smaller ones.

Jordan: The ion is fixed at an axial site; the water dipoles rotate and the 32 carbonyls deflect. We are doing some modelling to get a sense of what happens to the structure if we put in smaller ions. However, we will probably have to permit the ion translational freedom to treat this aspect of double occupancy.

Woolley: Did you only look at the case where you put the second ion at the end of the delocalized binding site nearest the first ion binding site?

Jordan: Yes. We're not quite sure how to do the calculation when we put the ion at the further end of the delocalized binding site in the potassium channel, because it's not clear whether we should incorporate another water molecule or not. We are going to do the calculations with and without, because there's a large space and it is reasonable to think that a second water molecule is present.

Roux: We have done some free energy calculations of doubly occupied gramicidin channels (Roux et al 1995). We computed the relative free energy between the different monovalent cations, and we found that there is a size dependence, i.e. it is relatively more favourable for large ions. The rationale behind these calculations was to find out where the free energy comes from, and we found that it partly comes from the single filing of water, which favours larger ions. This is because the single file is floppy in the case of a large ion, and therefore you don't pay a high price when you put the second one in. In the case of lithium, or other small ions, the single file is ordered and significantly perturbed by the second ion. This is also true in an ionic solution. The activity coefficients of lithium chloride increase more rapidly than those of caesium chloride. Having found this, we removed the channel and the bulk and constrained two ions at 18 Å apart with a single file of waters in between (in a cylinder). We found the opposite trend. This implies that the hydrogen bonding in the water channel was affecting how the water structure reacts to the presence of the second ion. I'm not saying we should not look at simplified models, but one should be aware that there are all sorts of small structural distortions, collisions and compression factors that affect the energetics.

Jordan: I agree. There are some large differences between what's going on in the potassium channel where the ions are close together and what's going on in a gramicidin-like system in which ions are separated. We're not talking about 3–4 kcal differences, we're talking about 20 kcal.

Roux: This implies that the selectivity filter part of the potassium channel must be an attractive well, otherwise the two ions would not be there.

Cross: Does anyone have a reasonable estimate for how much of the charge from a monovalent cation is distributed over nearest-neighbour waters?

Roux: We have done some preliminary calculations for carbonyls. We find that when one ion is interacting with one carbonyl, about 5% of the ion charge may be transferred to the oxygen (Roux & Karplus 1995). I don't recall having done the calculation for water, but my impression is that it could be on the same order of magnitude.

Busath: Ab initio calculations between calcium and water demonstrate more charge separation, i.e. 15–20%.

Roux: But one must remember that these initial calculations of charge transfer are sensitive to the *ab initio* basis sets that one uses.

Jakobsson: The amount of transfer is going to be larger for a small ion than for a large ion because the electric field at the surface is larger.

Roux: But a small ion such as lithium might not want to lose its electronic cloud as much as sodium.

Smart: Is the ion occupancy observed in the KcsA K^+ channel similar to the expected occupancy of the channel in a lipid bilayer?

Jordan: There are electrophysiological data which suggest that the channel is operating in a free ion mode.

Jakobsson: Ted Begenisich and co-workers have determined Ussing flux ratio exponents a bit less or a bit more than 3 in a variety of K channels (the most recent paper in the series is Stampe et al 1998), and there are three crystallographic ions in the KcsA X-ray structure, so there is good agreement between the structural data and the flux data in this regard.

Busath: That's been shown for the Shaker channel and other voltage-gated potassium channels but not for this particular potassium channel, KcsA.

Jakobsson: But this potassium channel has high homology to the Shaker channel. MacKinnon and co-workers made a convincing demonstration that the structural motif is essentially the same, by site-directed mutations that changed KcsA residues to Shaker counterparts, and thus conferred Shaker-like, toxin-binding properties to KcsA (MacKinnon et al 1998).

References

MacKinnon R, Cohen SL, Kuo A, Lee A, Chalt BT 1998 Structural conservation in prokaryotic and eukaryotic potassium channels. Science 280:106–109

Roux B, Karplus M 1995 Potential energy function for cations–peptides interaction: an *ab initio* study. J Comp Chem 16:690–704

Roux B, Prod'hom B, Karplus M 1995 Ion transport in the gramicidin channel: molecular dynamics study of single and double occupancy. Biophys J 68:876–892

Stampe P, Arreola J, Pérez-Cornejo P, Begenisich T 1998 Nonindependent K^+ movement through the pore in IRK1 potassium channels. J Gen Physiol 112:475–484

Peptide influences on lipids

J. Antoinette Killian, Sven Morein, Patrick C.A. van der Wel*, Maurits R. R. de Planque, Denise V. Greathouse* and Roger E. Koeppe II*

*Department of Biochemistry of Membranes, University of Utrecht, Padualaan 8, 5384 CH Utrecht, The Netherlands, and *Department of Chemistry and Biochemistry, University of Arkansas, Fayetteville, Arkansas 72701, USA*

Abstract. The extent to which the length of the membrane-spanning part of intrinsic membrane proteins matches the hydrophobic thickness of the lipid bilayer may be an important factor in determining membrane structure and function. To gain insight into the consequences of hydrophobic mismatch on a molecular level, we have carried out systematic studies on well-defined peptide–lipid complexes. As model peptides we have chosen gramicidin A and a series of artificial hydrophobic α-helical transmembrane peptides that resemble the gramicidin channel. These peptides consist of a hydrophobic stretch of alternating leucine and alanine residues with variable length, flanked by tryptophan residues. Using wide-line NMR techniques, we have investigated the interaction of these peptides with the bilayer-forming diacyl phosphatidylcholines and with phospholipids which by themselves have a tendency to form non-bilayer structures. We have shown that hydrophobic mismatch leads to systematic changes of the bilayer thickness and that it can even change the macroscopic organization of the lipids. The type of lipid organization induced by the peptides and the efficiency of the various processes depend on the properties of the lipids and on the precise extent of mismatch.

1999 Gramicidin and related ion channel-forming peptides. Wiley, Chichester (Novartis Foundation Symposium 225) p 170–187

Membrane proteins are embedded in the lipid bilayer with their hydrophobic parts in contact with the acyl chain region. In general, the length of these hydrophobic regions will be approximately equal to the hydrophobic bilayer thickness. The extent of hydrophobic matching between proteins and lipids can be important for protein structure and function, as well as for properties of the lipid bilayer (for recent review, see Killian 1998).

The importance of hydrophobic matching for protein structure and function was shown, for example, for the gramicidin channel. It was found that the bilayer thickness influences the equilibrium between different conformations of gramicidin (Mobashery et al 1997). However, the channel conformation itself appears to be remarkably stable and independent of mismatch (Katsaras et al 1992, Cornell et al 1989), perhaps due to the unusual β-helical structure of the peptide. For α-helical membrane proteins or other ion channel-forming peptides the situation may be different.

In principle, there are several ways in which a mismatch could alter protein structure, as illustrated in Fig. 1. When the hydrophobic part of a transmembrane protein is either too large or too small to match the hydrophobic bilayer thickness, the protein might oligomerize in the membrane to minimize the exposed hydrophobic area. Transmembrane helices that are too long could tilt or kink to reduce their effective hydrophobic length, whereas peptides that are too short might not incorporate and instead adopt a surface localization or aggregate outside the bilayer. The proteins might also adapt their conformation, and for instance go from an α-helix to the longer 3_{10} helix.

The extent of matching may also influence properties of the lipids, such as chain packing or phase behaviour. Lipids could modulate the membrane thickness by stretching or disordering their acyl chains, or they could even completely disrupt

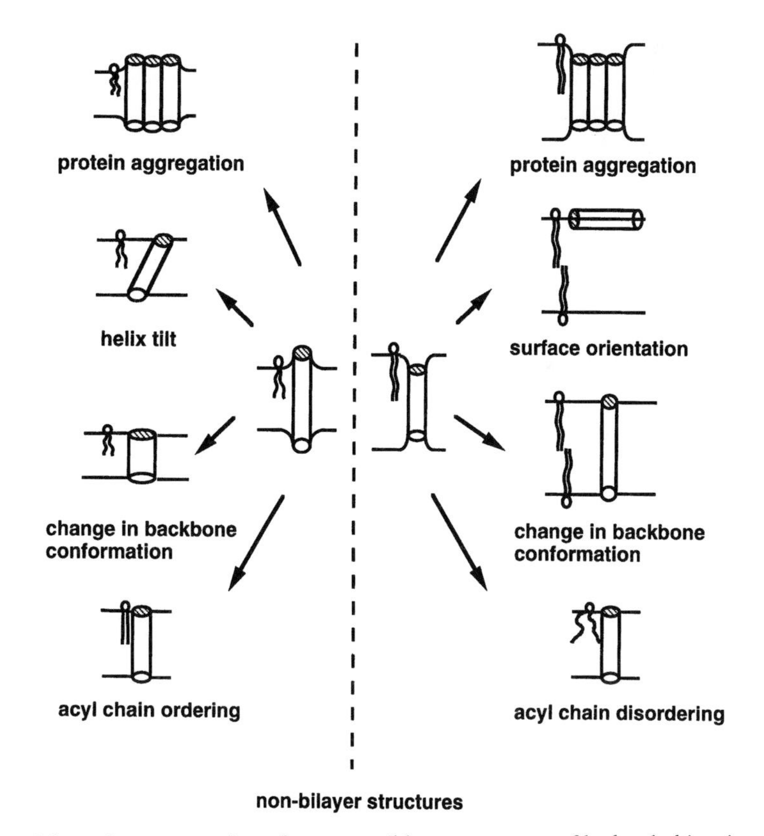

FIG. 1. Schematic representation of some possible consequences of hydrophobic mismatch. Any combinations of these possibilities may also occur.

the bilayer to assemble into another type of aggregate. These effects of mismatch on lipid structure and organization will be the focus of this chapter.

Design of peptides

To gain insight into the possible consequences of hydrophobic mismatch for lipid structure and organization, systematic studies have been carried out using well-defined peptide–lipid complexes consisting of synthetic lipids and model transmembrane peptides. The gramicidin channel was used as a model peptide, because it is transmembrane and has a well-known structure. However, it may not be representative for α-helical transmembrane peptides. Therefore, a family of peptides was also designed and synthesized (see Table 1) that mimic the gramicidin channel, but are α-helical. These peptides have a hydrophobic core of variable length of alternating leucine and alanine, flanked by tryptophans at each end. This also mimics the situation found for the membrane-spanning part of many membrane proteins, in which tryptophans are highly enriched at the membrane–water interface (e.g. Doyle et al 1998, Schiffer et al 1992, Ostermeier et al 1996, Landolt-Marticorena et al 1993). Based on their amino acid composition, the peptides are named WALP*n*, with *n* being the total number of amino acids.

Choice of lipids

Many membranes contain a large amount of lipids which by themselves do not form bilayers. The most common non-lamellar structures for lipids isolated from biological membranes are the hexagonal H_{II} phase and cubic phases (for reviews see Lindblom & Rilfors 1989, Seddon 1990). The lipid organization representative for these phases is schematically shown in Fig. 2, together with the corresponding ^{31}P

TABLE 1 Amino acid sequence of gramicidin and some of the WALP peptides

Peptide	*Amino acid sequence*[a]	*Total length (Å)*
WALP16	f-AWWLALALALALAWWA-e	26.5[c]
WALP23	f-AWWLALALALALALALALWWA-e	36.0[c]
WALP31	f-AWWLALALALALALALALALALALALWWA-e	48.0[c]
Gramicidin A	f-VGALAVVVWLWLWLW-e[b]	26[d]

[a]In all groups, f represents a formyl group, and e represents an ethanolamine.
[b]Underlined amino acids are D-residues.
[c]Assuming that the peptide forms an ideal α-helix with a length of 1.5 Å per amino acid with the C-terminal ethanolamine included as one additional amino acid.
[d]From a model of the gramicidin channel in the $\beta^{6.3}$ helical conformation.

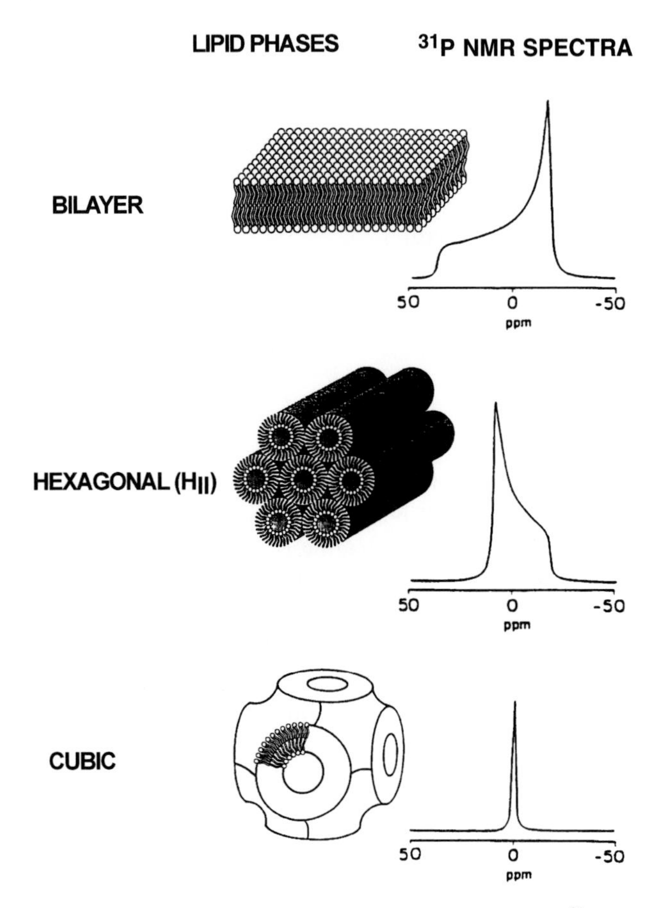

FIG. 2. Lipid bilayer and non-bilayer phases with corresponding ^{31}P NMR spectra.

NMR spectra (see below). In the H_{II} phase, the phospholipids are arranged in long tubes with their headgroups surrounding a narrow aqueous channel. Cubic phases consist of interwoven networks of aqueous channels and curved lipid bilayers. The lipid organization shown in Fig. 2 is one example. In phase diagrams cubic phases usually can be found in-between the phase boundaries for a bilayer and for a hexagonal H_{II} phase.

The phase preference of lipids can be understood on the basis of their effective shape (Israelachvili et al 1977, Cullis & de Kruijff 1979) or intrinsic radius of curvature (Gruner 1989). According to the 'shape concept', phosphatidylcholine has a more or less cylindrical shape and prefers organization in bilayers. Phosphatidylethanolamine has a much smaller effective size of the headgroup and prefers organization in structures with a high curvature, such as the H_{II} phase. Mixtures of both types of lipids may form a cubic phase.

To gain information on the effects of hydrophobic mismatch on lipid structure and organization, we have studied the interaction of gramicidin and WALP peptides of varying length with diacyl phosphatidylcholine, which is a typical bilayer-forming lipid, and with lipid systems which by themselves have a tendency to form non-lamellar structures.

Methods

^{31}P NMR spectroscopy

Wide-line ^{31}P NMR spectroscopy is a convenient technique to discriminate between organization of phospholipids in bilayer and in non-lamellar structures (Cullis & de Kruijff 1979, Lindblom & Rilfors 1989, Seelig 1978), as shown in Fig. 2. When lipids are organized in a bilayer, fast axial reorientation of the lipids about their long axis results in partial averaging of the chemical shift anisotropy (CSA), giving rise to characteristic spectra with a low field shoulder and a high field peak. For the H_{II} phase, the additional motion of the lipids about the tubes results in a ^{31}P NMR spectrum with a reversed asymmetry, and a further reduced CSA. In cubic phases, the lipids undergo complete motional averaging, leading to an isotropic ^{31}P NMR signal. It should be noted, however, that other arrangements of lipids are possible which result in complete averaging of the CSA and therefore give rise to an isotropic ^{31}P NMR signal. Examples are micelles and small vesicles.

^{2}H NMR spectroscopy

When using lipids with perdeuterated acyl chains, it is possible to analyse lipid acyl chain order by wide-line ^{2}H NMR techniques (Seelig & Seelig 1974). From order parameter profiles along the acyl chains, the average bilayer thickness can be derived in model membrane systems in the absence and presence of different peptides (see de Planque et al 1998).

Effects of mismatch on lipid organization in diacyl phosphatidylcholine model membranes

The effects of hydrophobic mismatch on lipid phase behaviour and acyl chain order were first studied in phosphatidylcholine model membranes. This lipid on its own will not organize in non-lamellar structures, but will form a stable bilayer.

Previously it was shown that gramicidin induces the formation of an H_{II} phase in phosphatidylcholine bilayers with acyl chains of 16 C-atoms or more at high molar peptide/lipid ratios typically of about 1/10 (Van Echteld et al 1982, Watnick et al

1990). However, the formation of non-bilayer phases in these lipids is not a general consequence of mismatch. The uncharged, hydrophobic α-helical polypeptide P15 (Boc-[L-Ala-Aib-L-Ala-Aib-L-Ala]$_3$-OMe, where Aib is aminoisobutyric acid), which is slightly shorter than gramicidin, was unable to affect lipid organization under the same conditions (Aranda et al 1987).

The question is, what are the specific properties of the gramicidin channel that are responsible for its effect on lipid organization? Studies on WALP peptides of different lengths, which mimic gramicidin but have an α-helical conformation, may answer this question.

Killian et al (1996) found that WALP16, which has a similar length as gramicidin (Table 1), also has the same effect on lipid phase behaviour as gramicidin. Furthermore, the effect of the peptides appeared to be mismatch dependent. The longest peptides did not affect lipid organization, slightly smaller peptides induced an isotropic phase and even smaller peptides induced an H_{II} phase.

When the hydrophobic thickness of the lipid bilayer was compared with the total peptide length for all samples measured, it was found that sharp 'mismatch boundaries' exist between the formation of a bilayer and an isotropic phase, and between the formation of an isotropic and an H_{II} phase (Fig. 3). The nature of this

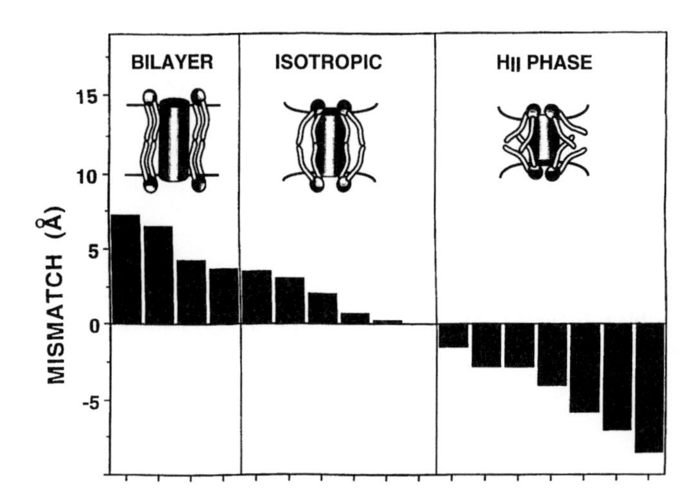

FIG. 3. Lipid phase preference as a function of hydrophobic mismatch. WALP peptides were incorporated into phosphatidylcholine bilayers with different thicknesses. The mismatch is arbitrarily defined as the difference between the total length of the peptides, assuming an idealized α-helix, and the hydrophobic thickness of the lipid bilayer. Each bar represents a different peptide/lipid mixture. The areas marked isotropic and H_{II} phase indicate spectra of which part or all of the lipids are present in an isotropic or H_{II} phase, respectively. For details see Killian et al (1996).

isotropic phase has not been defined, but it may be a 'molten cubic' phase (Lindblom & Rilfors 1989).

Why then do gramicidin and the WALP peptides behave differently from P15? One obvious possibility is that the presence of interfacially localized tryptophans is responsible. Tryptophans have a strong preference for the lipid–water interface (Wimley & White 1996, Persson et al 1998). In the case of P15, there are no interfacial residues that would effectively inhibit interactions between adjacent peptides, and peptide aggregation might be a more likely response to mismatch than formation of non-lamellar phases.

The effects described above were characteristic for negative mismatch, i.e. the peptide being too short to fit into the bilayer. The question remains what the consequences of mismatch will be when peptides are too long. Similarly, one might wonder what happens at lower peptide/lipid ratios, when there is not sufficient peptide to sustain a non-lamellar phase and the lipids are forced to remain in a bilayer. Do the peptides aggregate, or does the system react in another way, for instance by affecting the bilayer thickness?

Effects of mismatch on lipid chain order in phosphatidylcholine bilayers

To investigate effects of mismatch on lipid bilayer thickness, peptides with varying length were incorporated in deuterated phosphatidylcholine bilayers of varying thickness, in a low molar ratio of 1/30 peptide to lipid. The effects on average hydrophobic bilayer thickness were calculated from order parameter profiles of the lipid acyl chains in the absence and presence of peptide. These values are presented in Table 2.

TABLE 2 Estimated hydrophobic bilayer thicknesses (Å) of pure lipid, and changes in hydrophobic thickness relative to the pure lipids, for systems at a 1/30 molar ratio of peptide to lipid

Bilayer	*di-C12:0-PC*	*di-C14:0-PC*	*di-C16:0-PC*	*di-C18:0-PC*
Pure lipid	19.9	22.5	25.4	28.5
+WALP16	+0.6	+0.4	+0.0	−0.4
+WALP17	+1.0	+0.4	+0.1	−0.1
+WALP19	+1.4	+0.6	+0.2	+0.0
+Gramicidin	+1.5	+1.0	+0.4	−0.2

Measurements were performed at 10 °C above the main phase transition temperature of the pure lipids. For details see de Planque et al (1998).
PC, phosphatidylcholine.

In the absence of peptide, the bilayer thickness increases with the number of carbon atoms in the acyl chains. When WALP peptides are incorporated, in all lipid systems there is a small, but systematic, effect of peptide length on bilayer thickness. In dilauroyl phosphatidylcholine (di-C12:0-PC) lipids, all three peptides cause a stretching of the lipids, the longest peptide being the most efficient. In the longest lipid, distearoyl phosphatidylcholine (di-C18:0-PC), the incorporation of WALP19 has no effect, and the shorter WALP16 and WALP17 peptides cause a slight decrease in thickness relative to the pure lipid bilayer. Thus, WALP analogues with different hydrophobic length in the same lipid cause a length-dependent disturbance of the average bilayer thickness. Also, when comparing the effects of one particular WALP peptide in lipids of different length, a systematic decrease from stretching to shrinking upon increasing chain length is observed.

The effects of the peptides on bilayer thickness are small compared to the differences in mismatch, suggesting that either a partial mismatch is allowed, or that other counteractive moves, as shown for instance in Fig. 1, take place in addition to lipid stretching and shrinking.

With gramicidin, a similar systematic decrease from stretching to shrinking is observed as for the WALP peptides. Although not the longest peptide (Table 1), the gramicidin dimer exerts the largest perturbing influence in most of the lipid systems. This may be attributed to the different structural features of gramicidin and the WALP peptides. The amino acid side chains of gramicidin in general are rather bulky and have relatively small spatial separations (Fig. 4). The side chains of the α-helical WALP peptides with alternating leucine and alanine are less tightly packed. Therefore, gramicidin would be more rigid, and a possible explanation for the observed difference is that gramicidin is less able to adapt to a mismatch, e.g. by tilting or bending, thereby inducing a larger lipid response.

Effect of mismatch on lipid organization of dielaidoyl phosphatidylethanolamine (di-C18:1_t-PE)

Next, the effects of mismatch were investigated on the phase behaviour of lipid systems that on their own have a tendency to form non-bilayer structures, such as dielaidoyl phosphatidylethanolamine (di-C18:1_t-PE). This lipid undergoes a temperature-dependent bilayer to H_{II} phase transition, which has been well characterized and which occurs at an experimentally convenient temperature of about 60 °C (Killian & de Kruijff 1985).

It had previously been shown that incorporation of gramicidin promotes the formation of an H_{II} phase in di-C18:1_t-PE (Van Echteld et al 1981). At a molar ratio of 1/50, a lowering of the transition was observed with about 10 °C, indicating that little energy is required just to lower this phase transition. The

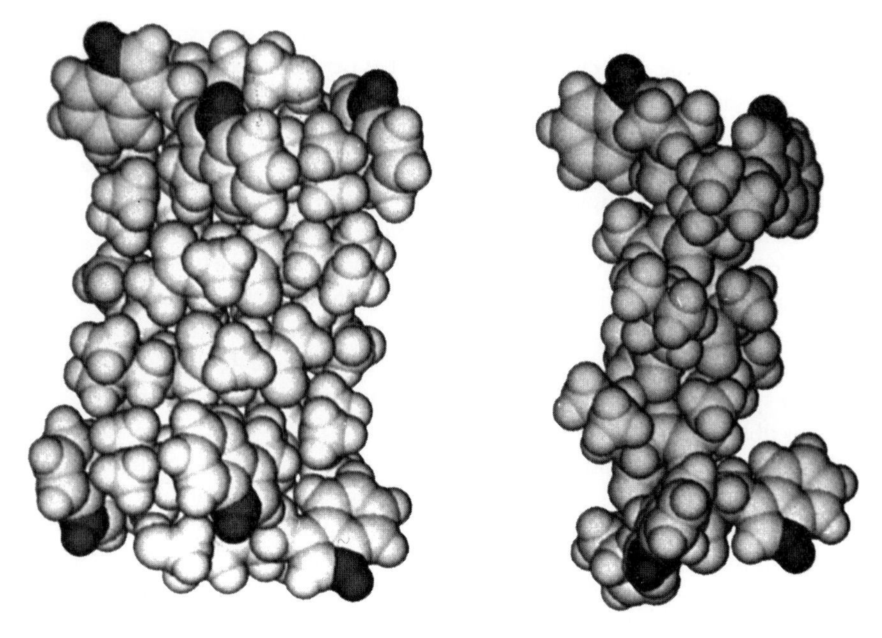

FIG. 4. Side view of WALP17 (right) and gramicidin (left). Tryptophans are visible as protruding rings near the termini of both peptides, with indole NH-groups darkened. For details see de Planque et al (1998).

lipid also easily can form isotropic phases, as was shown in the presence of low concentrations of alamethicin and gramicidin S (Keller et al 1996, Prenner et al 1997).

The questions to be answered now are whether and how the effects on phase behaviour depend on mismatch, and whether relatively long peptides can stabilize the bilayer. For this purpose WALP peptides with a large range of different lengths were incorporated in di-C18:1_t-PE and the phase behaviour was analysed as a function of temperature.

Using wide-line ^{31}P NMR, we found that the WALP peptides affect the lamellar to non-lamellar phase transition in a mismatch-dependent way. As shown in Fig. 5A, the shortest peptides (WALP14–19) induce a significant amount of H_{II} phase at 40 °C, similar to the effect previously observed for gramicidin. Slightly longer peptides (WALP19–27) induce an isotropic phase, suggesting that these peptides are too short to fit into an H_{II} phase, but that they fit well in an isotropic phase. A still longer peptide (WALP31), of which the length might even exceed the bilayer thickness, did not induce a significant amount of isotropic phase. Also at temperatures around the bilayer to H_{II} phase transition temperature, WALP31 did not affect lipid phase behaviour (not shown).

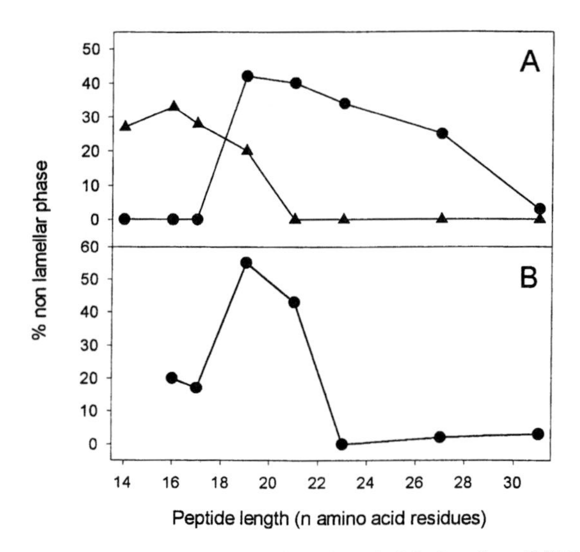

FIG. 5. Percentages of H_{II} (▲) and isotropic phase (●) induced at 40 °C by different length peptides in dielaidoyl phosphatidylethanolamine (di-C18:1_t-PE) at a 1/50 molar ratio of peptide to lipid (A) and in mixtures (7/3 molar ratio) of dioleoyl phosphatidylethanolamine/dioleoyl phosphatidylglycerol (di-C18:1_c-PE/di-C18:1_c-PG) at a 1/500 molar ratio of peptide to lipid (B). Both lipid systems adopt a pure lamellar phase at this temperature in the absence of peptide.

It is concluded that the mismatch dependence of the phase behaviour in di-C18:1_t-PE is similar to that in the phosphatidylcholine systems. Longer peptides do not appear to stabilize the bilayer, suggesting that such peptides perhaps will tilt or aggregate.

Effect of mismatch on lipid organization of mixtures of dioleoyl phosphatidylethanolamine (di-C18:1_c-PE) and dioleoyl phosphatidylglycerol (di-C18:1_c-PG)

Finally, the effects of mismatch were investigated in mixtures (7/3 molar ratio) of dioleoyl phosphatidylethanolamine (di-C18:1_c-PE) and dioleoyl phosphatidylglycerol (di-C18:1_c-PG). This mixture mimics the lipid composition and phase behaviour of *Escherichia coli* lipids and undergoes a temperature-dependent transition from a bilayer to an isotropic phase at about 60 °C.

Previously, it was shown in a similar lipid mixture that gramicidin significantly lowers the bilayer to isotropic transition temperature at low peptide/lipid ratios of 1/200 (Killian et al 1990). Again, this raises the question of whether this is a mismatch-dependent effect, and if so, whether relatively long peptides can stabilize the bilayer.

A series of WALP peptides were incorporated in di-C18:1_c-PE/di-C18:1_c-PG mixtures at different peptide/lipid ratios. All peptides were found to lower the bilayer to isotropic phase transition temperature. However, this effect was not simply mismatch related. As shown in Fig. 5B, at 40 °C and at a 1/500 molar ratio of peptide to lipid, WALP19 was the most efficient in promoting the formation of an isotropic phase. In general, the efficiency decreased with increasing peptide length, and the shorter WALP16 and WALP17 peptides were less efficient.

These results fit well with the results obtained for di-C18:1_t-PE. As shown by comparison of Figs 5A and 5B, the peptides that induce the highest percentage of isotropic phase in di-C18:1_t-PE are the most efficient at promoting the formation of an isotropic phase in di-C18:1_c-PE/di-C18:1_c-PG mixtures. This suggests that the shortest peptides, those that promote H_{II} phase formation in di-C18:1_t-PE, are too short to fit well into an isotropic phase. However, these peptides cannot induce an H_{II} phase because they are not present at a sufficiently high concentration. Instead, the peptides may aggregate or adopt a surface localization.

In this lipid system, no evidence was obtained for stabilization of the bilayer in the presence of the longer WALP27 or WALP31 peptides. Again, a possible explanation is that it is easy for a longer peptide to accommodate in the bilayer because it can tilt or kink. In that case, one might expect that larger and more rigid proteins might be able to stabilize the bilayer. An alternative explanation is that small concentrations of peptides lower the transition temperature by promoting the formation of required intermediates, but that there is no such mechanism to induce an increase in phase transition temperature.

Concluding remarks

We have shown that gramicidin and WALP peptides can induce non-lamellar structures in phosphatidylcholine bilayers, and promote formation of non-lamellar structures in bilayers of di-C18:1_t-PE and in di-C18:1_c-PE/di-C18:1_c-PG mixtures. The nature of the induced phase appears to be determined by the precise extent of mismatch, provided that sufficient peptide is present to support the non-lamellar phase. Structural features of the peptides may be less important, since gramicidin and WALP peptides of the same total length had similar effects. Yet, not all hydrophobic peptides can induce non-lamellar phases under mismatch conditions. We postulate that the anchoring residues, next to the hydrophobic transmembrane segment, play a crucial role in determining the consequences of mismatch. For example, tryptophans could inhibit peptide–peptide interactions, via their preference for interaction with the lipid–water interface, or charged amino acid residues could prevent peptide aggregation directly due to unfavourable intermolecular electrostatic interactions.

For effects on acyl chain order, other properties of proteins such as size or rigidity may play an additional role, as suggested by the observed differences between gramicidin and WALP peptides.

Implications for membrane structure and function

In biological membranes, lipids that are prone to forming non-lamellar structures will always be mixed with bilayer-preferring lipids, leading to an overall bilayer organization that is essential for the barrier function of biological membranes. Yet, the presence of non-bilayer lipids must be important for structural and/or functional properties of biomembranes because the balance between bilayer and non-bilayer lipids is strictly regulated (Rietveld et al 1993, Morein et al 1996). One possibility is that these non-bilayer lipids allow the local and transient formation of non-lamellar structures, which could be important for processes such as membrane fusion, exo- or endocytosis, or protein translocation (see de Kruijff 1987, 1997a,b). Another possibility is that by influencing lipid packing they can affect the conformational state or aggregational behaviour of membrane proteins, and thus influence protein function (de Kruijff 1997a). Striking examples are alamethicin (Keller et al 1993) and gramicidin (Lundbaek et al 1997), for which channel activity was found to be dependent on the tendency of lipids to form non-lamellar phases.

The results of this study show that hydrophobic mismatch can strongly affect lipid phase behaviour. Therefore, hydrophobic mismatch might influence these membrane processes that require the presence of non-bilayer lipids. In addition, hydrophobic mismatch can affect lipid packing in the membrane and thereby change the physical properties of the lipid bilayer.

Acknowledgements

This work was supported by The Netherlands Division of Chemical Sciences (CW) with financial aid from The Netherlands Organization for Scientific Research (NWO), by National Institutes of Health grant GM 34968 (to REK and DVG), by EU TMR Network Grant ERBFMRX CT96 0004 (to SM) and by NATO Grant CRG 950357.

References

Aranda FJ, Killian JA, de Kruijff B 1987 Importance of tryptophans of gramicidin for its lipid structure modulating activity in lysophosphatidylcholine and phosphatidylethanolamine model membranes. A comparative study employing gramicidin analogs and a synthetic α-helical hydrophobic polypeptide. Biochim Biophys Acta 901:217–228

Cornell BA, Separovic F, Thomas DE, Atkins AR, Smith R 1989 Effect of acyl chain length on the structure and motion of gramicidin A in lipid bilayers. Biochim Biophys Acta 985:229–232

Cullis PR, de Kruijff B 1979 Lipid polymorphism and the functional roles of lipids in biological membranes. Biochim Biophys Acta 559:399–420

de Kruijff B 1987 Polymorphic regulation of membrane lipid composition. Nature 329:587–588
de Kruijff B 1997a Biomembranes. Lipids beyond the bilayer. Nature 386:129–130
de Kruijff B 1997b Lipid polymorphism and biomembrane function. Curr Opin Chem Biol 1:564–569
de Planque MRR, Greathouse DV, Koeppe RE II, Schäfer H, Marsh D, Killian JA 1998 Influence of lipid/peptide hydrophobic mismatch on the thickness of diacylphosphatidylcholine bilayers. A ^{2}H NMR and ESR study using designed transmembrane α-helical peptides and gramicidin A. Biochemistry 37:9333–9345
Doyle DA, Cabral JM, Pfuentzer RA et al 1998 The structure of the potassium channel: molecular basis of K^+ conduction and selectivity. Science 280:69–77
Gruner SM 1989 Stability of lyotropic phases with curved interfaces. J Phys Chem 93:7562–7570
Israelachvili JN, Mitchell DJ, Ninham BW 1977 Theory of self-assembly of lipid bilayers and vesicles. Biochim Biophys Acta 470:185–201
Katsaras J, Prosser RS, Stinson RH, Davis JH 1992 Constant helical pitch of the gramicidin channel in phospholipid bilayers. Biophys J 61:827–830
Keller SL, Bezrukov SM, Gruner SM, Tate MW, Vodyanoy I, Parsegian VA 1993 Probability of alamethicin conductance states varies with nonlamellar tendency of bilayer phospholipids. Biophys J 65:23–27
Keller SL, Gruner SM, Gawrisch K 1996 Small concentrations of alamethicin induce a cubic phase in bulk phosphatidylethanolamine mixtures. Biochim Biophys Acta 1278:241–246
Killian JA 1998 Hydrophobic mismatch between proteins and lipids in membranes. Biochim Biophys Acta 1376:401–415
Killian JA, de Kruijff B 1985 Thermodynamic, motional, and structural aspects of gramicidin-induced hexagonal H_{II} phase formation in phosphatidylethanolamine. Biochemistry 24:7881–7890
Killian JA, de Jong AMPh, Bijvelt J, Verkleij AJ, de Kruijff B 1990 Induction of non-bilayer structures by functional signal peptides. EMBO J 9:815–819
Killian JA, Salemink I, de Planque MRR, Lindblom G, Koeppe RE II, Greathouse DV 1996 Induction of non-bilayer structures in diacylphosphatidylcholine model membranes by transmembrane α-helical peptides: importance of hydrophobic mismatch and proposed role of tryptophans. Biochemistry 35:1037–1045
Landolt-Marticorena C, Williams KA, Deber CM, Reithmeier RAF 1993 Non-random distribution of amino acids in the transmembrane segments of human type I single span membrane proteins. J Mol Biol 229:602–608
Lindblom G, Rilfors L 1989 Cubic phases and isotropic structures formed by membrane lipids. Possible biological relevance. Biochim Biophys Acta 988:221–256
Lundbaek JA, Maer AM, Andersen OS 1997 Lipid bilayer electrostatic energy, curvature stress, and assembly of gramicidin channels. Biochemistry 36:5695–5701
Mobashery N, Nielsen C, Andersen OS 1997 The conformational preference of gramicidin channels is a function of lipid bilayer thickness. FEBS Lett 412:15–20
Morein S, Andersson A-S, Rilfors L, Lindblom G 1996 Wild-type *Escherichia coli* cells regulate the membrane lipid composition in a 'window' between gel and non-lamellar structures. J Biol Chem 271:6801–6809
Ostermeier C, Iwata S, Michel H 1996 Cytochrome c oxidase. Curr Opin Struct Biol 6:460–466
Persson S, Killian JA, Lindblom G 1998 Molecular ordering of interfacially localized tryptophan analogs in ester– and ether–lipid bilayers studied by ^{2}H NMR. Biophys J 75:1365–1371
Prenner EJ, Lewis RN, Neuman KC et al 1997 Non-lamellar phases induced by the interaction of gramicidin S with lipid bilayers. A possible relationship to membrane-disrupting activity. Biochemistry 36:7906–7916

Rietveld AG, Killian JA, Dowhan W, de Kruijff B 1993 Polymorphic regulation of membrane phospholipid composition in *Escherichia coli.* J Biol Chem 268:12427–12433
Schiffer M, Chang C-H, Stevens FJ 1992 The functions of tryptophan residues in membrane proteins. Protein Eng 5:213–214
Seddon JM 1990 Structure of the inverted hexagonal (H_{II}) phase, and non-lamellar phase transitions of lipids. Biochim Biophys Acta 1031:1–69
Seelig A, Seelig J 1974 The dynamic structure of fatty acyl chains in a phospholipid bilayer measured by deuterium magnetic resonance. Biochemistry 13:4839–4845
Seelig J 1978 ^{31}P nuclear magnetic resonance and the head group structure of phospholipids in membranes. Biochim Biophys Acta 515:105–140
Van Echteld CJA, Van Stigt R, de Kruijff B, Leunissen-Bijvelt J, Verkleij AJ, de Gier J 1981 Gramicidin promotes formation of the hexagonal H_{II} phase in aqueous dispersions of phosphatidylethanolamine and phosphatidylcholine. Biochim Biophys Acta 648:287–291
Van Echteld CJA, de Kruijff B, Verkleij AJ, Leunissen-Bijvelt J, De Gier J 1982 Gramicidin induces the formation of non-bilayer structures in diacylphosphatidylcholine model membranes in a fatty acid chain length-dependent way. Biochim Biophys Acta 692:126–138
Watnick PI, Chan SI, Dea P 1990 Hydrophobic mismatch in gramicidin A′/lecithin systems. Biochemistry 29:6215–6221
Wimley WC, White SH 1996 Experimentally determined hydrophobicity scale for proteins at membrane interfaces. Nat Struct Biol 3:842–848

DISCUSSION

Ring: Is it possible to calculate the interaction energies of the peptides with the lipid from the changes in phase transition temperature?

Killian: It is probably possible, but I haven't tried it.

Ring: It would be interesting to look at all of the phases, and in particular the cubic phase you showed. If you look at the deformation of the membrane caused by gramicidin, each monolayer curves down as you approach the dimple but if you look in the 'horizontal' direction the curvature is in such a direction that these two curvatures at right angles give rise to a saddle. We don't know the size of that saddle splay (bending) coefficient, although we expect it to be of the same order as the normal, symmetrical splay coefficient. If you could calculate the deformation energy from the changes in transition temperature, it may be possible to use a cubic phase to calculate the saddle splay coefficient.

Huang: But you don't know how gramicidin and lipids are packed in an isotropic phase.

Killian: We have characterized the dioleoyl phosphatidylethanolamine/dioleoyl phosphatidylglycerol (di-C18:1_c-PE/di-C18:1_c-PG) mixtures by X-ray diffraction and shown that they are cubic. But we haven't characterized this in the presence of peptides. However, the samples become optically clear and viscous as a function of temperature, and the phase behaviour is in-between that of the bilayer and the H_{II} phase. Both observations are what you would expect of a cubic phase.

Heitz: Does the phase transition depend on the initial state of the lipid, i.e. whether it is a gel or liquid crystal?

Killian: The transition is not easily reversible, as is common for cubic phases. You have to go back to the gel state to retain the bilayer, but once you have a bilayer the transition temperature does not depend on whether it is in a gel or in a fluid phase.

Heitz: Have you tried to use a mixture of lipids to see if an adjustment to the surrounding lipid shifts the phase transition?

Killian: Do you mean mixtures of lipids with differing acyl chain lengths?

Heitz: Yes.

Killian: No, we haven't done that yet. We did some experiments to look at the effect on the gel to liquid crystalline phase transition temperature. We have introduced peptides of different lengths into dipalmitoyl phosphatidylcholine (di-C16:0-PC) bilayers, and at a low peptide/lipid ratio we see a systematic shift downwards, i.e. the shorter the peptide the more the phase transition shifts downwards.

Separovic: You showed the ^{2}H order parameter data that you used to calculate the change in lipid thickness, which were all generated using a lipid/peptide ratio of 30. Is it possible, since gramicidin is a dimer, that there is a lipid ratio of 60?

Killian: We thought of this possibility. We compared the effects of the gramicidin dimer with that of WALP, assuming that this peptide was present as either monomer or as a dimer. We then assumed that the effects on bilayer thickness were due only to the surrounding lipids. We found that even when WALP was present as a dimer, which would give it about the same dimensions as the gramicidin channel, gramicidin still had the largest effect on bilayer thickness.

Davis: You showed how the peptide fits into the hexagonal phase, which made me wonder whether the dimensions in the hexagonal phase correspond, so that you would expect the peptide to cross the lipid regions the way you drew it. Is this a realistic picture?

Killian: Yes, it is realistic. In collaboration with T. Pott (CNRS, Bordeaux) we did some X-ray diffraction studies of the different peptides in dielaidoyl phosphatidylethanolamine (di-C18:1_t-PE) and we found that at 40 °C the tube diameter increased linearly and with about the same amount as the length of the incorporated peptide. This fits with the model of the peptides in the H_{II} phase.

Hinton: What happens when you replace the tryptophans with lysines?

Killian: WALP16 induces an H_{II} phase when incorporated in dioleoyl phosphatidylcholine (di-C18:1_c-PC) model systems. When we replaced the tryptophans in this peptide with lysines, the lipids remained in a bilayer. We found that this was due to the high aqueous solubility of the peptide, so we synthesized a larger peptide that had a long hydrophobic part, and we incorporated this peptide into thicker bilayers. We then found that these lysine peptides also produced a hexagonal phase, so it seems to be a general effect. There was one difference, however, in that it behaved like a shorter WALP peptide. We concluded that the tryptophans are probably located deeper in the membrane, and that the lysines are higher up near the lipid headgroups, which would suggest that it is necessary to look at mismatch relative to the anchoring position at the interface.

Hinton: We just did a structure, binding and transport study on gramicidin A in which we replaced Trp15 with glycine, and we were surprised to find that the structure was similar to gramicidin A except that ethanolamine at position 16 has greater motion. However, the binding is much weaker and the transport is much lower compared to gramicidin A. Have you done any studies with completely deuterated lipids to see whether you still observe phase transitions? Some gramicidin analogues will insert into protonated SDS micelles, but if you want to do structural studies you have to put it in deuterated SDS, and then you find that it doesn't insert well into the membrane.

Killian: Sometimes this is due to the purity of the deuterated SDS.

Hinton: Yes, but we also wondered whether the subtle interactions between these lipids and the peptides are dependent on the amount of deuteration.

Killian: We did do deuterium NMR studies, but we didn't do a detailed study of whether the deuterated lipids behaved differently, for example by checking the phase transition temperature. But there are indications that deuterated lipids behave slightly different in this respect.

Stein: Has anyone seen other non-bilayer membrane structures in cellular material? Because one would think that they would be apparent by negative staining.

Killian: There are many non-bilayer lipids in biological membranes. The idea is that for certain membrane processes, e.g. membrane fusion or cell division, the bilayer has to be disrupted, and that non-bilayer lipids are required for this disruption. Non-bilayer structures themselves most likely do not play an important role in cellular functioning.

Stein: Non-bilayer lipids may also be involved in vesicular trafficking when endocytic vesicles come down from the cell membrane and the long-chain fatty acids separate into the ordered, globular pieces, and the disordered fatty acids into the tubes.

Jakobsson: We did a simulation of gramicidin in excess dimyristoyl phosphatidylcholine (DMPC), and we saw that the first layer of lipid around gramicidin was thickened by several ångströms, whereas the rest of the lipid was completely unchanged. We had 15 lipids in each leaflet, and the first leaflet changed by 4–5Å, whereas the rest were unchanged. This averages out to a thickening of about 1Å, which is consistent with your results.

Separovic: I would like to ask about the stabilizing and destabilizing effects. In the di-C18:1_c-PE/di-C18:1_c-PG mixtures you observed non-bilayer phases, and in real membranes there are mixtures of lipids and mixtures of proteins. What happens if you use a mixture of different length peptides? Would this stabilize the bilayer?

Killian: It would be unlikely, because we found that longer peptides do not stabilize the bilayer in di-C18:1_t-PE systems or in di-C18:1_c-PE/di-C18:1_c-PG mixtures.

Hladky: You mentioned that the haemolysis caused by gramicidin may involve membrane disruption. However, at least for low or moderate concentrations of gramicidin, the haemolysis can be completely prevented by impermeant ions; for example, if you replace sodium with choline or sucrose, you don't get haemolysis unless a high concentration of gramicidin is present.

Killian: But this also results in osmotic stabilization of the cells.

Hladky: The point I'm making is that it is a specific cation permeability that makes the cells osmotically unstable, and not a gross membrane disruption. I don't know if a gross membrane disruption occurs for erythrocytes at high gramicidin concentrations, but the haemolysis that limited the clinical use of gramicidin was seen at much lower concentrations of gramicidin.

Cornell: Is it fair to say that that if a lipid has a shorter chain or if a peptide has a longer sequence, then the bilayer structure is promoted?

Killian: We asked ourselves this question, and we generated these longer peptides and incorporated them in lipids that by themselves have a tendency to form non-lamellar structures. However, we didn't see any evidence for bilayer stabilization or promotion of the bilayer structure. We put WALP31, which we thought should be wider than the hydrophobic bilayer, in the di-C18:1_c-PE/di-C18:1_c-PG mixtures, and even at a 1:25 ratio the transition occurs at the same temperature as for pure lipids.

Cornell: But does the phase remain lamellar?

Killian: No, it goes from lamellar to cubic or isotropic at about 60 °C. At 40 °C a lamellar structure was maintained for the longer chain peptides, but not for the shorter ones.

Cornell: So, at the high temperature, a longer peptide and a shorter bilayer promotes a lamellar structure. In other words, for a particular length of peptide, if you have a shorter hydrocarbon in the lipid you can sustain more peptide and still keep the lamellar structure.

Jakobsson: Am I correct in saying that the short peptide destabilizes the membrane, and the long one doesn't make much difference?

Killian: Yes, that is correct. A likely explanation is that a relatively long peptide can adjust to the lipid length because it can tilt, and therefore it does not have much effect.

Davis: Regarding the question of how different length peptides interact with different thicknesses of lipid, what you should expect to see is that a longer peptide will preferentially partition into the lower temperature gel phase. It doesn't stabilize the lamellar phase, but it lowers the base transition into the fluid phase by preferentially partitioning at the phase transition into the gel phase. Shorter peptides prefer a fluid phase because the fluid phase is thinner.

Stein: Why should this be a specific effect of tryptophan, and not serine or threonine?

Sansom: Serine and threonine in the hydrophobic phase can accommodate its hydrogen bond by hydrogen bonding to the carbonyl backbone (Gray & Matthews 1984), whereas tryptophan cannot be folded back to satisfy its hydrogen bond.

Koeppe: Steve White would say that the situation is more complicated than this, and that the size, shape, rigidity and extensive aromatic system in the indole ring would all argue against it going into the bilayer centre and for tryptophan going towards the interface (Yau et al 1998).

Roux: There may also be other reasons related to protein assembly and insertion. For example, haemolysine has residues with aromatic side chains only on one side. This is probably related to the insertion mechanism.

Wallace: A number of crystal structures have been determined for membrane proteins, and almost all of them have a significant tryptophan asymmetry (Wallace & Janes 1999).

Roux: It is not clear why phenylalanine is observed at the interface because it cannot form a hydrogen bond. Observed propensity may not always be related to thermodynamics. There could be functional, as well as evolutionary, reasons.

Rosenbusch: All the phenolic groups of tyrosines point upwards, with the hydroxyls in the aqueous phase, whereas phenylalanine points in the reverse direction. The distance between aromatic groups exclude π–π interactions between the rings.

Sansom: This would tend to argue that the hydrogen bonding may be crucial.

References

Gray TM, Matthews BM 1984 Intrahelical hydrogen bonding of serine, threonine and cysteine residues within α-helices and its relevance to membrane-bound proteins. J Mol Biol 175: 75–81

Wallace BA, Janes RW 1999 Tryptophan in membrane proteins. Adv Exp Med Biol, in press

Yau W-M, Wimley WC, Gawrisch K, White SH 1998 The preference of tryptophan for membrane interfaces. Biochemistry 37:14713–14718

Peptide–lipid interactions and mechanisms of antimicrobial peptides

Huey W. Huang

Physics Department, Rice University, Houston, TX 77251-1892, USA

Abstract. Hydrophobic matching, in which transmembrane proteins cause the surrounding lipid bilayer to adjust its hydrocarbon thickness to match the length of the hydrophobic surface of the protein, is a commonly accepted idea. To test this idea, gramicidin was embedded in dilauroyl phosphatidylcholine (DLPC) and dimyristoyl phosphatidylcholine (DMPC) bilayers at the molar ratio 1:10. The bilayer thickness (PtP) was measured by X-ray lamellar diffraction. In the fluid phase near full hydration, PtP is 30.8 Å for pure DLPC, 32.1 Å for DLPC/gramicidin mixture, 35.3 Å for pure DMPC and 32.7 Å for a DMPC/gramicidin mixture. Gramicidin apparently stretches DLPC bilayers and thins DMPC bilayers toward a common thickness as expected by hydrophobic matching. Gramicidin pair correlations were measured by X-ray in-plane scattering. In the fluid phase, the gramicidin–gramicidin nearest-neighbour separation is 26.8 Å in DLPC bilayers but shortens to 23.3 Å in DMPC bilayers, thus confirming the conjecture that when proteins are embedded in a membrane, hydrophobic matching creates a strain field in the lipid bilayer that in turn gives rise to a membrane-mediated attractive potential between proteins. These results were analysed with an elasticity theory of membrane deformation. The same principle explains the 'concentration-gating' mechanism of pore formation by antimicrobial peptides via the membrane-thinning effect. Concentration-gated pore formation and membrane thinning by alamethicin and magainin have been observed.

1999 Gramicidin and related ion channel-forming peptides. Wiley, Chichester (Novartis Foundation Symposium 225) p 188–206

It is well known that amphiphilic peptides are membrane active. An amphiphilic peptide tends to associate its hydrophobic surface with that of a lipid bilayer. This is known as the hydrophobic effect. However, given the hydrophobic matching condition, amphiphilic helical peptides can associate with a lipid bilayer in many different ways. They can, for example, adsorb to the bilayer surface, form barrel-stave or toroidal pores, surround lipid bilayer disks or form other more complex lipid–peptide assemblies. The goals of studying peptide–lipid interactions are to: (1) find out experimentally which of the above actually takes place; and (2) understand theoretically why the peptide behaves that way. Since many of naturally produced amphiphilic peptides are immunogenic antimicrobials used in host defences throughout the animal kingdom, including humans, this is a subject of interest to both biology and medicine.

The question of how a peptide associates with a lipid bilayer is a problem of energetics. The energy of hydrophobic surface–surface interactions is dominated by the effect of water exclusion, i.e. the hydrophobic effect (Tanford 1980 [p 1–4], Israelachvili 1992 [p 128–135]). For a given peptide, this energy should be roughly the same in different peptide–lipid configurations, as it is measured by the area of hydrophobic matching surface. A crucial point is that any form of peptide association will likely involve a membrane deformation, and the energy of membrane deformation is an important part of the total energy that determines which form of association is most likely to occur. In this chapter I will report the use of gramicidin to investigate the energetics of membrane deformation, and I will suggest a concentration-gating mechanism for antimicrobial peptides based on their membrane-thinning effect.

Experiments with gramicidin

To study protein-induced membrane deformation, one needs to know the protein structure. Gramicidin is perhaps the best structurally characterized membrane protein. Elliott et al (1983) showed that the length of the hydrophobic surface of the gramicidin channel, h_G, is about 22 Å. If a gramicidin channel is embedded in a lipid bilayer, the thickness of the hydrocarbon region, h, is likely to be different from h_G. We know that lipid bilayers are about 400-fold more deformable than globular proteins, but does the lipid bilayer deform in order to achieve hydrophobic matching? For this to occur, the energy of the bilayer deformation must be smaller than the energy of hydrophobic mismatch. I will calculate these energies to determine under what conditions this is true. But first let's examine the experimental results (Harroun et al 1999a).

Hydrated lipid and gramicidin mixtures at the molar ratio 10:1 were prepared into oriented multilayers. The samples were equilibrated in a temperature–humidity chamber. We used circular dichroism analysis and other evidence to show that gramicidin was in the $\beta^{6.3}$ helical, dimeric channel form. The same sample was measured for the membrane thickness and the gramicidin pair correlation distribution.

Thickness of fluid membrane has been measured in many different ways. Each method, though, has its own complications. For the diffraction method, the main problem is the bilayer's undulatory fluctuations (Caillé 1972, Nagle et al 1996) that damp out high Bragg orders and distort the remaining diffraction signals. Our strategy for overcoming this difficulty consists of: (1) obtaining high ordered diffraction patterns by dehydration that diminishes the fluctuations; (2) measuring only the peak-to-peak separation (PtP, equivalent to the phosphate to phosphate distance across the bilayer), which is not affected by small undulation fluctuations; and (3) estimating the thickness (PtP) of fully hydrated membranes by extrapolation. This method has been demonstrated in detail (Wu et al 1995, He et al 1996a), including the particularly difficult region just above the main transition temperature (Chen et al 1997), and is supported by the fluctuation analysis of Nagle et al (1996). Figure 1 shows the PtP of dimyristoyl phosphatidylcholine (DMPC) bilayers with and without gramicidin at various temperatures as a function of the repeat spacing, *d*. In Table 1 we show the fluid phase PtP of DMPC and dilauroyl phosphatidylcholine (DLPC) bilayers near 98% RH, with and without gramicidin. The thickness of pure DLPC bilayers was measured by Olah (1990), and confirmed by Chen et al (1997); and the thickness of DLPC/gramicidin (10:1) bilayers was measured by Olah et al (1991). It is important to point out that all these results were obtained by the same diffraction method and by the same data reduction procedure, so that even if there are systematic errors, the relative changes of the membrane thickness are still reliable. Pure DMPC bilayers are 4.5 Å thicker than pure DLPC bilayers. But when the lipids contain gramicidin at a 10:1 ratio, the thicknesses of both of them approach a common value and become within 0.6 Å of each other.

These results support the estimate of the hydrophobic surface of gramicidin channel ($h_G \approx 22$ Å) by Elliott et al (1983). To show this, one has to deduce the thickness, h, of the hydrocarbon region of lipid bilayer from the PtP. For clarity we denote h_o the thickness in pure lipid, h_b the thickness at the lipid–gramicidin boundary, and $\bar{h}$ the average thickness in a bilayer containing gramicidin.

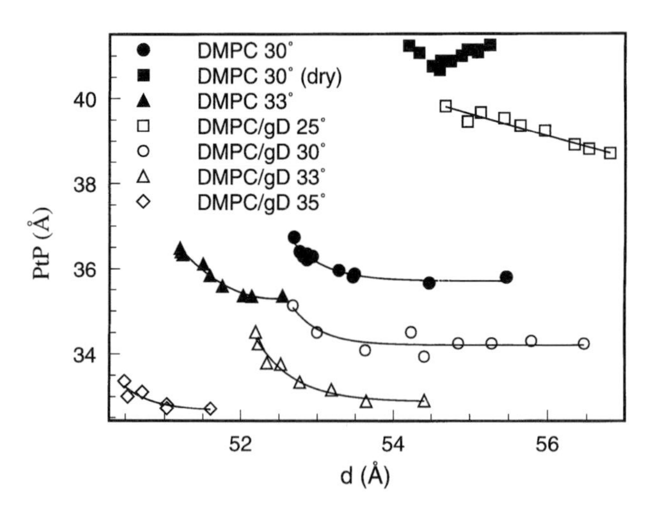

FIG. 1. Peak-to-peak (PtP) distance of the electron density profile as a function of repeat spacing, d. DMPC, dimyristoyl phosphatidylcholine; gD, gramicidin.

TABLE 1 Thickness (peak-to-peak; PtP) of dilauroyl phosphatidylcholine (DLPC) and dimyristoyl phosphatidylcholine (DMPC) bilayers at 98% relative humidity

Bilayer	*Gramicidin/lipid molar ratio*	*PtP (Å)*	*Temperature (°C)*
DLPC	0	30.8	20
DLPC	1/10	32.1	20
DMPC	0	35.3	33
DMPC	1/10	32.7	35

Hydrophobic matching implies $h_b \approx h_G$. However, $\bar{h}$ is in general different from h_b because the bilayer tends to restore itself to its unperturbed thickness, h_o, away from the boundary. X-ray crystallography (Hauser et al 1981), deuterium NMR (Seelig & Seelig 1980), neutron diffraction (Büldt et al 1978) and X-ray diffraction (Chen et al 1999) suggest that the glycerol region from the phosphate to the beginning of the hydrocarbon region is about the same in the gel and L_α phases (McIntosh & Simon 1986, Nagle et al 1996). This distance is about 5 Å (Nagle et al 1996, Chen et al 1999). Using this, we estimate $h_o \sim$ PtP$-$10 Å for pure lipids, and $\bar{h} \sim$ PtP$-$10 Å for bilayers containing gramicidin. For DLPC bilayers containing gramicidin, PtP = 32.1 Å gives $\bar{h}$ = 22.1Å, which is essentially equal to the estimated h_G as expected by hydrophobic matching. In this case $\bar{h} \sim h_b \sim h_G$ is reasonable because the lipid's natural hydrocarbon thickness ($h_o \sim 20.8$ Å) is sufficiently close to h_G. For DMPC bilayers, h_o ($\sim$25.3 Å) is sufficiently larger than h_G, so we expect $\bar{h}$ to be somewhat larger than $h_b \sim h_G$, as shown by the measurement $\bar{h} = 22.7$ Å. The energy cost of hydrophobic matching is proportional to $(h_o - h_G)^2$ (Huang 1986, 1995), so the strain energy in DMPC/gramicidin bilayers is much larger than in DLPC/gramicidin bilayers. We expect the strain field in the deformed bilayer to create an attractive membrane-mediated potential between gramicidin channels. Therefore, we expect the gramicidin–gramicidin nearest-neighbour separation to be smaller in DMPC than in DLPC bilayers.

The gramicidin–gramicidin nearest-neighbour separation was measured by X-ray in-plane scattering (Fig. 2). The gramicidin correlation peaks appear in the range $q \sim 0.2$–0.5 Å^{-1}. We want to compare DLPC/gramicidin and DMPC/gramicidin bilayers in the fluid phase (the lower two curves). DMPC/gramicidin bilayers below 27°C are in the gel phase, in which gramicidin forms aggregates (these data will not be discussed here). Qualitatively, the nearest-neighbour separation is inversely proportional to the q of the peak position. Therefore, the data clearly show a smaller nearest-neighbour separation in DMPC relative to DLPC bilayers. The precise interpretation of the scattering curve uses the scattering theory:

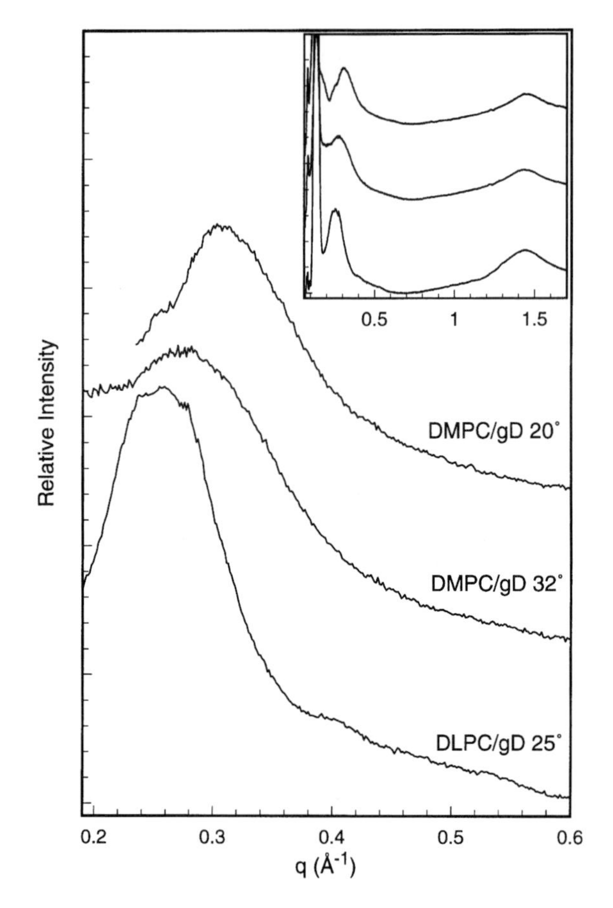

FIG. 2. X-ray in-plane scattering curves. The inset shows the raw data of dimyristoyl phosphatidylcholine (DMPC)/gramicidin (gD) (10:1) at 20 °C; DMPC/gramicidin (10:1) at 32 °C; and dilauroyl phosphatidylcholine (DLPC)/gramicidin (10:1) at 25 °C. The main frame shows the in-plane scattering curves after removing the background (pure lipid on the substrate). These are in-plane scattering curves of gramicidin.

$$I(q)/I_e = N|F(q)|^2 S(q), \tag{1}$$

where I_e is the scattering intensity by a single free electron; N is the number of channels; $F(q)$ is the scattering amplitude by an individual channel, called the form factor; $S(q)$ is the structure factor given by:

$$S(q) = 1 + \int [n(r) - \bar{n}] J_0(qr) 2\pi r dr. \tag{2}$$

$n(r)2\pi r dr$ is the average number of channels within ring of radius r and width dr, centred at an arbitrarily chosen channel; $\bar{n}$ is the mean number density of channels; and $J_0(qr)$ is the 0th order Bessel function of qr. The radial distribution function $2\pi r n(r)$ can be obtained by the Bessel transform:

$$2\pi r n(r) = 2\pi r \bar{n} + r \int [S(q) - 1] J_0(qr) q dq. \tag{3}$$

The form factor $F(q)$ of the gramicidin channel is well defined—it is almost entirely determined by the cylindrical peptide backbone of the channel. This is because only the rigid part of the molecule will contribute to the ensemble averaged form factor. The contribution of the side chains is for the most part motionally averaged to zero for the lack of correlations. The form factor generated by a molecular dynamics simulation (Woolf & Roux 1996), averaged over 250 molecular structures of the gramicidin channel, is close to the form factor of the backbone alone. Thus, the structure factor $S(q)$ can be reliably obtained from $I(q)/|F(q)|^2$ within a normalization factor. Then, the Bessel transform shown above gives the radial distribution function $2\pi r n(r)$. From the first maximum of this function we obtained the most probable nearest-neighbour separations: 26.8 Å in DLPC and 23.3 Å in DMPC bilayers.

Quantitative analysis

Some time ago, I proposed the following free energy of membrane thickness deformation to explain the effect of membrane thickness on gramicidin channel lifetime (Huang 1986).

$$f D = \frac{h_o B}{2} \left(\frac{D}{h_0}\right)^2 + \frac{\gamma}{4} (\nabla D)^2 + \frac{K_c}{8} (\nabla^2 D)^2, \tag{4}$$

where D is the deviation of the hydrocarbon thickness from its unperturbed value h_0, ∇ and ∇^2 are the in-plane gradient and Laplacian, $1/B$ is the thickness compressibility, γ is one-half of the bilayer tension coefficient, and K_c is the bending rigidity of the bilayer (Helfrich 1973). For local deformation, such as that induced by protein insertion, the tension contributes $<5\%$ of the total energy. This was recently demonstrated by the gramicidin channel kinetics under tension (Goulian et al 1998). Neglecting the tension term, we arrived at the Euler–Langrange equation:

$$\nabla^4 D + \lambda^{-4} D = 0, \tag{5}$$

which has only one length scale $\lambda = (h_o K_c/4B)^{1/4} \sim 13$ Å. We believe that the continuum theory can describe the energetics of membrane deformation even on the molecular scale. As long as there is a well-defined, time averaged molecular position for the interfaces, $D(x,y)$ can be defined. However, it is not clear whether the elasticity coefficients measured on the macroscopic scale are the same on the molecular scale. The solution of Equation 5, with appropriate boundary conditions, describes the effect of proteins on the membrane thickness profile. The boundary conditions are the values of D and the slope of D at the lipid–protein contact, designated as D_o and s, respectively. If the boundary conditions are known, D is determined and then the total energy of thickness deformation $F_M = \int f_D dxdy$ can be calculated. In our problem, D_o, the magnitude of hydrophobic mismatch, can be measured, but s is unknown. We can calculate F_M as a function of s. If we also know the boundary energy $E_{bd}(s)$, we can determine the boundary value s as a solution of $F_M'(s) + E_{bd}'(s) = 0$. It is important to note that one cannot neglect E_{bd}, because both F_M and E_{bd} are originated primarily from the hydrophobic effect—$F_M'(s)$ and $E_{bd}'(s)$ could be of the same order of magnitude. Not knowing $E_{bd}(s)$, we have no choice but to treat s as a parameter to be determined by comparison with experimental results.

Our samples were in the fluid phase. Gramicidin channels were in diffusive motion in the plane of the bilayer. We assumed a distribution of the channels and solved Equation 5 numerically for given D_o and s (Fig. 3), and calculated $F_M = \int f_D dxdy$. We then ran Monte Carlo simulations with the channels moving randomly without overlapping (the hardcore condition) and according to the Boltzmann probability in proportion to $\exp(-F_M/k_B T)$. After the simulation reached a steady state, we ran 2000 Monte Carlo cycles to represent 2000 channel distributions in thermal equilibrium. We calculated two quantities to compare with the experiments (Harroun et al 1999b): the average membrane thinning $<D>$ and the shift of the peak position q_{max} in $S(q)$ relative to the $S(q)$ of 'free hard disks', with no channel–channel interactions other than hardcore exclusion. The shift of q_{max} to a higher value in q represents the effect of membrane-mediated attractions between channels. Figure 4 shows the result of the simulations with the simulation parameters given in Table 2. D_o was chosen close to the value 3.3 Å,

---

FIG. 3. Examples of simulations. (A) A simulation of free hard disks. Circles represent gramicidin channels (or hard disks) 18 Å in diameter. There are no interactions between the channels other than the hard-core exclusion. (B) A simulation of channels (hard disks) in a membrane under the influence of Equation 4. Membrane deformation induces interactions between channels in addition to the hard-core exclusion. The area outside the circles is the membrane shown in a density plot. Dark areas represent height depression. There are many more pairs of channels close to each other in panel B than in panel A, due to membrane-mediated attractions in the former. (C) Enlargement of a small patch of (B). The contours indicate the membrane deformation $u(x,y)$.

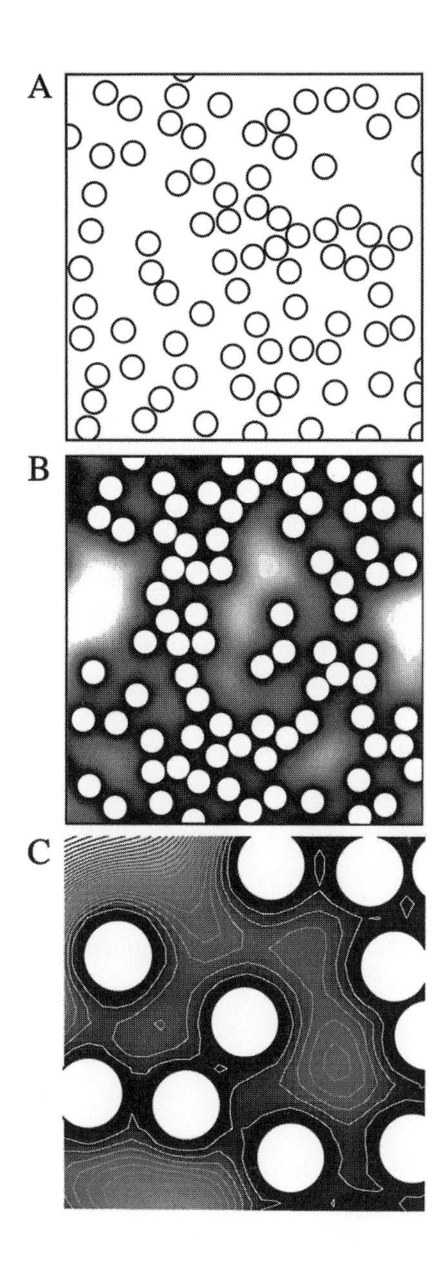
A
B
C

suggested by the thickness measurement. B and K_c were chosen close to their experimental values (see Harroun et al 1999b). We conclude that, for a reasonable choice of parameters: $D_o \sim (3.0$ to $3.4\,Å)$, $\lambda \sim (8.5$ to $12.6\,Å)$, $s \sim (-0.10$ to $-0.07)$, the result of gramicidin in DMPC bilayers can be explained with the elasticity theory given in Equation 4.

Conditions for hydrophobic matching

Let's now go back to the question: when do we expect hydrophobic matching to occur? Let's imagine that hydrophobic matching did not occur where a gramicidin channel is inserted and called it the mismatched condition, and compare it with the matching condition. The relevant energies for comparison are: for the mismatched condition, the mismatch energy $E(mis)$ and the binding energy between the gramicidin surface and first-shell lipids in their unstressed configurations E_{bd1}, and for the matching condition, the deformation energy F_M (per channel), and the binding energy between the gramicidin surface and first-shell lipids in their stressed configurations E_{bd2}. We do not know if E_{bd1} and E_{bd2} are different. We assume that their difference, if any, is small. Thus, for hydrophobic matching to occur the energy cost of membrane deformation F_M must be less than the energy cost of hydrophobic mismatch $E(mis)$. $E(mis)$ is estimated by the free energy change for transfer from organic solvent to water of non-polar residues: $16.7\,erg/cm^2$ (Chothia 1974). The external diameter of a gramicidin channel is $18\,Å$, so the area of mismatch is $18\pi D_0 Å^2$. This gives $E(mis) \approx 9.4 \times 10^{-14} D_o erg \approx 2.3 D_o k_B T$. In DMPC bilayers, $E(mis) = 7.6 k_B T$. In Huang (1986), an analytical expression for the deformation energy due to the insertion of a single gramicidin channel was derived according to the free energy Equation 4. Combining Equations IV.17 and IV.19 (of Huang 1986), and using the experimental values for various constants, we arrived at a total deformation energy caused by a single channel in an infinite membrane:

$$F_M = \int f_D dx dy \approx 1.8 \times 10^{-14} (D_o)^2 erg \approx 0.43 (D_o)^2 k_B T. \tag{6}$$

In DMPC bilayers, $F_M \approx 4.7 k_B T$. In comparison, the simulation results showed that at 1:10 peptide/lipid ratio F_M per channel is $\lesssim k_B T$. Since the deformation energy is proportional to D_o^2, whereas the energy cost of mismatch is proportional to D_o, one expects hydrophobic matching to occur for small D_o, $< 5.3\,Å$ according to the above estimates. For larger mismatches, some slippage or incomplete matching is expected to occur.

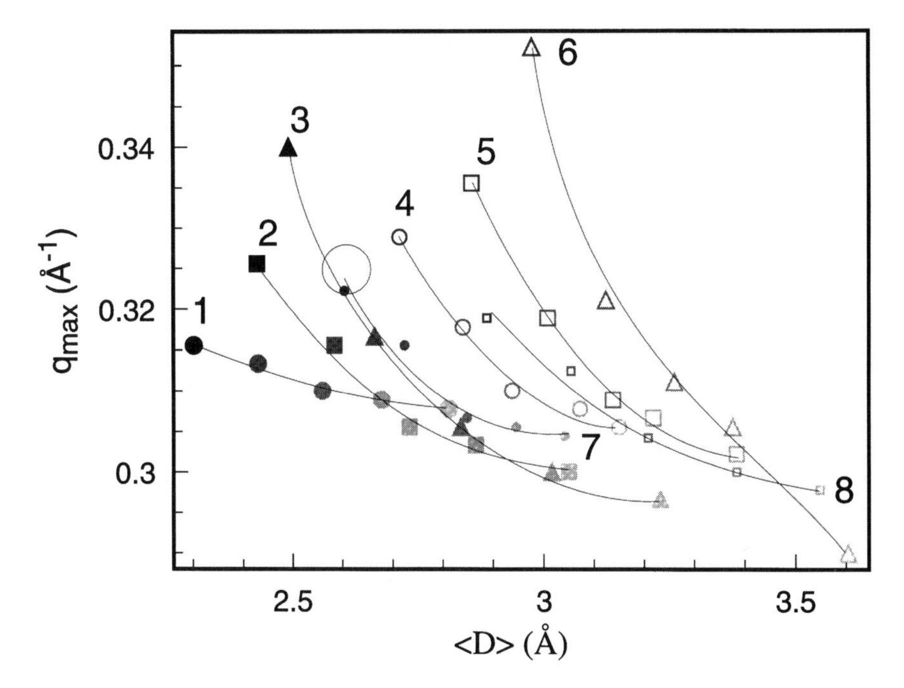

FIG. 4. The results of all simulations listed in Table 2 plotted as q_{max} versus $<D>$. Each curve (marked by the simulation number) is the result of the simulations for a chosen set of parameters (Table 2) as a function of s. Shading is used to indicate different values of s: from the darkest (-0.1) to the lightest $(+0.1)$. The experimental results of $<D>$ and the shift in q_{max} are in the large circled area. Different symbols are used to distinguish different sets of simulations.

TABLE 2 Simulation parameters

Simulation	D_o (Å)	K_c (10^{-12}erg)	B (10^8erg/cm^3)	λ(Å)
1	3.5	0.4	5.0	8.45
2	3.5	1.0	5.0	10.62
3	3.5	2.0	5.0	12.63
4	3.0	0.4	5.0	8.45
5	3.0	1.0	5.0	10.62
6	3.0	2.0	5.0	12.63
7	3.37	0.4	5.0	8.45
8	3.5	0.5	2.5	10.62

Dimyristoyl phosphatidylcholine (DMPC) hydrocarbon thickness in pure lipid $(h_0) = 25.3$ Å; channel diameter = 18 Å; number of channels = 81; the fraction of the total area occupied by the channels = 0.29; mesh size = 3.6 Å; $s = -0.1, -0.05, 0, +0.05, +0.1$.

Concentration-gating mechanism of antimicrobial peptides

To illustrate the importance of membrane deformation energy, we will now examine the molecular mechanisms of antimicrobial peptides. Gene-encoded antimicrobial peptides are ubiquitous components of host defences in mammals, birds, amphibia, insects and plants. Almost all of the peptides investigated in detail damage by first binding and then inserting into the microbial lipid membrane, thereby altering membrane permeability and impairing internal homeostasis (Martin et al 1995). We have studied the amphibian peptide magainin and two other α-helical peptides alamethicin and melittin. All of them adsorb on the membrane surface with helical axes parallel to the plane of the membrane at low peptide to lipid ratios (P/L). Perpendicular orientation and pore formation were detected only when P/L exceeds a threshold value, P/L*. In our experiments the helical orientation was measured by oriented circular dichroism (Huang & Wu 1991, Ludtke et al 1994) and pore formation was detected by neutron diffraction (He et al 1995, 1996b, Ludtke et al 1996, Yang et al 1998). The threshold P/L* varies with both peptide and lipid composition. In general, the larger the headgroup to chains area ratio, H/C, the smaller the P/L* (Heller et al 1997). For example, alamethicin is inserted in DMPC bilayers in the entire detectable range of P/L (low P/L*), but in diphytanoyl phosphatidylcholine bilayers P/L* is about 1/40. The existence of a threshold P/L for pore formation is crucial to the function of these peptides — we call it the concentration-gating mechanism. The host defence antimicrobial peptides are cationic. They are attracted to the negatively charged cell membranes of microbes, so the concentrations of the bound peptides are likely to exceed the P/L*. In contrast, the concentration-gating mechanism protects eukaryotic cells from the lytic effects of the peptides because these cells have neutral membrane surfaces, so the bound peptides tend not to exceed the threshold concentration.

What gives rise to the threshold concentrations? One possible explanation is the membrane-thinning effect. We have found that when the peptides are bound to the membrane surface, the membrane thickness decreases in proportion to P/L (Wu et al 1995, Ludtke et al 1995). The hydrocarbon region of a lipid bilayer is deformable but incompressible. The adsorption of the peptide in the headgroup region increases the area of the polar region. Therefore, the cross-sections of the chains must increase and their thickness decreases by volume conservation. The magnitude of thinning corresponds to the area extension caused by addition of peptides to the headgroup region (Wu et al 1995, Ludtke et al 1995). The energy cost of membrane thinning is part of the energy of peptide adsorption to the membrane surface. Using Equation 4, one can show that the energy of membrane thinning is proportional to the square of the thinning, and is hence proportional to the square of P/L (Huang 1995). This explains why, as P/L increases, the energy of surface adsorption will eventually exceed the energy of pore formation. When these two energies are equal the concentration is P/L*. Peptides form pores if P/L exceeds P/L*.

Acknowledgements

This work was supported by NIH grant GM55203, and by the Robert A. Welch Foundation.

References

Büldt G, Gally HU, Seelig A, Seelig J, Zaccai G 1978 Neutron diffraction studies on selectively deuterated phospholipid bilayers. Nature 271:182–184

Caillé A 1972 Remarques sur la diffusion des rayons X dans les smectiques. C R Acad Sci Serie B 274:891–893

Chen FY, Hung WC, Huang HW 1997 Critical swelling of phospholipid bilayers. Phys Rev Lett 79:4026–4029

Chen FY, Hung WC, Huang HW 1999 Hydration dependence of phospholipids, in preparation

Chothia C 1974 Hydrophobic bonding and accessible surface area in proteins. Nature 248: 338–339

Elliott JR, Needham D, Dilger JR, Hayden DA 1983 The effects of bilayer thickness and tension on gramicidin single-channel lifetime. Biochim Biophys Acta 735:95–103

Goulian M, Mesquita ON, Fygenson DK, Nielsen C, Andersen OS, Libchaber A 1998 Gramicidin channel kinetics under tension. Biophys J 74:328–337

Harroun TA, Heller WT, Weiss TM, Yang L, Huang HW 1999a Experimental evidence for hydrophobic matching and membrane-mediated interactions in lipid bilayers containing gramicidin. Biophys J 76:937–945

Harroun TA, Heller WT, Weiss TM, Yang L, Huang HW 1999b Theoretical analysis of hydrophobic matching and membrane-mediated interactions in lipid bilayers containing gramicidin. Biophys J 76:3176–3185

Hauser H, Pascher I, Pearson RH, Sundell S 1981 Preferred conformation and molecular packing of phosphatidylethanolamine and phosphatidylcholine. Biochim Biophys Acta 650:21–51

He K, Ludtke SJ, Worcester DL, Huang HW 1995 Antimicrobial peptide pores in membranes detected by neutron in-plane scattering. Biochemistry 34:15614–15618

He K, Ludtke SJ, Heller WT, Huang HW 1996a Mechanism of alamethicin insertion into lipid bilayers, Biophys J 71:2669–2679

He K, Ludtke SJ, Worcester DL, Huang HW 1996b Neutron scattering in the plane of membrane: structure of alamethicin pores. Biophys J 70:2659–2666

Helfrich W 1973 Elastic properties of lipid bilayers: theory and possible experiments. Z Naturforsch Sect C Biosci 28:693–703

Heller WT, He K, Ludtke SJ, Harroun TA, Huang HW 1997 Effect of changing the size of lipid headgroup on peptide insertion into membrane. Biophys J 73:239–244

Huang HW 1986 Deformation free energy of bilayer membrane and its effect on gramicidin channel lifetime. Biophys J 50:1061–1070

Huang HW 1995 Elasticity of lipid bilayer interacting with amphiphilic helical peptides. J Phys II France 5:1427–1431

Huang HW, Wu Y 1991 Lipid–alamethicin interactions influence alamethicin orientation. Biophys J 60:1079–1087

Israelachvili J 1992 Intermolecular and surface forces, 2nd edn. Academic Press, New York

Ludtke SJ, He K, Wu Y, Huang HW 1994 Cooperative membrane insertion of magainin correlated with its cytolytic activity. Biochim Biophys Acta 1190:181–184
Ludtke SJ, He K, Huang HW 1995 Membrane thinning caused by magainin 2. Biochemistry 34:16764–16769
Ludtke SL, He K, Heller WT, Harroun TA, Yang L, Huang HW 1996 Membrane pores induced by magainin. Biochemistry 35:13723–13728
Martin E, Ganz T, Lehrer RI 1995 Defensins and other endogenous peptide antibiotics of vertebrates. J Leukoc Biol 58:128–136
McIntosh TJ, Simon SA 1986 Area per molecule and distribution of water in fully hydrated dilauroylphosphatidylethanolamine bilayers. Biochemistry 25:4948–4952
Nagle JF, Zhang R, Tristram-Nagle S, Sun W, Petrache HI, Suter RM 1996 X-ray structure determination of fully hydrated Lα phase dipalmitoylphosphatidylcholine bilayers. Biophys J 70:1419–1431
Olah GA 1990 Tl^+ distribution in the gramicidin ion conducting channel determined by X-ray diffraction. PhD thesis, Rice University, Houston, TX, USA
Olah GA, Huang HW, Liu WH, Wu YL 1991 Location of ion binding sites in the gramicidin channel by X-ray diffraction. J Mol Biol 218:847–858
Seelig J, Seelig A 1980 Lipid conformation in model membranes and biological membranes. Q Rev Biophys 13:19–61
Tanford C 1980 The hydrophobic effect: formation of micelles and biological membranes. Wiley, New York
Woolf TB, Roux B 1996 Structure, energetics, and dynamics of lipid–protein interactions: a molecular dynamics study of the gramicidin A channel in a DMPC bilayer. Proteins 24:92–114
Wu Y, He K, Ludtke SJ, Huang HW 1995 X-ray diffraction study of lipid bilayer membrane interacting with amphiphilic helical peptides: diphytanoyl phosphatidylcholine with alamethicin at low concentrations. Biophys J 68:2361–2369
Yang L, Harroun TA, Heller WT, Weiss TM, Huang HW 1998 Neutron off-plane scattering of aligned membranes. I. Method of measurement. Biophys J 75:641–645

DISCUSSION

Woolley: Is the diameter of the alamethicin channel you report equal to the diameter of a relatively uniform channel aggregate (e.g. an octamer) or is it an average diameter for a range of channel types?

Huang: We initially thought there would be a relatively wide size distribution, but we were surprised to find that it was so narrow. I thought there would need to be a distribution of 7, 8 or 9 monomers to fit the data, but it turns out that the result was wider than the data. Therefore, we concluded that in dilauroyl phosphatidylcholine (DLPC) bilayers, the eight monomer channel is stable. We have checked with other lipid bilayers, and find that the channel has a different size (He et al 1996).

Sansom: You said that at low peptide/lipid ratios both alamethicin and magainin prefer to adopt a surface orientation. Is this dependent upon the lipid?

Huang: Yes. In a bilayer containing small headgroups and large chains, there is more room for the peptide in the headgroup region, so the threshold concentration for insertion is high. In a lipid like DLPC, the threshold concentration is so low that within the (spectroscopic) experimental range of concentration, alamethicin is always inserted in a transmembrane orientation.

Dempsey: It sounds as though you have 2D crystals in that system, and I was wondering whether they are likely to be amenable to image reconstruction electron microscopy analysis because this would give a higher resolution structure.

Huang: Yes, they are. I hope that someone will do the electron microscopy, but the sample preparation is difficult.

Bechinger: We found that the lipids are no longer oriented above 3% mole magainin. You used up to 10% magainin, so how did you make sure that your lipids were oriented?

Huang: We looked at them through a polarized microscope and analysed them by X-ray diffraction.

Davis: The level of hydration may be important. Lipids may orient more easily when there is less water present. What was your level of hydration?

Huang: We put the sample in a vacuum and pump out all the water, then we hydrate the sample at 100% relative humidity.

Bechinger: I would like to present some of our data relevant to this discussion. We have synthesized an amphipathic helical peptide similar to magainin in which the lysines have been replaced by histidines. Oriented ^{15}N solid-state NMR spectroscopy indicates that at pH 5.0 this peptide assumes an orientation along the surface of the bilayer, whereas at pH 7.0 it occurs in a transmembrane orientation (Bechinger 1996). A titration curve indicates that at pH 6.1 approximately the same amount of peptide occurs in transmembrane or in-plane orientations. In antibiotic assays the peptide is about two orders of magnitude more effective against *Escherichia coli* and other bacteria at pH 5.2 when compared to pH 7.3 (Vogt & Bechinger 1999). This strongly suggests that the in-plane peptide is the better antibiotic. We do not know if the antibiotic activity is caused by the same peptide configuration that also causes the channel-like properties of amphipathic peptides in electrophysiological recordings. It has always been assumed that this is the case, but there may be different mechanisms involved. For highly charged amphipathic peptides we suggest a model in which the detergent-like properties of these peptides cause defects in the bilayer (Bechinger 1997).

Dempsey: We have looked at pore formation in melittins linked together with a disulfide bond (Takei et al 1999). Under some conditions we find that discrete dimerization is rate limiting for pore formation. An amphipathic helix lying in a bilayer membrane is readily solvated, if you consider the hydrophobic interaction with the lipid in a monolayer, because the acyl chains are generally flexible and long enough to fill the space underneath the peptide. In contrast, for discrete dimers the lipids in the same monolayer can't fill the space underneath. This is a possible interpretation of the melittin dimerization results, where we think surface helix dimerization might trigger membrane disruption or pore formation.

Roux: How does this fit with the prostaglandin synthase structure of Picot et al (1994), which is an enormous structure that presumably lies flat on the surface of the membrane and contains four helices in the membrane?

Dempsey: There are other large parts of the protein on the outside of the membrane. We're looking at a particular situation in which there are only two amphipathic helices with nothing to prevent these from inserting further into the membrane.

Sansom: How far down into the membrane plane do these helices go? One tends to assume that because they're hydrophobic, they should be quite close to the mid-plane. Do we really know that? Or do they lie more loosely on the surface?

Roux: In the case of prostaglandin H2 synthase-1, the protein does not sit loosely on the surface because one has to use a fair amount of detergent to wash it off the membrane (Picot et al 1994).

Wallace: But there's no solid evidence that it penetrates.

Arseniev: If you use micelles as a model system you can do NMR spectroscopy to show that amphiphilic peptides have certain conformation in the bound state and you can obtain information about the positioning of the peptide in the micelle (Brown et al 1981). Our recent data (P. V. Dubovskii, H. Li, A. Takahashi, A. S. Arseniev & K. Akasaka, unpublished data 1998) revealed that E5 fusion peptide doesn't penetrate deeply into micelles but rather stays on the surface. About three-quarters of its amphiphilic helix is accessible to the bulk surface.

Jakobsson: I have a general point, but it is related to Huey Huang's talk. I wonder to what extent the hydrophobic thickness of a peptide like gramicidin is an inherent property that depends on the nature of the lipid. Because Huey, Avi Ring and I have all thought about how the membrane might be deformed around gramicidin. We all expected, based on data from glycerol membranes, that hydrophobic thickness was such that it would have caused a thinning in dimyristoyl phosphatidylcholine (DMPC) bilayers, and yet Benoît Roux's data (Woolf & Roux 1996) and ours (Chiu et al 1999a,b) suggest that gramicidin causes DMPC to thicken, so it looks as though gramicidin has a different hydrophobic length if it's in phosphatidylcholine membranes as opposed to being in glycerol membranes.

Ring: Our theory and your experiments were performed in the presence of solvent, whereas Huey Huang's and Benoît Roux's were in the absence of solvent, therefore I'm not sure whether Benoît Roux's data would be applicable to this more complicated situation. A large part of the deformation energy for the situation we looked at was in the boundary lipids, and so the steric conditions and the interactions between the phospholipid headgroups and the gramicidin ends is different in Benoît Roux's case than it is with glycerol monoolein (GMO). I can imagine that the polar head of a phospholipid is attracted and pulled out by gramicidin, or is pushed away by hydrophobic interactions, but it's difficult from our studies to conclude anything about the boundary lipids in DMPC bilayers.

Killian: Our results are different. A likely explanation is that our studies were done at a 1:30 ratio, whereas yours were done at 1:10. Watnick et al (1990) showed that deuterium NMR order parameter is a function of peptide concentration. They found that the order parameter increases with the peptide concentration up to 1:15, and then it decreases again. Therefore, something is happening as a function of concentration, and our experiments cannot be directly compared.

Huang: I would like to comment on the use of the Seelig formula. The deuterium NMR order parameter is related to the average of cosine squared of the angle between the bilayer normal and the CD bond. To calculate the membrane thickness, you need the projection of the C–C chain onto the bilayer normal, which is the cosine of the angle between the bilayer normal and the normal of the two CD bonds. In general, it is impossible to get the average of cosine from the order parameter, so to obtain the Seelig formula, it is necessary to restrict the hydrocarbon chain to a few most likely conformations. Then you can obtain the equation just like Seelig did. The result of applying this formula to a non-saturated chain is ambiguous, and it is also questionable for bilayers containing peptides.

Killian: I would like to repeat that our systems are not comparable, and I agree that we have to make certain assumptions when we apply this interpretation to systems in which there are peptides present.

Ring: One would expect from the models to see thinning in DMPC bilayers, but the molecular dynamics show a thickening. What is the origin of that thickening? Does it contradict elasticity theory or is it a different phenomenon?

Roux: Huey Huang described the results from a continuum elasticity theory for the bilayer deformations around embedded membrane-bound proteins and impurities. Questions have been raised about the spatial range of the local deformation around a protein, and I would like to add a few comments on the significance of such theoretical treatments. Statistical mechanical theories can be developed for all kinds of perturbations and deformations. Let us call $f(x)$ the membrane deformation at the point x. For example, f could be the membrane thickness, the bending, the stretching or the hydrocarbon density. Often, f is assimilated to a generalized order parameter. It is important to realize that in all such theories arise integrals of the following type:

$$I = \int dx' \chi(x - x') f(x') \qquad (1)$$

where χ is the susceptibility response function of the membrane associated with the 'order parameter' f. For example, I could be one term in the free energy of the system. Typically, the response function is expected to be of moderate to short range compared to the spatial extent of the deformation f. This implies that a Taylor series may be used to express $f(x')$ in terms of a local expansion around the point x:

$$f(x') = f(x) + \frac{df(x)}{dx}(x - x') + \frac{1}{2}\frac{d^2f(x)}{dx^2}(x - x')^2 + \ldots \tag{2}$$

This expansion is then inserted into the integral I, yielding

$$I = Af(x) + B\frac{d^2f(x)}{dx^2} \tag{3}$$

where the coefficients A and B are given by

$$A = \int dx\chi(x) \tag{4}$$

$$B = \frac{1}{2}\int dx\chi(x)x^2. \tag{5}$$

The integral of the linear term vanishes because the response function is usually assumed to be isotropic. Equation 3 is reminiscent of continuum theories, which usually have the form of a differential equation for some function $f(x)$. This derivation makes it clear that continuum theories of bilayer deformations assume that the spatial range of the response function $\chi(x)$ is much shorter than that of the order parameter $f(x)$. It is likely that the spatial range of the function $\chi(x)$ for a bilayer corresponds to at least the diameter of one or two phospholipid molecules, i.e. 10 to 20 Å. In conclusion, one must be aware of the limitations of using a continuum treatment to describe the local deformation in the vicinity of a membrane-bound protein. If the variations in the order parameter f are sudden and abrupt, e.g. over 1 to 5 Å, it is possible that the underlying assumptions of the continuum theory (a Taylor series expansion) are not satisfied. Thus, one must be careful not to overinterpret the results of such theories.

Ring: Those coefficients are not sufficient to cover the liquid crystal theory (Ring 1996). This theory was proposed Frank (1958) and was supported by de Gennes (1974), Helfrich (1973) and Huang (1986). The Evans theory doesn't cover local deformations, and neither do theories on erythrocyte deformations. There are many more terms than what you showed, and this is your interpretation of the theory.

Roux: It's not an interpretation, it is how you derive such continuum theories.

Ring: There are some classical examples where continuum theory has been used successfully down to molecular dimensions, e.g. in hydrodynamics. Also, Einstein's derivation of ionic mobilities is a classic example of how you use continuum theory down to atomic scale and get qualitatively good results. There were also three calculations in the 1930s and 1940s of the surface tension of small droplets of water consisting of no more than nine water molecules and the conclusion is that the differences to the macroscopic value is quantitative but not qualitative (Tolman 1949, Kirkwood & Buff 1949, Benson & Shuttleworth 1951).

Jakobsson: But the point isn't that those theories don't ever work at these dimensions, it is that if you see them fail you shouldn't be surprised.

References

Bechinger B 1996 Towards membrane protein design: pH-sensitive topology of histidine-containing polypeptides. J Mol Biol 263:768–775

Bechinger B 1997 Structure and functions of channel-forming peptides: magainins, cecropins, melittin and alamethicin. J Membr Biol 156:197–211

Benson GC, Shuttleworth R 1951 The surface energy of small nuclei. J Chem Phys 19:130–131

Brown LR, Bosch C, Wuthrich K 1981 Location and orientation relative to the micelle surface for glucagon in mixed micelles with dodecylphosphocholine. EPR and NMR studies. Biochim Biophys Acta 642:296–312

Chiu S-W, Subramaniam S, Jakobsson E 1999a Simulation study of a gramicidin/lipid bilayer system in excess water and lipid. I. Structure of the molecular complex. Biophys J 76: 1929–1938

Chiu S-W, Subramaniam S, Jakobsson E 1999b Simulation study of a gramicidin/lipid bilayer system in excess water and lipid. II. Rates and mechanisms of water transport. Biophys J 76:1939–1950

de Gennes PG 1974 The physics of liquid crystals. Clarendon Press, Oxford

Frank C 1958 I. Liquid crystals. On the theory of liquid cystals. Disc Faraday Soc 25:19–28

He K, Ludtke SJ, Worcester DL, Huang HW 1996 Neutron scattering in the plane of membranes: structure of alamethicin pores. Biophys J 70:2659–2666

Helfrich W 1973 Elastic properties of lipid bilayers: theory and possible experiments. Z Naturforsch Sect C J Biosci 28:693–703

Huang HW 1986 Deformation free energy of bilayer membrane and its effect on gramicidin channel lifetime. Biophys J 50:1061–1070

Kirkwood JG, Buff FP 1949 The statistical mechanical theory of surface tension. J Chem Phys 17:338–343

Picot D, Loll PJ, Garavito RM 1994 The X-ray crystal structure of the membrane protein prostaglandin H2 synthase-1. Nature 367:243–249

Ring A 1996 Gramicidin channel-induced lipid membrane deformation energy: influence of chain length and boundary conditions. Biochim Biophys Acta 1278:147–159

Takei J, Remenyi A, Dempsey CE 1999 Generalized bilayer perturbation from peptide helix dimerization at membrane surfaces: vesicle lysis induced by disulphide-dimerized melittin analogues. FEBS Lett 442:11–14

Tolman RC 1949 The effect of droplet size on surface tension. J Chem Phys 1949 17:333–337

Vogt TCB, Bechinger B 1999 The interactions of amphipathic helical peptide antibiotics with lipid bilayers: the effect of charges and pH. in prep

Watnick PI, Chan SI, Dea P 1990 Hydrophobic mismatch in gramicidin A′/lecithin systems. Biochemistry 29:6215–6221

Woolf TB, Roux B 1996 Structure, energetics, and dynamics of lipid–protein interactions: a molecular dynamics study of the gramicidin A channel in a DMPC bilayer. Proteins 24: 92–114

Folding patterns of membrane proteins: diversity and the limitations of their prediction

Jurg P. Rosenbusch

Department of Microbiology, Biozentrum, University of Basel, CH-4056 Basel, Switzerland

Abstract. Significantly more high resolution structures of membrane proteins, obtained either by X-ray analysis, electron crystallographic methods or, in the future, by NMR spectroscopy, will be required for reliable structure predictions. Aberrations from the motifs of α-helical bundles and β-barrels occur that are not easily identified by algorithms unless structural homologies exist in the data banks. The coexistence of secondary structure motifs, originally proposed for a neurotransmitter receptor, has now been confirmed for a bacterial iron-siderophore translocating protein (FhuA) by X-ray analysis to 2.7 Å resolution. This protein contains a plug domain that has both α- and β-structures.

1999 Gramicidin and related ion channel-forming peptides. Wiley, Chichester (Novartis Foundation Symposium 225) p 207–214

The protein sequence of bacteriorhodopsin (Khorana et al 1979, Ovchinnikov et al 1979) is in agreement with its structure, which was previously unravelled by electron crystallography (Henderson & Unwin 1975). It revealed a correlation between the seven transmembrane α-helices and the seven hydrophobic segments in the protein sequence, each consisting of some 20 residues. This correlation immediately suggested an algorithm that allowed such segments to be identified in protein data banks (Kyte & Doolittle 1982). Not only is this procedure expedient, but it can also be applied to many other membrane proteins, as confirmed by high resolution structure analyses (see below). This led to the notion that most, if not all, integral membrane proteins occur in an α-helical configuration.

At about the same time, another membrane protein, porin, was described that resides stably in bacterial envelopes and forms aqueous pores across the bilayer (Nakae & lshii 1978). Its function and structure were studied by single-channel recordings and electron crystallography, respectively (Schindler & Rosenbusch 1978, Engel et al 1985). Unlike bacteriorhodopsin, it exhibits a high degree of polarity (>50%), and spectroscopic studies revealed predominantly a β-structure (Rosenbusch 1974). Its protein sequence, however, does not contain a single stretch of 20 hydrophobic residues, suggesting that its folding pattern is different from the α-helical bundles described above. Using turn predictions based on *ab initio* or empirical methods, Paul & Rosenbusch (1985) proposed that its folding pattern comprised a β-sheet of 16 antiparallel β-strands. Although these methods are not precise, they do suggest that the polypeptide exists in an extended conformation with β-strands forming stretches of 8–10 residues, long enough to span a hydrophobic membrane core of ~30 Å. In such an arrangement, only 4–5 residues in alternating positions are exposed to a hydrophobic environment, with intervening residues undefined with regard to polarity. Such limited constraints do not allow unequivocal folding patterns to be predicted, as such segments also exist in soluble proteins. None the less, subsequent X-ray analysis of the OmpF porin structure to high resolution (2.5 Å) confirmed the proposed pattern (Cowan et al 1992). The structure revealed that aromatic residues, located at the lipid–water interfaces on either side of the membrane, may impose stronger constraints for β-structure predictions (Schirmer & Cowan 1993). These are, however, still minimal, and they limit attempts to identify transmembrane β-structures from a data bank.

Another question that must be raised is whether membrane domains always consist of either all-α or all-β structures, or whether folding patterns with mixed secondary structures exist. It is only recently that clear evidence for a membrane domain with mixed structures has been described (Locher et al 1998). This finding may be useful for testing the criteria and the reliability of the methods predicting membrane protein structures.

α-helical bundles

The high resolution structure of bacteriorhodopsin, investigated either by electron crystallography to ~3 Å (Grigorieff et al 1996, Kimura et al 1997) or by X-ray crystallography in lipidic cubic phases to <2.5 Å (Pebay-Peyroula et al 1997, Luecke et al 1998), confirmed the original structure of a seven α-helical bundle. That this fold occurs as a recurring motif in membrane proteins was extended to ~10 protein families (Ostermeier & Michel 1997), and recently also to aquaporin (Heymann et al 1998), a bacterial K^+ channel (Doyle et al 1998), and a mechanosensitive channel (Chang et al 1998). A particularly intriguing example is the structure of rhodopsin, which is related to the G protein-dependent receptor proteins and reveals structural homology to bacteriorhodopsin (Baldwin et al 1997). Interesting variants are the electrically gated specific ion channel-forming proteins. It was originally proposed that their folding, derived from the sequence of a Na^+ channel (Noda et al 1984), appeared to be a paradigm for channels of various ion selectivities. Its proposed structural motif, repeated four times, consisted of six α-helices per functional domain. However, mutational analyses identified that the functionally significant part is in a region originally excluded from the membrane domain (Yellen et al 1991). Recent X-ray analysis to 3.2 Å resolution revealed that the functionally critical site is an α-helix which is too short to span the membrane, and a non-periodic segment carrying the signature sequence. Both of these are located within membrane boundaries (Doyle et al 1998).

An analogous problem occurs in the ligand-gated nicotinic acetylcholine receptor protein. It was originally proposed that this protein contains four α-helices per subunit (Noda et al 1983), and this remains the consensus opinion despite spectroscopic investigations which show that nearly half of the backbone is found in a β-conformation (Mielke & Wallace 1988). An electron crystallographic study at a 9 Å resolution suggested that polypeptides in a β-sheet configuration are present within membrane boundaries and coexist with α-helices (Unwin 1993). This issue is not yet entirely clarified, and the outcome of this question will be significant.

β-barrels

The original proposal of a β-sheet folding pattern (Paul & Rosenbusch 1985) has been applied, with minor variations, to several bacterial proteins spanning the outer membrane. However, it was the high resolution structural determination of the porins (Cowan et al 1992, Weiss et al 1991) that demonstrated conclusively that β-sheets exist as barrels. For example, the membrane domain of a small protein, OmpA (19 kDa), was predicted correctly, using algorithms proposed for the porins (Paul & Rosenbusch 1985), to consist of eight transmembrane β-strands (Vogel & Jähnig 1986). Recently, this prediction was confirmed (Fig. 1a) by X-ray analysis at 2.5 Å resolution (Pautsch & Schulz 1998). The prediction was originally developed for the unspecific porin OmpF (37 kDa) from *Escherichia coli* (Paul & Rosenbusch 1985) and, together with spectroscopic evidence, it suggested that OmpF consists of ~16 transmembrane β-strands. This folding arrangement challenged the results of extensive topological studies (Klebba et al 1990) that postulated 19 membrane-spanning β-strands, although subsequent high resolution structural analysis showed 16 strands (Cowan et al 1992), illustrating the limitations of overly ambitious predictions. The reason for the discrepancy is that unexpectedly the largest loop in the molecule bends backwards into the channel to form a constriction (Fig. 1b), thus evading the predictions. This suggests the sobering thought that we can predict only what we know. Another example of this is that β-bulges in β-strands change the periodicity of the alternating membrane-exposed residues, thus giving rise to significant errors in the predictions of transmembrane β-strands (Schirmer & Cowan 1993).

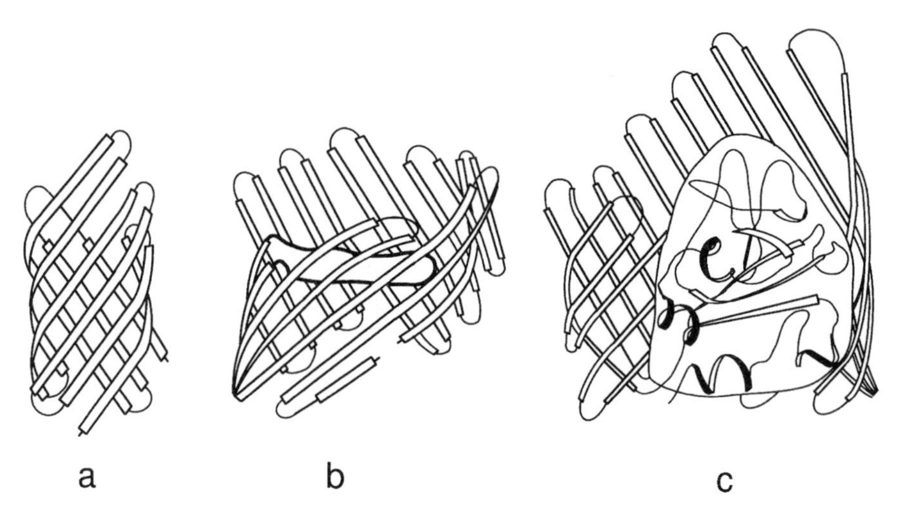

FIG. 1. Structure predictions of three β-barrel proteins in schematic representations. (a) N-terminal domain of OmpA (residues 1–171; 19 kDa) consists of eight antiparallel β-strands that form a closed barrel (Pautsch & Schulz 1998), as predicted by Vogel & Jähnig (1986). (b) Structure of the unspecific OmpF porin from *Escherichia coli* (37 kDa) with 16 β-strands (Cowan et al 1992) and a constriction loop (bold line) inside the channel. This constriction loop could not have been predicted (Klebba et al 1990). (c) The β-barrel of FhuA (78 kDa) is sliced open to visualize the plug (thin lines). The original study predicted, based on the constriction loop in unspecific porins, a barrel consisting of 32 β-strands and a constriction loop representing the channel-closing device (Koebnik & Braun 1993). High resolution analysis revealed a barrel with 22 β-strands (~60 kDa) and a plug (140 residues, represented in thin lines, with α-helices in bold), the sequence and the structure of which have no precedent in the respective data banks. This domain, located in the channel lumen, contains five α-helices and six β-strands, which prediction and spectroscopic methods could not have easily identified (Locher et al 1998).

Mixed structures

The difficulties of predictions are further illustrated by the structure of the ligand-gated, channel-forming, membrane-spanning protein FhuA (Locher et al 1998). This bacterial iron-siderophore translocating protein has a molecular mass of 78 kDa. Sequencing and extensive topological and functional studies (including analyses of protein fusions, deletion mutants, phage binding sites and channel state) suggested that it has a 32-stranded β-barrel structure, with one large loop bending backwards into the pore (Koebnik & Braun 1993). This prediction seemed a logical extension of the structure of OmpF, except that X-ray analysis at 2.7 Å resolution (Locher et al 1998) revealed a barrel structure containing 22 β-strands, with a lumen clogged by a plug (Fig. 1c). This plug (~15 kDa) undergoes complex conformational alterations that eventually allow ligand translocation (for details see Locher et al 1998). What is interesting in this context is that the plug contains five short α-helices in addition to a four-stranded β-sheet which is nearly parallel to the plane of the membrane (Fig. 1c). Therefore, the structure is composed of a mixture of α- and β-structures within a membrane domain, and could not have been predicted as no homologous structure is known.

Diagnostic tools to assess membrane protein structures

It is clear that spectroscopy is a most useful method to estimate secondary structure. Fourier-enhanced infrared spectroscopy is the method of choice, not only because it requires small quantities ($\sim$10 μg; Cowan & Rosenbusch 1994), but also because it can be performed irrespective of the state of aggregation of a protein. The question thus arises as to what methods, if any, can be employed to recognize mixed structures and assign them to membranous or extramembraneous domains. Proteolysis is not a viable method, since it may affect the structure markedly. Preferential deuteration involves sophisticated interpretations, and thus cannot be employed as a routine method. Spectroscopic methods are subject to further limitations, as it is not known whether they can differentiate reliably between α-helices and, for example, 3_{10} or Π helices. Albeit uncommon in soluble proteins, they cannot be excluded a priori in environments with low dielectric constants (2–8). The problem of whether different types of secondary structure motifs are frequently segregated in membrane-integral and extramembraneous domains, or whether they can form motifs such as those found in triosephosphate isomerase barrels, also requires further work. These considerations do not account, of course, for the insertion mechanisms during biosynthesis.

Conclusions

High resolution structures are required to understand the basis of functional properties of membrane proteins, and to calibrate the current prediction methods. Crystallizations from detergent solutions, limited thus far to about a dozen families of membrane proteins (versus over a thousand of families of soluble proteins), may imply that crystallizable membrane proteins are, based on their stability, a strongly biased selection, with many, or most, of those eluding high resolution research being too labile to form ordered three-dimensional crystals. This may imply that we have to revise our notions of folding patterns once we obtain more information. Aggregation, heterodispersity and thus failures in crystallization may be avoided by a different approach using lipidic mesophases (Landau & Rosenbusch 1996). The lateral pressure in bilayers as they occur in lipidic cubic phases (Marsh 1996) may be responsible for preserving the native state of membrane proteins during crystal growth, so the enclosure of membrane proteins in three-dimensional membrane matrices may stabilize labile proteins and may thus allow X-ray analysis (M. L. Chiu, P. Nollert, M. C. Loewen, H. Belrhali, E. Pebay-Peyroula, J. P. Rosenbusch & E. M. Landau, unpublished results 1999). This approach was successful for bacteriorhodopsin (Pebay-Peyroula et al 1997), and it may be successfully applied to other proteins in the future.

Acknowledgement

I thank the Swiss National Science Foundation for their support.

References

Baldwin JM, Schertler GF, Unger VM 1997 An alpha-carbon template for the transmembrane helices in the rhodopsin family of G-protein-coupled receptors. J Mol Biol 272:144–164

Chang G, Spencer RH, Lee AT, Barclay MT, Rees DC 1998 Structure of the MscL homolog from *Mycobacterium tuberculosis*: a gated mechanosensitive ion channel. Science 282:2220–2226

Cowan SW, Rosenbusch JP 1994 Folding pattern diversity of integral membrane proteins. Science 264:914–916

Cowan SW, Schirmer T, Rummel G et al 1992 Crystal structures explain functional properties of two *E. coli* porins. Nature 358:727–733

Doyle DA, Cabral JM, Pfuetzner RA et al 1998 The structure of the potassium channel: molecular basis of K^+ conduction and selectivity. Science 280:69–77

Engel A, Massalski A, Schindler H, Dorset DL, Rosenbusch JP 1985 Porin triplets merge into single outlets in *Escherichia coli* outer membranes. Nature 317:643–645

Grigorieff T, Ceska TA, Downing KH, Baldwin JM, Henderson R 1996 Electron-crystallographic refinement of the structure of bacteriorhodopsin. J Mol Biol 259:393–421

Henderson R, Unwin PNT 1975 Three-dimensional model of purple membrane obtained by electron microscopy. Nature 257:28–32

Heymann JB, Agre P, Engel A 1998 Progress on the structure and function of aquaporin 1. J Struct Biol 121:191–206

Khorana HG, Gerber GE, Herlihy WC et al 1979 Amino acid sequence of bacteriorhodopsin. Proc Natl Acad Sci USA 76:5046–5050

Kimura Y, Vassylyev DG, Miyazawa A et al 1997 Surface of bacteriorhodopsin revealed by high-resolution electron crystallography. Nature 389:206–211

Klebba PE, Benson SA, Bala S et al 1990 Determinants of OmpF porin antigenicity and structure. J Biol Chem 265:6800–6810

Koebnik R, Braun V 1993 Insertion derivatives containing segments of up to 16 amino acids identify surface- and periplasm-exposed regions of FhuA outer membrane receptor of *Escherichia coli* K-12. J Bacteriol 175:826–839

Kyte J, Doolittle RF 1982 A simple method for displaying the hydropathic character of a protein. J Mol Biol 157:105–132

Landau EM, Rosenbusch JP 1996 Lipidic cubic phases: a novel concept for the crystallization of membrane proteins. Proc Natl Acad Sci USA 93:14532–14535

Locher K, Rees B, Koebnik R et al 1998 Transmembrane signaling across the ligand-gated FhuA-receptor: crystal structures of the free and ferrichrome-bound states reveal allosteric changes. Cell 95:771–778

Luecke H, Richter HT, Lanyi JK 1998 Proton transfer pathways in bacteriorhodopsin at 2.3 Å resolution. Science 280:1934–1937
Marsh D 1996 Components of the lateral pressure in lipid bilayers deduced from H_{II} phase dimensions. Biochim Biophys Acta 1279:119–123
Mielke DL, Wallace BA 1988 Secondary structure analyses of the nicotinic acetylcholine receptor as a test of molecular models. J Biol Chem 263:3177–3182
Nakae T, Ishii J 1978 Transmembrane permeability channels in vesicles reconstituted from single species of porins from *Salmonella typhimurium*. J Bacteriol 133:1412–1418
Noda M, Takahashi H, Tanabe T et al 1983 Structural homology of *Torpedo californica* acetylcholine receptor subunits. Nature 302:528–532
Noda M, Shimizu S, Tanake T et al 1984 Primary structure of *Electrophorus electricus* sodium channel deduced from cDNA sequence. Nature 312:121–127
Ostermeier C, Michel H 1997 Crystallization of membrane proteins. Curr Opin Struct Biol 7:697–701
Ovchinnikov Y, Abdulaev NG, Feigina M, Kiselev AV, Lobanov NA 1979 The structural basis of the functioning of bacteriorhodopsin: an overview. FEBS Lett 100:219–224
Paul C, Rosenbusch JP 1985 Folding patterns of porin and bacteriorhodopsin. EMBO J 4:1593–1597
Pautsch A, Schulz GE 1998 Structure of the outer membrane protein A transmembrane domain. Nat Struct Biol 5:1013–1017
Pebay-Peyroula E, Rummel G, Rosenbusch JP, Landau E 1997 X-ray structure of bacteriorhodopsin at 2.5 Å from microcrystals grown in lipidic cubic phases. Science 277:1676–1681
Rosenbusch JP 1974 Characterization of the major envelope protein from *Escherichia coli*. Regular arrangement on the peptidoglycan and unusual dodecyl sulfate binding. J Biol Chem 249:8019–8029
Schindler H, Rosenbusch JP 1978 Matrix protein from *Escherichia coli* outer membranes forms voltage-controlled channels in lipid bilayers. Proc Natl Acad Sci USA 75:3751–3755
Schirmer T, Cowan SW 1993 Prediction of membrane-spanning β-strands and its application to maltoporin. Protein Sci 2:1361–1363
Unwin N 1993 Nicotinic acetylcholine receptor at 9 Å resolution. J Mol Biol 229:1101–1124
Vogel H, Jähnig F 1986 Models for the structure of outer-membrane proteins of *Escherichia coli* derived from Raman spectroscopy and prediction methods. J Mol Biol 190:191–199
Weiss MS, Abele U, Weckesser J, Welte W, Schiltz E, Schulz GE 1991 Molecular architecture and electrostatic properties of a bacterial porin. Science 254:1627–1630
Yellen G, Jurman ME, Abramson T, MacKinnon R 1991 Mutations affecting internal TEA blockade identify the probable pore-forming region of a K^+ channel. Science 251:939–942

DISCUSSION

Sansom: Do you have any suspicions that there might be membrane proteins in the sequence databases for non-bacterial, non-endosymbiotic membranes that could have β- or mixed β/α-motifs? Perhaps we haven't seen them because they cannot be readily detected by current prediction algorithms.

Rosenbusch: We all agree on the occupancy of α-motifs. Predictions of β-motifs are not so trivial, because by scanning a data bank, the constraints of the hydrophobic and aromatic residues are too loose, so you may find many false positives. In the porin sequence there is a conserved motif of about five amino acids at the tip of constriction loop, but this is specific for porins. What one may say is that our current search algorithms are too selective, or are not selective enough, and hence most problems that are not simple α-helical bundles elude identification.

Stein: As far as I understand from membrane proteins, the problem is reversed because there are huge protein families that have common motifs, e.g. the seven hydrophobic strand families. Are you suggesting that there are not seven hydrophobic strand families?

Rosenbusch: No. In the case of bacteriorhodopsin an α-helical bundle clearly exists, and the same α-helical bundle is found in G proteins. But whether other segments exist in membrane domains, such as in the bacterial potassium channel (Doyle et al 1998), is anyone's guess at this time.

Stein: Is the problem with members of the 12 membrane family that we can't crystallize them?

Rosenbusch: Yes, but a further word of caution is necessary because we don't know whether they have the number of helices that we believe they have. They may have domains, cores or plugs that we cannot yet predict, but this will become easier as we identify more homologues.

Sansom: It definitely is not as clear cut as one might like to imagine because the sequences of the ATP-binding cassette transporters, for example, do not all have the same canonized six or 12 transmembrane helix pattern.

Rosenbusch: It is clear that there are proteins of the α-helical and β-sheet type, but we don't know whether there are different constraints in stabilizing the membrane that affect their lability. We also don't know whether the core of the large membrane proteins such as the sodium channel should not be considered rather as a globular protein, as there is no immediate contact with membrane lipids. It is much more difficult to make predictions if there are no membrane constraints.

Sansom: To play devil's advocate, if the two-stage folding model (Popot & Engelman 1990) is correct, you can predict all-α membrane proteins because they are inserted into the membrane not as the whole assembly, but as single helices (i.e. folding domains). This may be why it works in cases where you wouldn't have imagined it to work, from simply looking at the final, folded structure.

Rosenbusch: It is not mutually exclusive. The partitioning of an α-helical hairpin is easy to visualize, but whether other segments are taken for a ride is difficult to predict.

References

Doyle DA, Cabral JM, Pfuetzner RA et al 1998 The structure of the potassium channel: molecular basis of potassium conductance and selectivity. Science 280:69–77

Popot JL, Engelman DM 1990 Membrane protein folding and oligomerization: the two-state model. Biochemistry 29:4031–4037

Molecular basis of the charge selectivity of nicotinic acetylcholine receptor and related ligand-gated ion channels

Pierre-Jean Corringer, Sonia Bertrand*, Jean-Luc Galzi[1], Anne Devillers-Thiéry, Jean-Pierre Changeux and Daniel Bertrand*

*Neurobiologie Moléculaire, Unité de recherche associée au Centre National de la Recherche Scientifique D1284, Institut Pasteur, 25 rue du Docteur Roux, 75724 Paris Cedex 15, France, *Département de Physiologie, Centre Médical Universitaire (Faculté de Médecine), 1211 Geneva 4, Switzerland*

Abstract. Nicotinic acetylcholine receptors are homo- or heteropentameric proteins belonging to the superfamily of receptor channels including the glycine and GABA-A receptors. Affinity labelling and mutagenesis experiments indicated that the M2 transmembrane segment of each subunit lines the ion channel and is coiled into an α-helix. Comparison of the M2 sequence of the cation-selective $\alpha 7$ nicotinic receptor to that of the anion-selective $\alpha 1$ glycine receptor identified amino acids involved in charge selectivity. Mutations of the $\alpha 7$ homo-oligomeric receptor within (or near) M2, namely E237A, V251T and a proline insertion P236′ were shown to convert the ionic selectivity of $\alpha 7$ from cationic to anionic. Systematic analysis of each of these three mutations supports the notion that the conversion of ionic selectivity results from a local structural reorganization of the 234-238 loop. The 234-238 coiled loop, previously shown to lie near the narrowest portion of the channel, is thus proposed to contribute directly to the charge selectivity filter. A possible functional analogy with the voltage-gated ion channels and related receptors is discussed.

1999 Gramicidin and related ion channel-forming peptides. Wiley, Chichester (Novartis Foundation Symposium 225) p 215–230

The superfamily of phylogenetically related ligand-gated ion channels, which includes receptors for acetylcholine/nicotine (nAChR), glycine (glyR), γ-aminobutyric acid ($GABA_AR$) and serotonin ($5\text{-}HT_3R$) have been intensively investigated for several decades (Galzi & Changeux 1995, Karlin & Akabas

[1]Present address: Unité propre de recherche CNRS 9050, Ecole Supérieure de Biologie de Strasbourg, boulevard Sebastien Brant 67400 Illkirch, France.

1995, Lindstrom 1996, Role & Berg 1996). These receptors are allosteric membrane proteins (Changeux & Edelstein 1998) that transduce the binding of specific agonists into the opening of selective channels through which ions passively diffuse across the cell membrane. The superfamily includes members where the ion channel is either selective to cations (nAChR, $5HT_3R$, permeable in particular to Na^+ and K^+, but also in some cases to Ca^{2+}) or to anions (GlyR, $GABA_AR$). These ionic selectivities are of crucial physiological importance, since typically cationic channels elicit excitatory responses, whereas anionic channels cause inhibition. The nAChRs are still the best characterized members of the superfamily. They result from the association of five homologous subunits, arranged with a pseudo-fivefold symmetry, each subunit crossing the membrane four times at the level of transmembrane segments M1, M2, M3 and M4. So far, five subunits coding for the muscle type receptor, α, β, γ, δ and ε, as well as 11 subunits coding for the neuronal receptors, α2-9 and β2-4 have been cloned (Le Novère & Changeux 1995). In this chapter we will briefly summarize our present understanding of the ion channel that has accumulated for the last decade, and we will present our recent results on the localization of its charge selectivity filter.

The ion channel is the target of a particular class of non-competitive inhibitors, referred to as channel blockers, that act through steric occlusion within the pore. Some amino acids contributing to the binding site of channel blockers have been identified by affinity labelling experiments (Giraudat et al 1986, Hucho et al 1986). In parallel, site directed mutagenesis experiments identified several amino acids which mutation altered the unitary conductance of the ion channel (Imoto et al 1986). These complementary approaches led to a model of functional organization where: (i) the ionic pathway is located along the axis of pseudo-symmetry of the protein, through the contribution of the M2 segment of each subunit; (ii) the ion channel is formed by a superimposition of chemically defined rings of homologous amino acids; (iii) the M2 segment is coiled into an α-helix (Galzi & Changeux 1995). Electron microscopy of the *Torpedo* receptor revealed, at the level of the membrane, five rods symmetrically arranged around the axis of rotational symmetry which were tentatively assigned to the M2 segments (Unwin 1995). Figure 1 summarizes these conclusions, taking the homopentameric α7 nAChR as a reference.

Conversion of the ionic selectivity of α7 nAChR from cationic to anionic

The nAChR superfamily possess the unique property to include both cationic (nAChR, $5HT_3$) and anionic (glycine and GABA-A) members, thus paving the way for the identification of the charge selectivity filter of the ion channel. Comparison of the M2 sequences and their flanking regions between the homo-oligomeric α7 nAChR and the glycine α1 receptor revealed several differences

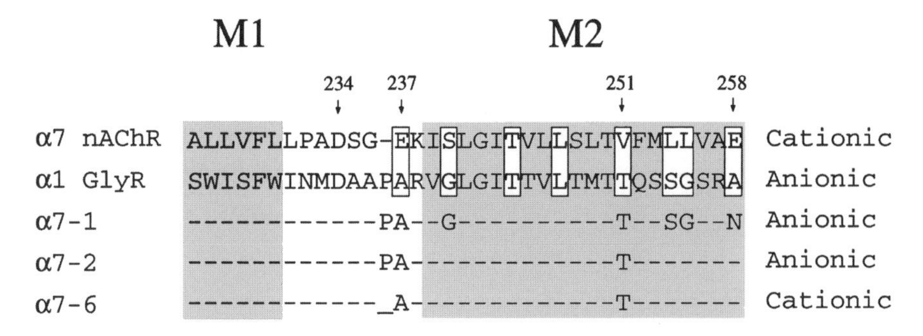

FIG. 1. Three mutations convert the α7 channel from cationic to anionic. Aligned sequences of the α7 nAChR and glycine α1 subunits. Grey boxes indicate the transmembrane M2 segment and part of the transmembrane M1 segment. Boxed residues are those believed to face the lumen of the ion channel, from affinity labelling and site directed mutagenesis experiments. Numbering is given according to the chick α7 subunit. Some mutants described by Galzi et al (1992) are shown, together with their ionic selectivity.

(Fig. 1). In a first step (Galzi et al 1992), the amino acids assumed to face the lumen of the ion channel were transferred from the glycine α1 receptor into the α7 receptor (Fig. 1). The resulting construct, α7-1, was found to possess an anionic channel gated by ACh. Then, all the amino acids that did not contribute to the conversion of ionic selectivity were eliminated. The minimal set of mutations sufficient to confer anionic selectivity was found located at the level of: (i) one ring of hydrophobic amino acids (V251T); and (ii) two rings of amino acids located at the N-terminal (cytoplasmic) end of the M2 transmembrane segment, i.e. mutation of the 'intermediate ring of charged residues' (E237A) and the insertion of a proline between positions 236 and 237, referred to here as P236′ (numbering according to chick α7 nAChR, see Fig. 1). It is noteworthy that the V251T and E237A mutations are fully compatible with a cationic channel (see α7-6 in Fig. 1), and the proline insertion by itself converts the selectivity of the ion channel (from α7-6 to α7-2).

Pleiotropic allosteric effects of mutations within (or near) M2

Mutations distributed within the ion channel are expected to affect primarily intrinsic channel properties, such as conductance and selectivity. This view is in harmony with the standard interpretation of the mutagenesis experiments according to which the physiological effect of a given mutation is determined by a local change of structure that takes place at the level of the mutated residue. However, in the course of our extensive studies of α7 nAChR, we observed that

TABLE 1 Mutational analysis of the charge selectivity filter of the α7 receptor

Type	*Sequence*	*Imax (nA) ACh-evoked*	*Erev (mV) ACh-evoked* *NaCl*	*Mannitol*	*Selectivity*
Position 251 analysis					
	234 ↓ 238 ↓ 251 ↓				
α7-2	ADSG**PA**KI———**T**	114±11	−35±2	18±4	Anionic
α7-15	ADSG**PA**KI———**D**	13±3	−48±7	23±2	Anionic
α7-16	ADSG**PA**KI———**K**	39±11	−43±1	0±2	Anionic
Insertion analysis					
α7-19	A**P**DSG**A**KI———**T**	230±61	−25±1	−45±7	Cationic
α7-20	AD**P**SG**A**KI———**T**	48±11	−25±2	39±3	Anionic
α7-2	ADSG**PA**KI———**T**	114±11	−35±2	18±4	Anionic
α7-21	ADSG**AP**KI———**T**	56±6	−18±1	45±2	Anionic
α7-22	ADSG**A**K**PI**———**T**	0	nd	nd	nd
α7-20	———**AKISLGPITVLL**—**T**	0	nd	nd	nd
234-238 loop analysis					
α7-33	A**A**SG**PA**KI———**T**———	58±20	−18±1	43±4	Anionic
α7-59	AD**AG**P**A**KI———**T**———	6±0.5	−18±6	35±8	Anionic
α7-63	ADS**APA**KI———**T**———	31±5	−18±2	43±3	Anionic
α7-31	ADSG**PAQIG**———**T**FM**SG**VA**N**	174±46	−36±4	53±4	Anionic

Bold residues correspond to mutations introduced in the α7 subunit, and underlined residues indicate insertions. Mean amplitudes (given in absolute values) were measured at a holding potential of −100 mV with 100 μM ACh. Reversal potentials were measured before and after substitution of 90% NaCl with mannitol. nd indicates not determined. These data are taken from Corringer et al (1999).

mutations within the M2 segment often result in pleiotropic phenotypes (Revah et al 1991, Bertrand et al 1992, Devillers-Thiéry et al 1992, Galzi et al 1996).

The V251T mutation has been shown to produce a gain-of-function phenotype. This phenotype is well illustrated by α7-2 and α7-6, which both harbour the V251T mutation and are characterized by (Corringer et al 1999): (i) a weak desensitization to the response to ACh; (ii) an approximately 100-fold decrease in EC_{50} for ACh (data not shown); and (iii) the occurrence of leak currents recorded in the absence of agonist. These currents correspond to spontaneous openings since part of them can be closed by the competitive α7 antagonist methyllycaconitine (MLA; Bertrand et al 1997). Interestingly, the proline insertion is shown here to dramatically increase the spontaneous versus ACh-evoked currents of the receptor. Furthermore, we found that the ionic selectivity of the spontaneous (MLA-inhibited) and ACh-evoked currents are identical. Indeed, Fig. 2 illustrates that exchanging external NaCl (control) by mannitol for α7-2 shifts the reversal potential (Erev) of both the ACh-evoked and MLA-inhibited leak currents towards positive values, from −37 to +15 mV, a result consistent with an anionic channel according to the Nernst relation (Galzi et al 1992). Similarly, we found that both the ACh-evoked and the MLA-inhibited

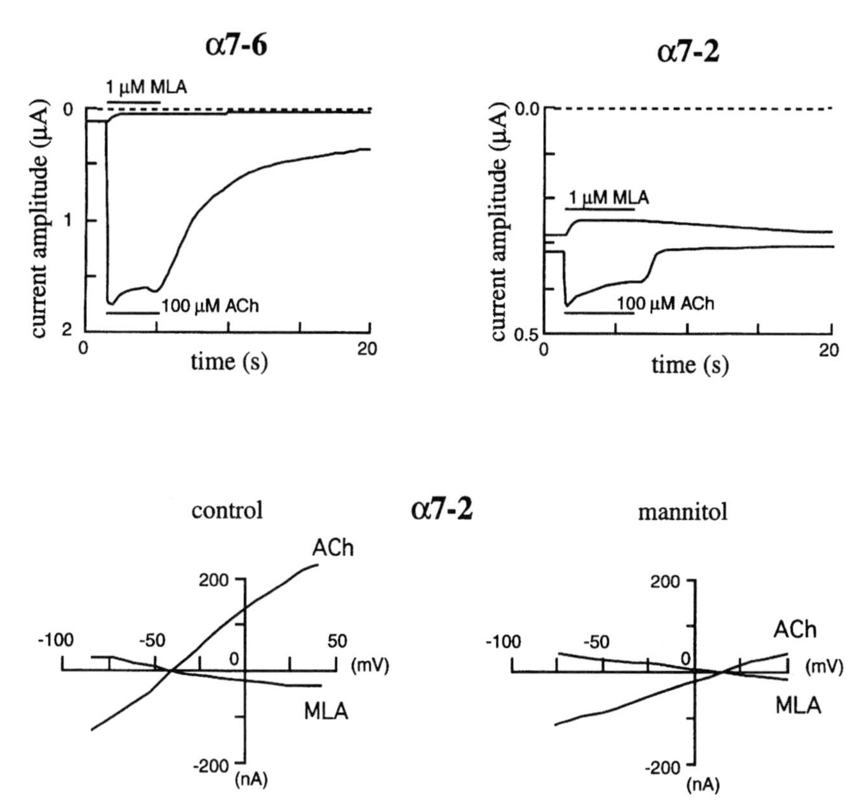

FIG. 2. Spontaneous openings of the α7-2 and α7-6 mutants. Upper panel: agonist- (ACh) and antagonist (methyllycaconitine; MLA)-evoked currents on oocytes expressing the α7-6 (E237A/V251T) (left panel) and α7-2 (P236′/E237A/V251T) (right panel) receptors. Application of saturating concentrations of ACh evokes a steady-state current with little desensitization for both receptors. The holding current in the absence of effectors is higher than the usual leak current of the oocytes (typically −30 to −60 nA), especially in the case of the α7-2 receptor. Application of a high concentration of MLA quickly reduces a significant fraction of this current (corresponding to MLA-inhibited currents). Lower panel: current–voltage (I–V) relationships for both the ACh-evoked (ACh 100 μM) and MLA-inhibited (MLA 1 μM) currents for the α7-2 receptor. Typical I–V relationships recorded first under control conditions and after exchanging 90% of external NaCl by mannitol are shown. Note that the ACh-evoked and MLA-inhibited currents display identical reversal potentials under both conditions. These data are taken from Corringer et al (1999).

leak currents of α7-6 are found to be cationic, since the corresponding Erev shift towards negative values upon NaCl substitution (not shown). Thus the MLA-inhibited and ACh-evoked currents display the same ionic selectivity, and the conversion of ionic selectivity from cationic to anionic equally affects the spontaneous and ACh-evoked currents (Corringer et al 1999).

Contribution of each mutation to the conversion of ionic selectivity

V251T

The presence of a threonine at 251 is required to obtain an anionic channel. To evaluate its contribution, we introduced at this position positively (V251K, α7-16) or negatively (V251D, α7-15) charged residues, along with the P236′ and E237A mutations. Both receptors were found to be anionic (Table 1; Corringer et al 1999). The observation that mutants V251D (α7-15) and V251K (α7-16) similarly display anionic selectivity, despite a possible electronic repulsion with aspartate, suggests that the side chains do not interact directly with the permeant chloride ion.

E237A

Elimination of one or two negative charges at a homologous position in the muscle-type nAChR was previously found to decrease dramatically the unitary conductance of the channel, suggesting that these negatively charged residues could contribute to the ionic pathway in cationic channels (Imoto et al 1988). The requirement of the complete elimination of the ring of charged residues to get an anionic channel could thus be consistent with either a direct or indirect contribution of this position to the anion selection.

P236′

A critical mutation for the conversion of ionic selectivity is the insertion of a proline at the cytoplasmic end of M2 (Galzi et al 1992). Several plausible mechanisms may underlie the conversion of the ionic selectivity caused by the proline insertion. An extra residue at this position could have conformational effects. The resulting conformational reorganization may either be limited to the neighbouring amino acids or propagate into the M2 segment which lines the ion pore. To evaluate these possibilities, the locus of insertion was systematically investigated in combination with the permissive E237A and V251T mutations (Corringer et al 1999).

We found that the proline could be inserted either at position 234′ (α7-20), 236′ (α7-2) and 237′ (α7-21) resulting in functional receptors displaying comparable current amplitudes and anionic selectivity. In contrast, the insertion of the proline upstream at position 233′ (α7-19) results in a cationic channel. On the other hand, for the insertion of the proline downstream to position 237′ (at 238′ [α7-22] and 242′ [α7-23]), no ACh-evoked nor leak currents could be detected. In addition, no receptor was found at the surface of oocytes injected with the α7-22 and α7-23 constructs by [^{125}I]α-Bgtx binding (data not shown). This indicates that insertion within the M2 segment alters the expression of the protein. The present analysis thus suggests that the structural reorganization resulting from the proline insertion is mainly restricted to the 234-238 loop (Corringer et al 1999). Indeed, the M2 segment is expected to be constrained by a cluster of tertiary interactions with the other transmembrane segments, whereas the 234-238 loop is predicted to adopt a coiled structure accessible to the solvent, allowing structural alteration.

Proposed location of the charged selectivity filter

Permeability studies have supported that both the muscle-type nAChR and GlyR ions channels have a minimal diameter of 6–7 Å (reviewed in Hille 1992). These rather wide pores may accommodate partially hydrated but not fully hydrated ions. This indicates that, at the level of the narrowest part of the channel, at least partial dehydration of the ion may occur via interaction with the channel walls. Several lines of evidence have located the narrowest portion of the channel at the cytoplasmic border of M2: (i) mapping of the binding site for channel blockers, which enter the ion channel but are too large to permeate, indicates that the narrowest part of the open channel is closer to the intracellular compartment than residues corresponding to α7T244 (Giraudat et al 1986); (ii) mutation of residues homologous to α7E237 and α7T244 in the muscle type receptor were found to critically affect the unitary conductance of the channel. In particular, the unitary conductance of the channel was found to be inversely related to the volume of the amino acid introduced at these positions (Imoto et al 1991, Cohen et al 1992, Villarroel & Sakmann 1992); and (iii) recently, the SCAM method has been used to identify amino acids involved in the gate of the ion channel (Wilson & Karlin 1998). By applying cysteine-reactive reagents both intracellularly or extracellularly together with reagents of different chemical structures, these authors located the gate between residues homologous to α7G236 and α7S240 of the muscle-type α1 subunit. Our results show that the critical proline has to be inserted within the 234-238 region to switch the ionic selectivity from cationic to anionic. Since the narrowest part of the ion channel is located within or at the vicinity of this region, it supports the conclusion that this segment directly interacts with the permeant ions and contributes at least in part to the charge selectivity filter of the ion channel (Corringer et al 1999).

The 234-238 loop is thus proposed to contribute defined chemical groups for the dehydration of specific ions. Systematic mutations of each residue within this loop revealed however that no particular side chain is critical for the anionic selectivity (see Table 1). In particular, the positively charged K238 can be mutated to Q without alteration of the ionic selectivity, showing that no charged residue contribute to the conversion of ionic selectivity from α7-2 to α7-6. The putative electropositive features at the level of the 234-238 loop may be contributed by hydrogen atoms of the hydroxyl group of the serine residue (α7-20, α7-2 and α7-21), and/or of the NH groups of the peptide backbone.

Functional analogy with voltage-gated potassium channels

The feature that a coiled loop contributes to the selectivity filter is reminiscent of the voltage-gated and glutamate-gated ion channel structures, where the selectivity filter consists of a set of so called P-loops that enter a widely open channel pore (MacKinnon 1995). The structure of the voltage-gated potassium channel from *Streptomyces lividans* has been recently resolved by X-ray analysis at 3.2 Å (Doyle et al 1998). It is composed of two functionally and structurally different regions. First, a transmembrane α-helical component, composed of the four hydrophobic inner α-helices acting as a water pore, through which ions diffuse in a fully hydrated form, thus lowering the energy barrier predicted for a non-hydrated ion in the low dielectric medium of the membrane. Second, a so-called 'pore region' component, which is involved in the specific dehydration of the permeant cations through the contribution of three residues, GYG, folded in an extended conformation. This component is located close to the extracellular compartment. Whereas no sequence homologies exist between the nAChR and the voltage-gated channel superfamilies, one may speculate that the ion channel of α7 displays a functional organization analogous to that of the potassium channel but with an inverted disposition: the selectivity filter being located at the cytoplasmic side of the α-helical region (Corringer et al 1999).

Recently, a secondary structure prediction was performed on a representative set of sequences from the ligand-gated ion channel superfamily using several third-generation algorithms (Le Novère et al 1999). According to this prediction, the synaptic part of M2 is coiled into an α-helix, whereas a small portion of the cytoplasmic part, from residues 239 to 243 would adopt a β-strand configuration. Furthermore, the cytoplasmic part of the M1 transmembrane segment is also predicted to adopt a β-strand configuration, suggesting that two β-strands are bordering the M1-M2 coiled loop. Figure 3 shows a speculative attempt to represent the charge selectivity filter of the channel, in the context of these secondary structure predictions and of the putative functional analogy with the potassium channel. At this stage, structural studies at atomic resolution of the

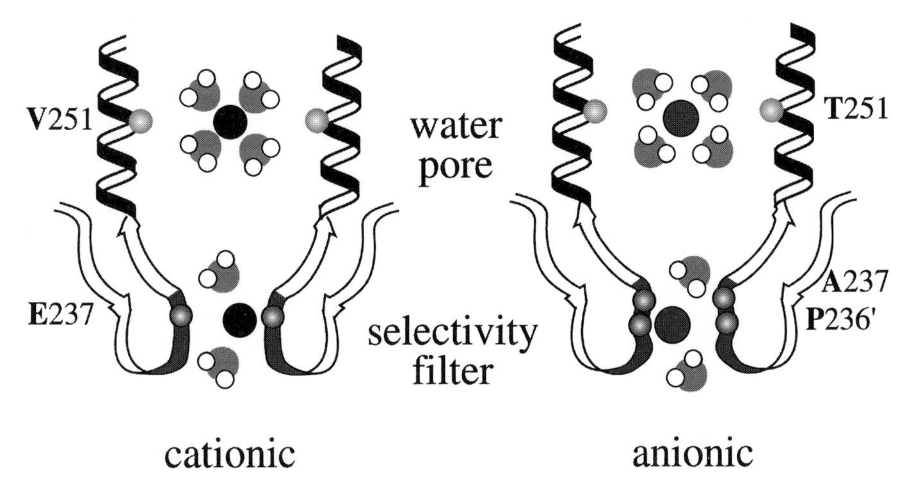

FIG. 3. Speculative representation of the functional architecture of the α7 ion channel. A portion of the transmembrane segments (including part of M1, the M1-M2 loop and M2) of only two subunits is shown here, to illustrate their possible contribution to the ion channel, which actually results from the contribution of the five subunits of the pentamer. The secondary structure proposed here for the M1, M2 and M1-M2 loop is taken from the predictions of Le Novère et al (1999). The two predicted β-strands respectively at the cytoplasmic border of M1 and M2 are tentatively shown here to interact with each other. This forms a β-hairpin-like structure that orientates the 234-238 loop towards the lumen of the channel. The left panel represents the cationic channel α7, and the right panel the anionic channel α7-2 where the three mutations V251T, E237A and P236′ are present. We propose that the 234-238 loop (shown in grey) contributes directly to the ion dehydration whereas, by analogy with the voltage-gated ion channel, the upper part of the channel acts as a 'water pore' where the cations (dark sphere) or anions (grey sphere) diffuse in a fully hydrated state. This figure is adapted from Corringer et al (1999).

nAChR channel are needed to further progress in the understanding of the ion channel function and its regulation by ACh.

References

Bertrand D, Devillers-Thiéry A, Revah F et al 1992 Unconventional pharmacology of a neuronal nicotinic receptor mutated in the channel domain. Proc Natl Acad Sci USA 89:1261–1265

Bertrand S, Devillers-Thiéry A, Palma E et al 1997 Paradoxical allosteric effects of competitive inhibitors on neuronal alpha7 nicotinic receptor mutants. Neuroreport 8:3591–3596

Changeux JP, Edelstein SJ 1998 Allosteric receptors after 30 years. Neuron 21:959–980

Cohen BN, Labarca C, Czyzyk L, Davidson N, Lester HA 1992 $Tris^+/Na^+$ permeability ratios of nicotinic acetylcholine receptors are reduced by mutations near the intracellular end of the M2 region. J Gen Physiol 99:545–572

Corringer PJ, Bertrand S, Galzi JL, Devillers-Thiéry A, Changeux JP, Bertrand D 1999 Mutational analysis of the charge selectivity filter of the α7 nicotinic acetylcholine receptor. Neuron 22:831–843

Devillers-Thiéry A, Galzi JL, Bertrand S, Changeux JP, Bertrand D 1992 Stratified organization of the nicotinic acetylcholine receptor channel. Neuroreport 3:1001–1004

Doyle DA, Cabral JM, Pfuetzner RA et al 1998 The structure of the potassium channel: molecular basis of K^+ conduction and selectivity. Science 280:69–77

Galzi JL, Changeux JP 1995 Neuronal nicotinic receptors: molecular organization and regulations. Neuropharmacology 34:563–582

Galzi JL, Devillers-Thiéry A, Hussy N, Bertrand S, Changeux JP, Bertrand D 1992 Mutations in the ion channel domain of a neuronal nicotinic receptor convert ion selectivity from cationic to anionic. Nature 359:500–505

Galzi JL, Edelstein SJ, Changeux JP 1996 The multiple phenotypes of allosteric receptor mutants. Proc Natl Acad Sci USA 93:1853–1858

Giraudat J, Dennis M, Heidmann T, Chang JY, Changeux JP 1986 Structure of the high-affinity site for noncompetitive blockers of the acetylcholine receptor: serine-262 of the delta subunit is labeled by [^{3}H]chlorpromazine. Proc Natl Acad Sci USA 83:2719–2723

Hille B 1992 Ionic channels of excitable membranes, 2nd edn. Sinauer Associates, Sunderland, MA

Hucho F, Oberthür W, Lottspeich F 1986 The ion channel of the nicotinic acetylcholine receptor is formed by the homologous helices M II of the receptor subunits. FEBS Lett 205:137–142

Imoto K, Methfessel C, Sakmann B et al 1986 Location of a delta-subunit region determining ion transport through the acetylcholine receptor channel. Nature 324:670–674

Imoto K, Busch C, Sackmann B et al 1988 Rings of negatively charged amino acids determine the acetylcholine receptor channel conductance. Nature 335:645–648

Imoto K, Konno T, Nakai J, Wang F, Mishina M, Numa S 1991 A ring of uncharged polar amino acids as a component of channel constriction in the nicotinic acetylcholine receptor. FEBS Lett 289:193–200

Karlin A, Akabas MH 1995 Toward a structural basis for the function of nicotinic acetylcholine receptors and their cousins. Neuron 15:1231–1244

Le Novère N, Changeux JP 1995 Molecular evolution of the nicotinic acetylcholine receptor: an example of multigene family in excitable cells. J Mol Evol 40:155–172

Le Novère N, Corringer PJ, Changeux JP 1999 Improved secondary structure predictions for a nicotinic receptor subunit. Incorporation of solvent accessibility and experimental data into a 2D representation. Biophys J 76:2329–2345

Lindstrom J 1996 Neuronal nicotinic acetylcholine receptors. In: Narahashi T (ed) Ion channels, vol 4. Plenum Press, New York, p 377–450

MacKinnon R 1995 Pore loops: an emerging theme in ion channel structure. Neuron 14:889–892

Revah F, Bertrand D, Galzi JL et al 1991 Mutations in the channel domain alter desensitization of a neuronal nicotinic receptor. Nature 353:846–849

Role LW, Berg DK 1996 Nicotinic receptors in the development and modulation of CNS synapses. Neuron 16:1077–1085

Unwin N 1995 Acetylcholine receptor channel imaged in the open state. Nature 373:37–43

Villarroel A, Sakmann B 1992 Threonine in the selectivity filter of the acetylcholine receptor channel. Biophys J 62:196–205

Wilson GG, Karlin A 1998 The location of the gate in the acetylcholine receptor. Neuron 20:1269–1281

DISCUSSION

Koeppe: When you made the mutations did you change only one subunit or all of the subunits?

Corringer: All of the subunits. These are homopentameric receptors, so when we make one mutation we have to mutate five residues.

Koeppe: Could you do the same experiment in receptors that have four different subunits?

Corringer: Yes, this could be done. However, in the γ-subunit of the muscle-type receptor there is an insertion in the wild-type γ-subunit and the receptor is cationic, so I would guess that five mutations are required to switch to anionic selectivity in the five-subunit receptor.

Woolley: Have you looked at isolated channels in reconstituted systems rather than in oocytes?

Corringer: The problem is that these are neuronal nicotinic acetylcholine receptors and it is difficult to obtain single-channel recordings from them because when you patch the receptor you activate it once and then it dies. Also, the unitary conductance of these anionic channels is low compared to wild-type channels; there is a decreased maximum amplitude of the current.

Cross: What is the charged state in that selectivity filter? Are there five charges in a homopentamer, for example?

Corringer: This is a good question, and it is related to Mark Sansom's presentation. We have looked only at the selectivity and we have no ideas about the state of the charge.

Cross: Do you know if it is symmetrical? If only one of them is charged, this would break the symmetry.

Eisenberg: This depends on the time-scale.

Corringer: We have also mutated every charged residue in this region in the chloride channel, and we always observe anionic selectivity, suggesting that charged residues do not contribute to anionic selectivity in this case, which contrasts with the cationic channel.

Cross: I would like to mention some preliminary results from our lab. Mark Sansom introduced the work on the M2 protein from the influenza A virus. This protein is a proton channel that is pH activated at low pH, and a histidine residue is thought to be involved in this pH activation. Our NMR spectroscopy data indicate that the four histidines are charged at low pH (i.e. pH 5.0).

Eisenberg: It is possible to work backwards and estimate the electrochemical potential of the electric field, i.e. the effect of pK depends on the electrical energy to take an ion from water down to that spot, and if you work backwards you will probably find that you're estimating a potential difference of about 100 mV.

Woolley: There is no applied field.

Eisenberg: But there are plenty of local fields.

Smart: It seems counterintuitive to have a pH-activated channel with a histidine that is charged when it is activated.

Bechinger: Are you looking at the monomer or the tetramer?

Cross: Probably the tetramer because the helix tilt is insensitive to bilayer thickness, but that's the only evidence we have.

Jordan: Have you modelled the proline mutations to see what happens to the size of the selectivity region?

Corringer: That's a good point. We haven't done any modelling because we don't have any structural data on this region. At the moment we are continuing with our mutagenesis studies, and then we will try to validate the model.

Busath: You said that you had to replace V251 with a hydrophilic residue in order to obtain tolerance to the proline insertion. What happens if you don't mutate it, or you replace it with a hydrophobic residue? Does the cation selectivity in that region conflict with anion selectivity in the proline region, or is there a folding defect?

Corringer: All I can say from these experiments is that we have zero current. There are two possible explanations. The first is that this mutation has been shown to represent a gain-of-function phenotype, characterized by increased maximal currents. This gain of function might be required to see the ionic conductance, which is weak. The second explanation is that there could be a reorganization of the upper part of the channel. This mutation is compatible with both cationic and anionic conductances, so the reorganization would permit the permeation of both anions and cations.

Jakobsson: What does calcium do to cation-selective channels?

Corringer: Calcium is 10-fold more permeable than sodium, but when the intermediate ring of charged residues is removed, calcium is no longer permeable. This suggests that the monovalent and divalent permeation mechanisms are different for this channel, and possibly also for other ion channels. Zhorov & Ananthanarayanan (1997) have done some modelling studies of calcium permeation, in which the calcium ion binds to two glutamate side chains. There are four glutamate side chains and two calcium ions in the synthetic calcium channel they studied, and this is the proposed basis of the calcium permeation mechanism.

Woolley: It is surprising that you have made structurally perturbing mutations, but you never see loss of selectivity, i.e. non-selective channels.

Corringer: Yes, the channels are either cationic or anionic. We never see intermediate phenotypes, which suggests that there is a radical structural reorganization that produces an all-or-nothing conversion of ionic selectivity.

Woolley: You are suggesting that peptide bonds are the source of anion selectivity. Is there a peptide bond orientation, in gramicidin for instance, that interacts favourably with anions?

Roux: The peptide bond does not solvate the anions very well, and anions are not well solvated in liquids. For instance, the transfer free energy of chloride between water and liquid formamide is 3.3 kcal/mol because the NH does not solvate an anion very well.

Stein: I would like to mention that the acetylcholine receptor is also permeable to small non-electrolytes, e.g. ethylene glycol and ethanolamine (Huang et al 1978). This could be useful for modelling.

Jakobsson: It seems to me that there's a good reason why any relatively small channel, i.e. smaller than a gap junction channel, should always select for either cations or anions, because there has to be a charge localization for desolvation. What I mean is, in order to overcome the solvation energy and remove hydration waters from a small ion, you've got to have a locus of charge in the channel structure strategically located at the narrow part of the channel. So whatever the sign of that charge, you'll select for the opposite charge of ion.

Corringer: Are there any models of chloride channels at atomic resolution?

Sansom: No, but the structures of weakly anion-selective porins are known at atomic resolution, and there are data on side chain mutations and small shifts in selectivity.

Ring: What are the current–voltage (I–V) curves of the switched channels compared to wild-type channels? The wild-type I–V curves are linear, which is compatible with a short selectivity region because it has a low field dependence. This could, however, be interpreted differently because it doesn't contradict that there may be a long region and a different mechanism.

Corringer: The I–V curves of switched channels are roughly linear. There don't seem to be any differences.

Cross: You mentioned the α-helical segments, and in the model you showed a β-sheet structure extending down to the selectivity filter, but are your data also consistent with an α-helix in this region?

Corringer: This is the secondary structure prediction, and such predictions have an accuracy of 70% for third-generation algorithms on soluble proteins. This prediction is related to the transmembrane segment, and we have few data on this, so we have to be cautious.

Cross: But the exposure of side chains from a β-sheet versus an α-helix is different, so do you favour a β-sheet in that region?

Corringer: No, it could be either an α-helix or β-sheet from these data.

Cornell: How far away from the surface of the lipid bilayer is the anionic selectivity determined? Is it within a few ångströms, or is it in the entrance to the pore?

Corringer: I only presented a hypothetical model, and I haven't looked at this. It is surrounded by M1, M3 and M4, and not just by the membrane, so the concept of the transmembrane is slightly different. The loop could also fall back inside the pore, as in the potassium channel.

Cornell: Do M3 and M4 have a role in this, or is it strictly M2?

Corringer: We've only looked at M2.

Wallace: You're dealing with an insertion of one amino acid and such an insert can rotate a molecule around its helical axis if it is part of the helix. Is it possible that you could be seeing different faces of the helix rotated into the pore region?

Corringer: This was our initial idea, so we made a number of mutations in M2 and in all cases this resulted in non-functional receptors. Karlin & Akabas (1995) then did cysteine scanning mutagenesis within the M2 region, and they found similar patterns of accessibility in cationic and anionic channels, suggesting that M2 has the same orientation in both types of channel.

Sansom: Is your working hypothesis that the pore lined by the M2 helices is wide, whereas the constriction is localized at the intracellular mouth and is due to the M1-M2 loop? This would fit with Unwin's electron microscopy data that suggest the helices are further apart than one would expect (Unwin 1995).

Corringer: Yes. Wilson & Karlin (1998) also looked at the receptor gate, and found it at the level of the loop. Therefore, the loop could be both the narrowest portion and the gate of the channel.

Smart: Has Unwin any evidence for electron density this far down?

Corringer: He has structure features at 7.5 Å resolution (Unwin 1996), but he has no published maps of the pore region at this resolution.

Smart: Presumably, he also sees the membrane boundaries and M2 within the membrane.

Wallace: It is not possible to define that from electron microscopy studies at this resolution. You would probably need at least a 3–4 Å resolution structure in order to be able to see those kinds of details.

Eisenberg: I have a comment on this need for resolution. The work on porins that Jurg Rosenbusch has done, and our work on the mutants, is in a sense discouraging, because all the mutations that Schirmer has made also change the structure enough so that if you put them into our one-dimensional model you get substantial changes in either selectivity or conductance (Cowan et al 1992, Jeanteur et al 1994, Lou et al 1996, Saint et al 1996a,b, Tang et al 1997). You can interpret what's going on, but you need to know the structure at the 1 Å resolution to predict the current, at least for porin. If you don't know it you're going to be wrong by a factor of two and you will obtain an incorrect reversal potential.

Sansom: In your porin mutant calculations, do you calculate the pK_a for every mutant?

Eisenberg: Of course. We have data at different pH values, but we don't have structures at different pH values. There are clear indications that there are significant structural changes that are significant, it's just that we have chosen not to pursue that line. However, we are aware of that issue, and the particular method we use to integrate the equations allows us to analyse the effect of pK_a — we insert the ionization equation into the iteration loop, and it converges rapidly.

Jakobsson: Regarding your comment about needing 1 Å resolution structures, I was thinking about all of us putting 3 Å resolution structures into the simulation programs in our computers and it really scares me. We all have a tendency to look at the beautiful pictures of the K channel which show everything in one place, and somehow we think we're going to infer functionality, but we do need to bear in mind the issue of resolution.

Wallace: The B factors, which are also known as the temperature factors, are a measure of the uncertainty at particular positions, and if you look at the coordinates of the K channel structure, you find that the B factors are high, which means that there's a great uncertainty at a number of positions. It's important to look at the B factors before you make too many decisions based on a crystal structure.

Eisenberg: The temperature is also important. These experiments are carried out at a low temperature, so you have to be extremely cautious about interpreting the localizations of the ions if you ultimately want to find out how they behave at the physiological temperature. They may be much more delocalized at the physiological temperature.

Cross: A further sobering comment on this is that we had a good structural model of gramicidin in 1971, and we're still arguing about function, selectivity, binding sites, etc.

Sansom: Isn't this going to be a challenge even when we have a higher resolution structure for the potassium channel or, for that matter, for any other gated channel?

Wallace: We don't even know which state the potassium channel is in the crystals.

Eisenberg: But when we talk about state, we're talking about something inferred from function, and the variable that controls permeation is the potential of mean force. The relevant conformation is the potential of mean force conformation, and it is easy to obtain large changes in the potential of mean force with trivial changes in structure. The model of gating that the community seems to be jumping on at the moment, i.e. the so-called 'tethered permeon' has small structural changes. I'm not saying that I believe this model, but small structural changes can result in 50–100 mV changes in the potential of mean force.

Smart: It is important to be able to make the distinction between the conductance state of a channel and the structure which produces this. For instance, we have been working (Smart et al 1999) on α-toxin (also known as α-haemolysin), and we have a wonderful crystal structure that is heptameric, but biochemical observations previous to the structure indicated that the major lipid-bound form is hexameric. It is significant that there are two conducting states. Our interpretation, which may be too naïve, is that these correspond to two different oligomerization states. Interestingly the channel never closes. A problem for this type of identification is that often the conducting form represents a minor proportion of the protein in the bilayer (we known this to be the case for α-toxin, and gramicidin is probably similar). Do you know how many molecules are expressed on the oocyte in your experiments?

Corringer: Yes, on the basis of bungarotoxin-binding experiments, we measured 0.1 to 0.5 fmol of receptor per oocyte.

Smart: How many of these are open when you do the conductance measurements?

Corringer: That's a more difficult question to answer.

Sansom: You can get some idea of this if you treat it as an allosteric model, and you have some estimates of the equilibrium constants between the different states. If a channel is gated, i.e. it spends most of the time closed, then the C↔O equilibrium constant is going to be such that the closed state is favoured. For example, when glycogen phosphorylase is activated, it doesn't mean that 100% of the protein molecules are in the active state, rather that *c*. 10% of them are in the active state. It took some time to characterize the active state crystallographically. Therefore, it's going to be a challenge to get crystallographic structures for channels in their open forms, unless we're lucky, because they are built to remain closed for most of the time.

References

Cowan SW, Schirmer T, Rummel G et al 1992 Crystal structures explain functional properties of two *E. coli* porins. Nature 358:727–733

Huang L-YM, Catterall WA, Ehrenstein G 1978 Selectivity of cations and nonelectrolyes for acetylcholine-activated channels in cultured muscle cells. J Gen Physiol 71:397–410

Jeanteur D, Schirmer T, Fourel D et al 1994 Structural and functional alterations of a colicin-resistant mutant of ompF porin from *Eschericia coli*. Proc Natl Acad Sci USA. 91:10675–10679

Karlin A, Akabas MH 1995 Toward a structural basis for the function of nicotinic acetylcholine receptors and their cousins. Neuron 15:1231–1244

Lou K-L, Saint N, Prilipov A et al 1996 Structural and functional characterization of ompf porin mutants selected for large pore size. I. Crystallographic analysis. J Biol Chem 271:20669–20675

Saint N, Lou K-L, Widmer C, Luckey M, Schirmer T, Rosenbusch JP 1996a Structural and functional characterization of ompf porin mutants selected for large pore size. II. Functional characterization. J Biol Chem 271:20676–20680

Saint N, Prilipov A, Hardmeyer A, Lou K-L, Schirmer T, Rosenbusch J 1996b Replacement of the sole hisdinyl residue in ompF porin from *E. coli* by threonine (H21T) does not affect channel structure and function. Biochem Biophys Res Comm 223:118–122

Smart OS, Coates GMP, Sansom MSP, Alder GM, Bashford CL 1999 Structure-based prediction of the conductance properties of ion channels. Faraday Disc 111:185–199

Tang J, Chen D, Saint N, Rosenbusch J, Eisenberg R 1997 Permeation through porin and its mutant G119D. Biophys J 72:A108

Unwin N 1995 Acetylcholine receptor channel imaged in the open state. Nature 373:37–43

Unwin N 1996 Projection structure of the nicotinic acetylcholine receptor: distinct conformation of the alpha subunit. J Mol Biol 257:586–596

Wilson GG, Karlin A 1998 The location of the gate in the acetylcholine receptor. Neuron 20:1269–1281

Zhorov BS, Ananthanarayanan VS 1997 Docking of verapamil in a synthetic calcium channel: formation of a ternary complex involving calcium ions. Arch Biochem Biophys 341:238–244

The gramicidin-based biosensor: a functioning nano-machine

B. A. Cornell, V. L. B. Braach-Maksvytis, L. G. King, P. D. J. Osman, B. Raguse, L. Wieczorek and R. J. Pace

Cooperative Research Centre for Molecular Engineering and Technology, 126 Greville Street, Chatswood, NSW 2067, Australia

Abstract. Biosensors combine a biological recognition mechanism with a physical transduction technique. In nature, the transduction mechanism for high sensitivity molecular detection is the modulation of the cell membrane ionic conductivity through specific ligand–receptor binding-induced switching of ion channels. This effects an inherent signal amplification of six to eight orders of magnitude, corresponding to the total ion flow arising from the single channel gating event. Here we describe the first reduction of this principle to a practical sensing device, which is a planar impedance element composed of a macroscopically supported synthetic bilayer membrane incorporating gramicidin ion channels. The membrane and an ionic reservoir are covalently attached to an evaporated gold surface. The channels have specific receptor groups attached (usually antibodies) that permit switching of gramicidin channels by analyte binding to the receptors. The device may then be made specific for the detection of a wide range of analytes, including proteins, drugs, hormones, antibodies, DNA, etc., currently in the 10^{-7}–10^{-13} M range. It also lends itself readily to microelectronic fabrication and signal transduction. By adjusting the surface density of the receptors/channel components during fabrication, the optimum sensitivity range of the device may be tuned over several orders of magnitude.

1999 Gramicidin and related ion channel-forming peptides. Wiley, Chichester (Novartis Foundation Symposium 225) p 231–254

Nanotechnology is a term widely used to describe an emerging discipline that specializes in machines and devices with functional components nanometres in dimension. Despite a substantial literature on nanotechnology (Special report 1991) its application to the fabrication of functional devices is in its infancy. Ion beam machining and deposition techniques have been used to generate passive features as small as 10 nm (Davies & Khamsehpour 1996). However, the minimum feature size achieved by leading-edge microelectronics technology is still an order of magnitude larger at ~ 0.1 μm. In this chapter we describe the first example of an engineered device with functional moving parts composed of individual molecules nanometres in diameter. The miniaturization of microchip technology is not necessarily the most practical route to engineering at these dimensions. Biopolymers and biological assemblies are commonplace examples of supramolecular arrays operating as nanomachines. Unlike the physical techniques used by the microelectronics industry, biological systems assemble by a process of energy minimization coded by the structure and chemical reactivity of the component molecules. This process is broadly described as self assembly (Philp & Stoddart 1996).

Many functions in biology are accessible to mimicry using synthetic compounds and the process of self assembly. Biological sensory systems function by registering changes in the electrical conductivity of specialized cell membranes. In particular, olfaction in mammals and pheromone detection in insects depend upon the binding of an analyte to a targeted receptor protein causing a change in the electrical conductivity of an ion channel (Avrone & Rospars 1995). An important feature of this mechanism is its massive inherent amplification with the detection of a single molecule potentially triggering the passage of up to a million ions per second across an otherwise impermeable membrane. A well-documented example of this mechanism is the acetylcholine-triggered cation channel involved in cross-synaptic nerve conduction (Reiken et al 1996). The present design (Fig. 1) mimics this nanoscale machinery but uses robust synthetic molecular components and a novel molecular architecture adaptable to a wide range of receptor-based sensing applications. The approach employs a tethered lipid bilayer membrane and a population of receptor–ion channel complexes that switch their conduction when the targeted analyte binds to the receptor. We call the completed device an ion channel switch (ICS) biosensor.

The ion channel switch biosensor

Tethered membrane

Many attempts to make a robust lipid bilayer membrane with the properties of a black lipid membrane (BLM) have been reported. These have principally focused on physisorbing or chemically attaching a layer of hydrocarbon to a silicon (Heysel et al 1995), hydrogel (Lu et al 1996) or metal (Steinem et al 1996, Bain et al 1989) surface and then fusing lipid vesicles onto the layer to form a bilayer membrane. An important element in the stability of the present device is the use of a high concentration of ether-linked, bilayer membrane-spanning lipids in the formation of the membrane. A precedent in nature for this approach is the membrane-spanning lipid that occurs in archaebacteria (Dante et al 1995). These are bacteria that have evolved to survive in environments normally thought to be too hostile for biological systems. Archaebacteria are able to withstand temperatures in excess of 100 °C and pH variations from 2 to 11 (Stetter 1996). Using ether-linked, membrane-spanning lipids in the tethered bilayer results in a

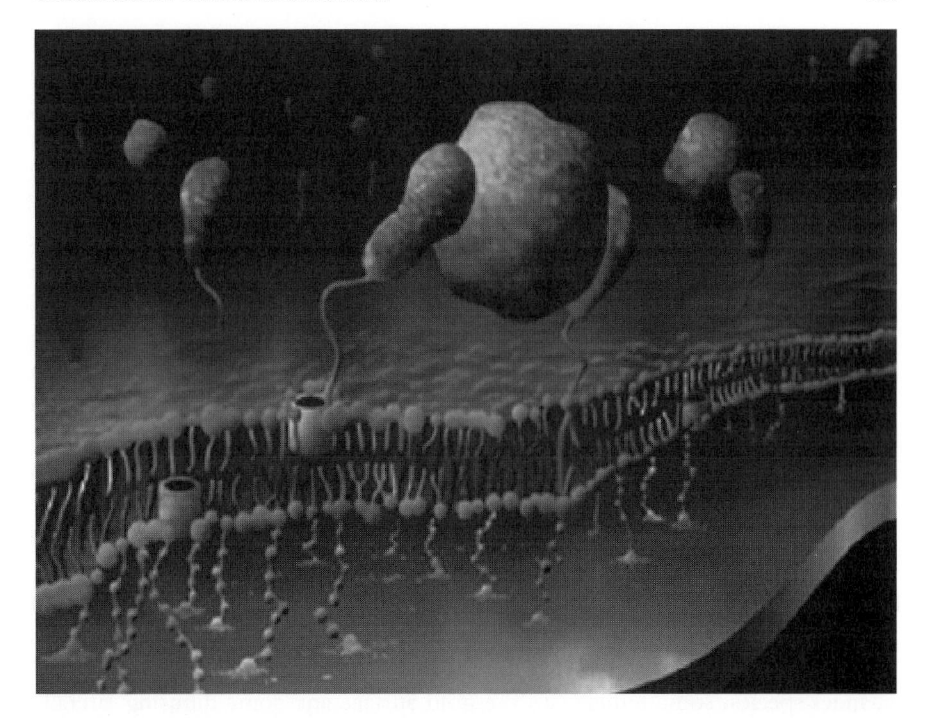

FIG. 1. A graphical representation of the key elements of the ion channel switch biosensor. A membrane is tethered to an electrode using gold–sulfur chemistry. A feature of the membrane is an ionic reservoir between the membrane and the electrode. This permits the measurement of ion fluxes across the membrane through an electrical connection to the electrode and to a reference wire in the solution. Tethered and mobile ion channels are incorporated into the inner and outer layers of the membrane, respectively. Antibody fragments are attached to the mobile, outer layer channel population and to a fraction of the tethered membrane-spanning lipids. The introduction of analyte causes these two antibody populations to cross-link, preventing the formation of ionically conducting channel dimers between the mobile and immobile channels, and resulting in a reduction in the electrical conductance of the membrane.

membrane structure which is able to resist challenges that would normally disrupt a liposome or hydrolyse the acyl chain attachments. Tethering a fraction of the membrane lipids also results in the stabilization of a lamellar phase for the non-tethered lipids and a resistance to insertion of additional material into the pre-formed membrane assembly. Tethered membranes may be formed that possess excellent stability over many months.

Gated ion channel

The molecular mechanisms of naturally occurring channels are poorly understood. At this level of detail, a literal mimicry of nature is no longer practicable and a degree of nanoscale engineering becomes necessary. Many attempts to engineer receptor-based gated ion channels have been reported. Mechanisms range from anti-channel antibodies that disrupt ion transport (Bufler et al 1996) to molecular plugs that block the channel entrance (Lopatin et al 1995). All mechanisms so far proposed have had a limited range of application and require re-engineering for each new analyte. In biology, the receptor and channel functions are normally combined in one supramolecular protein assembly. In this chapter we describe a synthetic receptor–channel complex assembled by attaching antibodies, using a streptavidin–biotin linker, to the ethanolamine moiety of the well-characterized bacterial ion channel, gramicidin A (Koeppe & Andersen 1996). The gating mechanism for this complex draws on the properties of gramicidin A within BLMs. Gramicidin monomers diffuse within the individual monolayers of a BLM. The flow of ions through gramicidin only occurs when two non-conducting monomers align to form a conducting dimer. The gating mechanism proposed here depends upon altering the dimer population by cross-linking the channel-attached antibodies to an incoming analyte. A change in the number of dimers is reported by a change in the electrical admittance of the membrane.

A stable membrane incorporating such a structure can be assembled on a smooth gold surface using a combination of sulfur–gold chemistry and physisorption. This is shown schematically in Fig.2a. The membrane comprises amphiphiles and channel species, some tethered to the gold surface and some diffusing laterally within the plane of the membrane. By using sulfur-containing lipids with extended domains of differing polarity a structure can be formed which is reminiscent of a biological membrane enclosing a hydrated compartment that may be electrically accessed by the gold electrode. This compartment provides a reservoir into which ions may flow when the membrane channels are conductive. The arrival of analyte cross-links the antibodies attached to the mobile outer layer channels to those attached to the membrane-spanning lipid tethers. Because of the low density of channels within the membrane, this anchors them distant, on average, from their immobilized inner layer channel partners. Gramicidin dimer

FIG. 2. Schematic of the ion channel switch biosensor. (a) Two-site sandwich assay. Immobilized ion channels (G_T), synthetic archaebacterial membrane-spanning lipids (MSLs) and double length reservoir half membrane-spanning phytanyl lipids (DLP) are attached to the gold surface via polar tetraethylene oxide linkers and sulfur–gold bonds. Polar spacer molecules (MAAD) are directly attached to the gold surface using the same sulfur–gold chemistry. Mobile half membrane-spanning lipids (DPEPC/GDPE) and mobile ion channels (G_α) complete the membrane. The mobile ion channels are biotinylated and coupled to biotinylated antibody fragments (F_{ab}') using a streptavidin (SA) intermediate. This means of coupling may be used across a range of antibodies. Biotinylated fragments are used in preference to biotinylated whole antibodies as they provide a beneficial orientation of the binding site on the F_{ab}'. The removal of the F_c portion of the antibody also eliminates a class of non-specific interactions

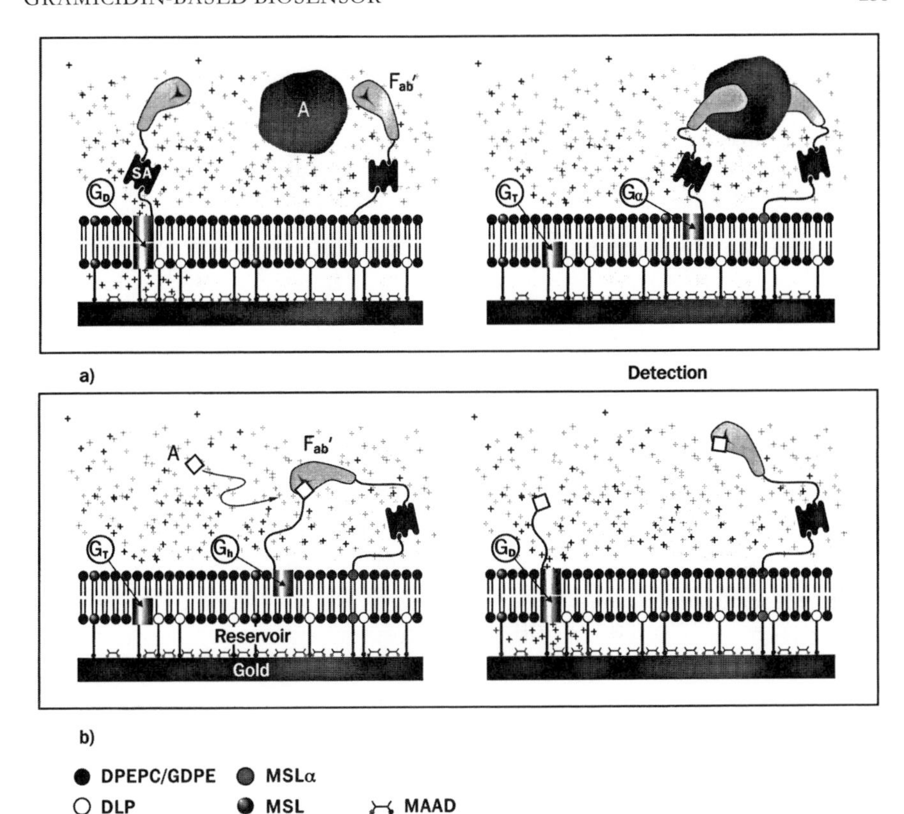

that arise in patient serum. Some of the membrane-spanning lipids (MSL_α) possess biotin tethered F_{ab}'. In the absence of analyte (A) the mobile ion channels diffuse within the outer monolayer of the tethered membrane, intermittently forming conducting dimers (G_D). These dimers permit ions within the bathing saline solution to cross the membrane, producing a high electrical conductance between the gold electrode and a reference wire in contact with the bathing solution. The addition of the targeted analyte cross links the F_{ab}' on the MSL_α and G_α and forms complexes that tether the G_α distant from their immobilized inner layer partners. This prevents the formation of channel dimers and lowers the electrical conductivity of the membrane. (b) Competitive assay. For the detection of small analytes with only a single binding site, a competitive mechanism is employed. A similar membrane is formed to that used for large analyte detection but containing a different population of mobile outer layer ion channels, (G_h). These channels are flexibly linked to haptens that match the targeted analyte. The membrane is rinsed with a streptavidin solution after which an appropriate biotinylated, hapten-specific F_{ab}' is added, forming complexes between the MSL_α and the G_h. The G_h is thus tethered distant from its immobilized inner layer partners, G_T, preventing the formation of dimers and lowering the electrical conductance of the membrane. The sensor is stored in this state until the addition of analyte competes with the hapten for the F_{ab}', liberating the channel and resulting in an increase in the membrane conductance.

conduction is thus prevented and the admittance of the membrane decreases. We describe this form of ion channel aggregation as lateral segregation. This switching mechanism is generic and may be adapted to many applications and analytes with no change in basic design. Examples of the gramicidin analogues and sulfur-containing lipids that act as the immobilized half of the membrane are shown in

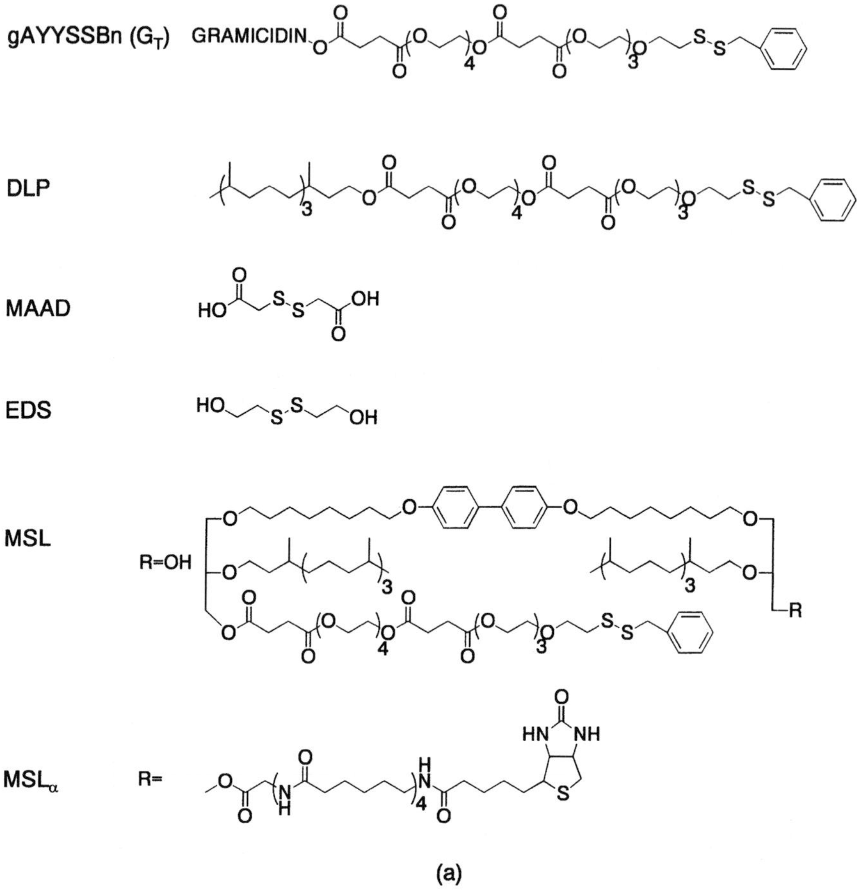

gAYYSSBn (G_T) GRAMICIDIN

DLP

MAAD

EDS

MSL R=OH

MSL_α R=

(a)

FIG. 3. Components used in the ion channel switch biosensor assembly: (a) Immobilized components. The immobilized components of the membrane comprise a mixture of tethered gramicidin, gAYYSSB$_n$ (G_T), double length reservoir half membrane-spanning phytanyl lipids, (DLPs), and full membrane-spanning lipids, (MSLs). The surface density of these tethered species is controlled by dilution with the small hydrophilic mercaptoacetic acid disulfide (MAAD). A fraction of the tethered membrane-spanning lipid, MSL_α, is biotinylated and carries an antibody fragment. These compounds are based on a common benzyl disulfide

Fig. 3a. These compounds possess matching sulfur-containing and reservoir-forming segments. A fraction of the membrane-spanning lipid carries flexibly linked antibody groups that act as the anchor points in the lateral segregation mechanism. The complementary membrane-forming mobile species are shown in Fig. 3b. For the detection of analytes too small to allow two binding antibodies, a competitive version of the switch has been engineered. Figure 2b shows the complete assembly and Fig. 3b a further mobile gramicidin analogue with a flexible linker attaching a hapten, shown here as digoxin.

Comparison with ELISA

It is instructive to contrast the operation of the ICS biosensor with that of the well known ELISA. An ELISA is based on a sequence of reactions. First analyte is captured by a layer of antibodies attached to the surface of a plastic well. A

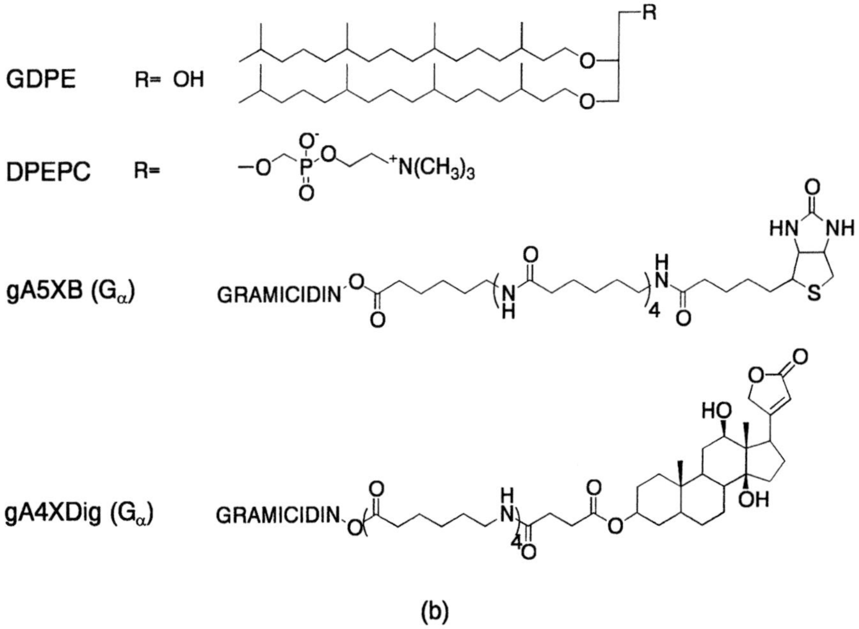

GDPE R= OH

DPEPC R= $-O-CH_2-CH_2-O-P(=O)(O^-)-O-CH_2-CH_2-{}^+N(CH_3)_3$

gA5XB (G_α) GRAMICIDIN

gA4XDig (G_α) GRAMICIDIN

(b)

attachment moiety and an ethylene glycol chain. The thickness of the reservoir region is determined by the length ($\sim$4 nm) of the ethylene glycol chain segment. (b) Mobile components. The major mobile component is a 70:30 mole ratio mix of diphytanyl ether phosphatidylcholine (DPEPC)/ glycerodiphytanylether (GDPE) which is mixed with a small fraction of mobile ion channels (G_α). For large analyte detection, the channel species employed is gA5XB. For small analyte detection, the mobile channel is the hapten linked gA4Xdig.

second, or reporter antibody, is then added which binds to a further site on the analyte. The second antibody, which has attached to it an enzyme probe, is used to report on the amount of bound analyte. An optical readout is obtained via the enzyme probe, converting a separately added substrate into a coloured product.

The operation of the ICS biosensor shares many common elements with ELISA but is engineered to incorporate all of the steps into a single machine that is considerably faster and more convenient. One obvious advantage is that it responds immediately upon the addition of analyte and requires none of the washing or multiple reagent additions of the ELISA. This arises from the capture

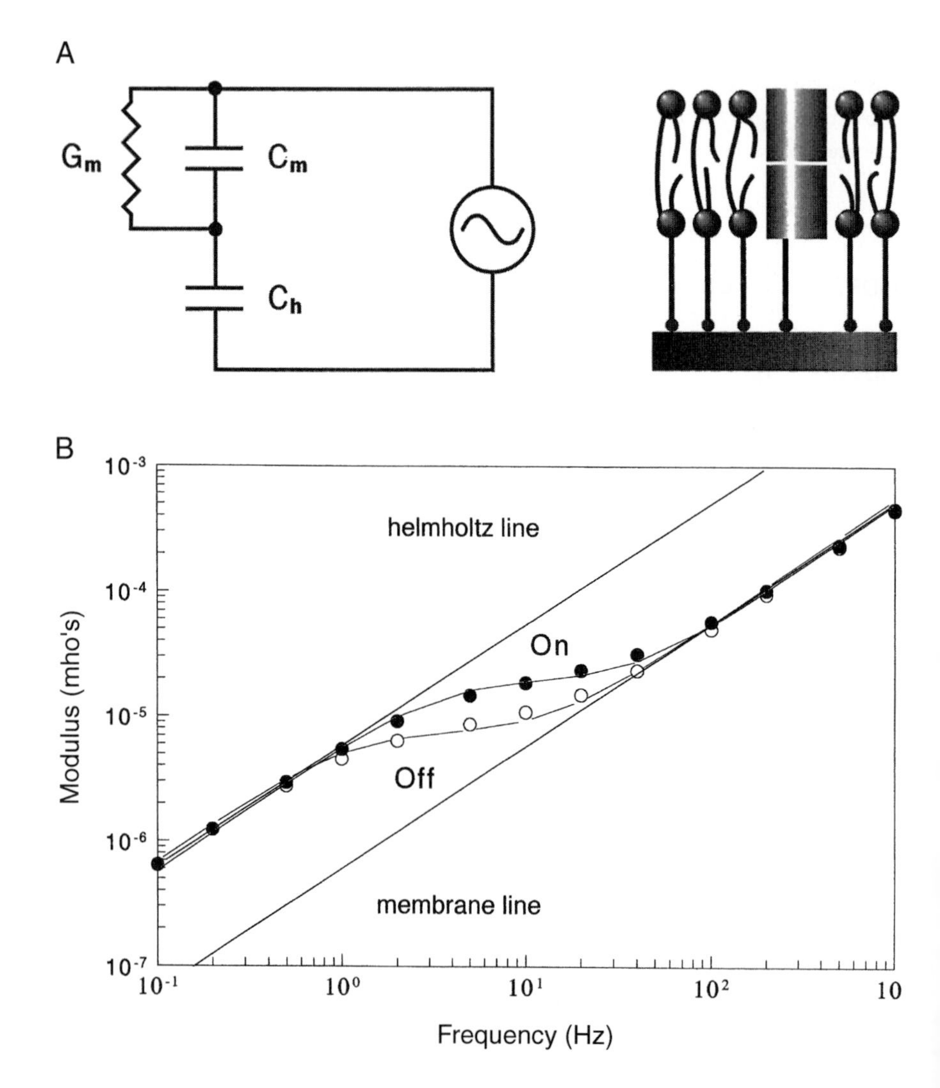

and reporter antibodies being preassembled at the membrane surface. The capture antibodies are attached to the tethered membrane-spanning lipid and the reporter antibodies to the mobile outer layer channels. A more fundamental difference between the biosensor and the ELISA arises from the outer layer membrane of the biosensor being a two dimensional liquid crystal. This permits the reporter antibodies to scan an area of order 1 μm^2 in less than 5 min. Each mobile channel thus has access to 10^2–10^3 more capture antibodies than if the gating mechanism were triggered simply by binding to the channel-attached antibodies. The speed and sensitivity of the response of the sensor is improved in direct proportion to the number of binding sites accessible to the mobile channels. This property is effectively a molecular analogue of an electronic multiplexer.

Design parameters

Electrical properties of tethered membrane. The integrity and fluidity of the tethered membrane may be probed by incorporating the ionophore valinomycin. Valinomycin selectively transports K^+ ions across the membrane. The conductance at high [K^+] demonstrates that the lipid chains are sufficiently fluid both to dissolve the ionophore and to permit the conformational changes required for complexation of the K^+ ion. With low [K^+], the high impedance at low

---

FIG. 4. Characterization of the tethered membrane. (A) Electrical equivalent circuit. The electrical equivalent circuit of a tethered membrane containing the K^+ selective ionophore valinomycin. For an excitation of typically 50 mV the equivalent circuit may be approximated as an effective Helmholtz capacitance, C_h, in series with the capacitance of the membrane, C_m, which is bypassed by the conductance, G_m, due to the ion flux transported by the ionophore. C_h and C_m values of approximately 7 uF/cm^2 and 0.6 uF/cm^2, respectively, are deduced from the spectrum. In the ion channel switch configuration the same equivalent circuit applies. The G_m element is then dominated by the channel conductance. (B) Impedance spectrum. For the capacitive elements in (A), sweeping the frequency (v) causes the impedance ($Z = Y^{-1}$) to vary according to the relationship $Z = 1/2\pi vC$. Analysing the relationship on $\log_{10}/\log_{10}$ axes yields a linear relationship with a slope of -1 and an intercept at $1/2\pi$ (Hz) of $-\log_{10}C$. For a non-conductive membrane to which no ionophore has been added, $C = C_m C_h/(C_m + C_h)$. The subscripts 'm' and 'h' refer to the membrane and effective Helmholtz layers, respectively. Introducing a high K^+ ion concentration creates a short circuit to C_m and results in a similar capacitive response as a function of frequency but with an intercept at $1/2\pi$ (Hz) of C_h. At intermediate K^+ ion concentrations the frequency-independent conductance, G_m, of the valinomycin is evident, which, if superimposed on the capacitive responses gives a characteristic impedance profile seen above. By observing the frequency at which the impedance profile crosses from C_m-dominated to C_h-dominated behaviour, it is possible to follow the change in G_m with a change in the [K^+]. This crossing point is conveniently identified by frequency, v, at which the phase relationship, ϕ, between the excitation and the resultant current is minimized. Traces are shown here for different [K^+] possessing phase minima at frequencies $v1$ and $v2$. The approach applies equally well for measurements of the channel dimer concentration, $[D] \propto Z^{-1}{}_{\phi min} = Y_{\phi\,min}$.

frequency (<1 Hz) indicates that the membrane seals against significant ionic leakage. The impedance at 1 kHz under these conditions allows an estimate of the thickness of the membrane. A good approximation to the equivalent electrical circuit of the tethered membrane in physiological saline is shown in Fig. 4A. A small alternating potential is applied between the gold electrode and a reference electrode in ohmic contact with the electrolyte solution. The resulting impedance spectrum is shown in Fig. 4B for two different levels of $[K^+]$. From the phase relationship between the excitation potential and the current flow we derive the admittance at minimum phase, $Y_{\phi min}$. It is possible to titrate $Y_{\phi min}$ against $[K^+]$ over two to three decades of concentration. The electrical properties of a membrane containing the ICS may be measured in the same manner. However, for large analytes the sensor response is more conveniently determined from the maximum rate of change of $Y_{\phi min}$. To correct for variations in electrode size, we normalize this to the admittance at the time of the sample addition, $Y_{\phi min\ t=0}$. The normalized maximum slope, $-(dY_{\phi min}/dt)_{max}/Y_{\phi min\ t=0}(s^{-1})$ can be shown by numerical simulation to be a linear function of the concentration of analyte over more than six decades of detectable admittance change. The maximum slope can typically be attained after only a 10% change in $Y_{\phi min}$, and it provides about three to four decades of dynamic range for only two decades change in $Y_{\phi min}$.

Receptor-binding kinetics. Assuming, for the moment, adequate mass transport of analyte to the membrane, the behaviour of the system may be summarized by a family of coupled equations. The most significant of these are given below.

$$[MSL\alpha] + [A] \underset{k_A^{-1}}{\overset{k_{3D}}{\leftrightarrow}} [MSL\alpha\ {*}A] \qquad \text{where } K_{3D} = k_{3D}/k_A^{-1} \tag{1}$$

$$[G\alpha] + [G_T] \underset{k_6^{-1}}{\overset{k_{2d}}{\leftrightarrow}} [G_D] \qquad \text{where } K_d = k_{2d}/k_G^{-1} \tag{2}$$

$$[MSL\alpha{*}A] + [G\alpha] \underset{k_A^{-1}}{\overset{k_{2D}}{\leftrightarrow}} [MSL\alpha{*}A{*}G\alpha] \qquad \text{where } K_{2D} = k_{2D}/k_A^{-1}, \tag{3}$$

where $[MSL_\alpha]$, $[A]$ and $[MSL_\alpha{*}A]$ are the surface density of the membrane-spanning lipid tethered antibodies, the 3D analyte concentration and the surface density of their complexes, respectively. The concentrations $[G_\alpha]$, $[G_T]$ and $[G_D]$ refer to the mobile gramicidin–antibody monomers in the outer layer, the tethered gramicidin in the inner layer and the conducting gramicidin dimers, respectively. The rate constants k_{3D}, k_{2d}, k_{2D}, $k^{-1}{}_A$ and $k^{-1}{}_G$ are the 3D and 2D forward and reverse reaction rates, and K_{3D}, K_d and K_{2D} are the equilibrium constants for interactions (1), (2) and (3). Once analyte has been captured, the mobile gramicidins cross-link via 2D diffusion to form the tertiary complex, $MSL_\alpha{}^*A^*G_\alpha$.

For small analyte detection $[G_\alpha]$ in expression (2) is replaced by $[G_h]$, the concentration of mobile gramicidin–hapten as shown in (4). Competition for the tethered antibodies between the analyte and the hapten on the mobile gramicidins establishes an equilibrium that determines the concentration of the complex $MSL_\alpha{}^*G_h$.

$$[MSL_\alpha] + [G_h] \underset{k_A^{-1\,h}}{\overset{k_{2D}^{h}}{\leftrightarrow}} [MSL_\alpha{}^*G_h]. \tag{4}$$

The final measure is the number of gramicidin dimers, which is proportional to the membrane admittance at the minimum phase, $Y_{\phi\,\min}$.

$$Y_{\phi\,\min} \propto [D]. \tag{5}$$

The 2D reactions will generally be faster than the corresponding 3D rates, and therefore it is anticipated that within the analyte concentration range of interest, the 3D processes will be rate limiting. However, at high concentrations of analyte 2D processes will intrude. When (1) is limiting, the flux of analyte captured and detected at the membrane surface is given by $[MSL_\alpha]\,k_{3D}\,[A]$ molecules/cm^2 per s. When (3) is limiting the response is independent of [A]. When (1) and (3) are similar, analyte capture by the MSL_α and its cross-linking to gramicidin occur at comparable rates. This produces a response proportional to $[A]^{1/2}$. The transition between these regimes may be shown to depend on a dimensionless parameter, θ, given by $[G_\alpha]^2_{t=0}k_{2D}/(2[MSL_\alpha]k_{3D}[A])$. When θ approaches two the response becomes proportional to $[A]^{1/2}$. A further limiting condition is the lifetime of the dimeric channel (2). Under these conditions the response time becomes analyte independent and is limited by $1/k_G^{-1}$. Based on expressions (1)–(4), and incorporating mass transport kinetics, we have developed a detailed numerical simulation of the device for large and small analyte detection. Figures 6 and 7 show a superposition of the experimental and modelled responses.

The sensor performance may be adapted to a range of applications. For $[G_T]$ and $[G_\alpha]$ in the range 10^{10}–10^8 molecules/cm^2, the ratio $[MSL_\alpha]/[G_\alpha]_{t=0}$ 'amplifies' the apparent capture rate of analyte from what would otherwise be the simple first-order rate constant, $k_{3D}[A]$ to $k_{3D}[A][MSL_\alpha]/[G_\alpha]_{t=0}$. The maximum amplification in this configuration approaches 10^3. This indicates a potential for quantitative detection of analyte at sub-picomolar concentrations in practical measurement periods. At higher analyte concentrations, in the range

10^{-9}–10^{-6} M, the amplification may be adjusted downwards by lowering the $[MSL_\alpha]/[G_\alpha]_{t=0}$ ratio.

When operating in competitive mode for small analyte detection, either a kinetic or equilibrium measure may be employed. With the competitive system the 'amplification' effect described above is not available. However, it is still possible to adjust the sensitivity of the device over a considerable range by manipulation of component surface densities. Using either equilibrium or kinetic measures the fractional gating response ΔY is given by:

$$\Delta Y \sim (K_{3D}[A] + 1)/(K^{h}_{2D}[MSL_\alpha] + K_{3D}[A] + 1). \tag{6}$$

Experimentally, K^{h}_{2D} approximately equals $10^{-11}-10^{-9}$ (molecules/cm^2)$^{-1}$. Similarly, K_{3D} typically equals $10^{10}-10^{11}$ M^{-1} for commercially available antibodies. This means that for a particular antibody, K^{h}_{2D} $[MSL_\alpha]$ may be varied by two to three orders of magnitude, giving a tuneable range of analyte sensitivity. For an individual configuration the practical dynamic range is ~2 orders of magnitude of analyte concentration.

For both two-site and competitive assays, the conductance per channel remains constant as the membrane area is reduced. The membrane leakage however, decreases in proportion to area. This means that with smaller electrodes (<30 μm

FIG. 5. Response to thyroid-stimulating hormone (TSH). A tethered membrane 0.16 cm^2 in area is formed on a freshly evaporated gold film. The gold film is 100 nm thick over a 5 nm bonding layer of Cr on a clean glass microscope slide. The slide is immersed in an ethanol solution of the immobile lipid layer mixture comprising 76 μM DLP (double length reservoir half membrane-spanning phytanyl lipid), 15 nM MSL-4XB, 150 nM MSLOH, 9 nM gA-YY-SS-Bn and 35 μM MAAD (mercaptoacetic acid disulfide). Following immersion at 20 °C for 1 h the slide is removed and rinsed thoroughly in ethanol, air dried and clamped to a series of 4 mm ID teflon cylinders which define a 200 μl volume above the coated gold surface. A further 5 μl of mobile layer mix in ethanol solution is added containing 5 nM gA-5XB, and a 10 mM (30:70) mix of glycerodiphytanylether (GDPE)/diphytanyl ether phosphatidylcholine (DPEPC). Self-assembly occurs on thoroughly rinsing the slide with approximately 500 μl of phosphate-buffered saline (PBS), spontaneously inducing the formation of a tethered bilayer membrane. Following a further three rinses with PBS to eliminate any residual excess membrane lipid, 5 μl of 40 nM streptavidin in PBS is added. Following incubation for 10 min the supernatant is thoroughly rinsed with PBS and a further 5 μl added of 1 μM b-F_{ab}' in PBS. Following incubation for 10 min the residual material is eliminated with a further three rinses with PBS. The sensors are now available for use. The impedance is measured at 10 Hz using an excitation amplitude of 50 mV and an offset of −300 mV applied to the electrode. A different population of biotinylated F_{ab}' were used in each case: (▼) b-antiferritin, (▲) b-antiα-TSH, (○) b-antiβ-TSH, and (+) a 50:50 mix of b-antiα-TSH and b-antiβ-TSH. At the arrow, 2 nM TSH is added. The mixed population of b-F_{ab}' elicited a substantial response on the addition of TSH. The antiβ-TSH and antiα-TSH yielded either a small or negligible response. This shows the lateral segregation gating mechanism is dominant requiring a complementary F_{ab}' on each of the MSL_α and G_α to elicit a response.

diameter) it should become possible to resolve the ion flux arising from individual channels. Operation within this regime promises to improve the sensitivity of the device by many orders of magnitude.

Reservoir characteristics. Single channel conductivity measurements (G_y) of gramicidin within BLMs have been shown to be dependent on the species of monovalent cation being transported, i.e. $G_{Cs}(6) > G_K(2) > G_{Na}(1)$ (Myers & Haydon 1972). The conductance of the tethered membrane is insensitive to ion type, indicating that it is not limited by the conductivity of the channel. The present limit to conductance is thought to be the reservoir. When fully extended, the hydrophilic reservoir region is sufficiently restricted that all ions are within a few Debye lengths of either the metal surface of the electrode or the underside of the membrane. Also the reservoir contains >10% w/v ethylene oxide tethers and is therefore different to a bulk electrolyte solution.

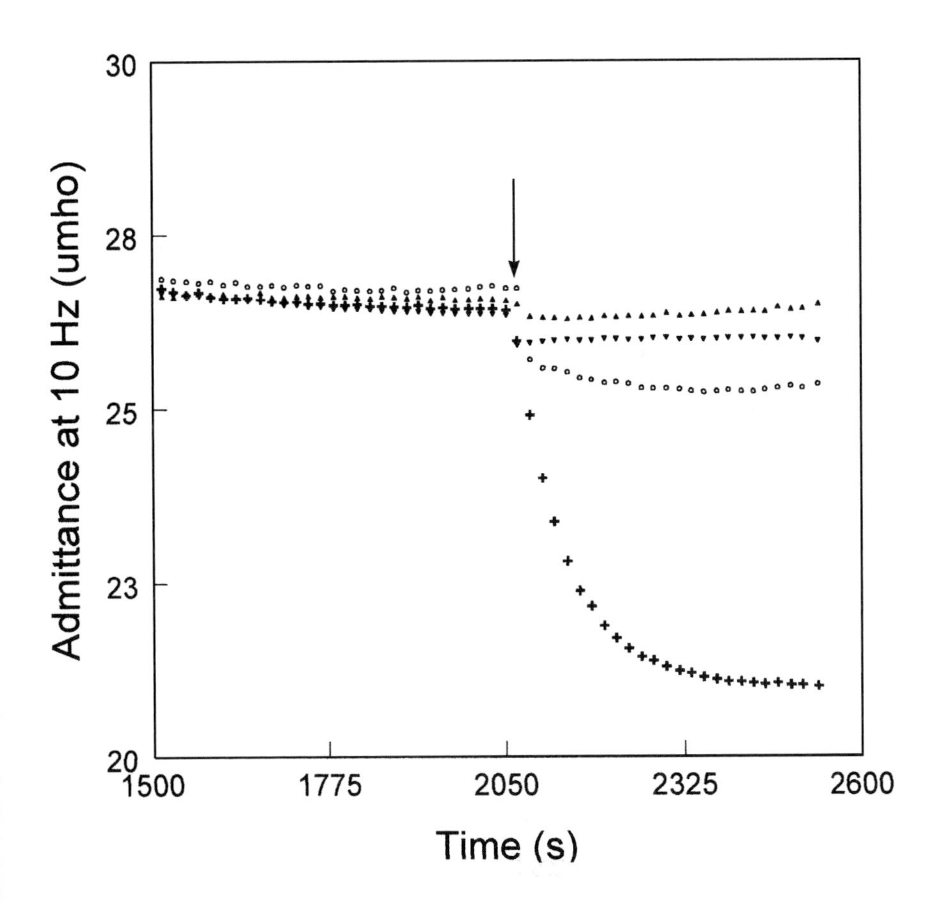

A finite element analysis of the electrical response of the reservoir has been undertaken. The reservoir is treated as a homogeneous ionically conducting medium with a variable specific conductivity. In order to simulate the experimental device behaviour, it is necessary to use a specific conductivity for the reservoir region about 10^4-fold lower than that of bulk saline at the same total ionic concentration. The membrane conductance remains proportional to the number of conducting channel dimers [D], whilst the effective conductivity per channel is substantially less than its usual value. A channel limited conductance is achieved when the reservoir conductivity is above a threshold greater than the present example but, because of the linker groups, this will always be less than the specific conductivity of bulk water. Hydrogel, ion-conducting polymers or even solid electrolyte phases of suitable thickness may be considered as alternative reservoir materials.

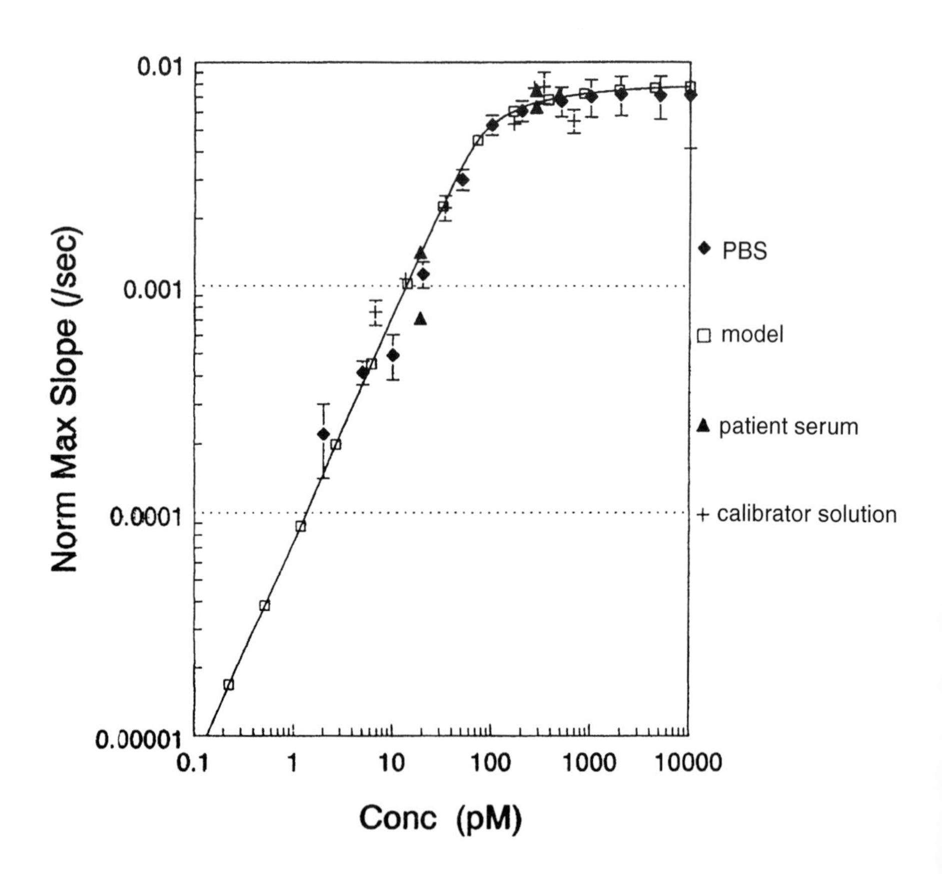

Examples of performance

Two-site or 'sandwich' assay

Thyroid-stimulating hormone. Thyroid-stimulating hormone (TSH) provides a useful model to demonstrate the response of the ICS biosensor when challenged by a large analyte. With a molecular mass of 28 kDa, this protein comprises an α and β subunit to which it is possible to bind complementary F_{ab}'s. A matched pair of antibodies is used, each one sensitive to different non-overlapping epitopic sites on TSH. The response to TSH is shown in Fig. 5. If the membrane surface is prepared with antiferritin, antiα-TSH or antiβ-TSH F_{ab}' exclusively, the addition of TSH results in a negligible change in admittance. If, however, an equal mixture of antiα-TSH and antiβ-TSH F_{ab}' is bound, the addition of TSH elicits a dramatic change in membrane admittance. The dependence of the admittance response on the F_{ab}' composition is wholly consistent with the lateral segregation design. Because of the approximate symmetry of the cross-linking event it is not critical how the two classes of F_{ab}' are distributed between the tethered lipids and the mobile channels. Assembly may be achieved by adding an equimolar mixture of

FIG. 6. Ferritin titration. A similar membrane is prepared to that described in the caption to Fig. 5 except that $[MSL_{\alpha}] = 5.5$ nM. In this example b-antiferritin F_{ab}' fragments are used as the receptor. The normalized maximum slope $-(dY_{\phi\,min}/dt)_{max}/Y_{\phi min\ t=0}$ (s^{-1}) is plotted as a function of ferritin concentration in phosphate-buffered saline (PBS), patient sera and serum calibrators. Controls assembled with b-antitheophylline F_{ab}' fragments show no response. All measurements are performed at 30 °C. Superimposed on the data is the predicted response curve based on estimates of the various kinetic and concentration parameters in the membrane. These are: $k_{3D}[MSL\alpha] = 7.5\times10^{15}\ M^{-1}s^{-1}$ molecules/cm^2; $k_A^{-1} = 10^{-3}\ s^{-1}$ $k_{2D} = 10^{-7}\ cm^2$ molecule$^{-1}\ s^{-1}$; $k_{2d} = 1.25\times10^{-11}\ cm^2$ molecule$^{-1}s^{-1}$, $k^{-1}{}_G = 1.25\times10^{-2}$ s; $[G_{\alpha}] = 1\times10^8$ molecules cm^{-2}; and $[G_T] = 3\times10^{10}$ molecules cm^{-2}. The simulation is for a static non-flowing analyte solution. The nominal value of $[MSL_{\alpha}] = 1.5\times10^{10}$ molecules cm^{-2} results in a $k_{3D} = 5\times10^5\ M^{-1}\ s^{-1}$. Whilst this is a reasonable value for the antibody 'on' rate it may be elevated by multiple epitopic sites on the ferritin and by an elevated effective $[MSL_{\alpha}]$ site density. The $k_{2D} = 1\times10^{-7}\ cm^2$ molecule$^{-1}\ s^{-1}$ supports the expectation that the 2D kinetics are rapid. The dimer lifetime, $(k^{-1}{}_G = 1.25\times10^{-2}\ s^{-1})^{-1}$, is consistent with the lifetime of gramicidin dimers in solvent-free black lipid membranes (BLMs; Sawyer et al 1990). It is $k_{\bar{G}}^{1}$ that causes the response to limit at analyte concentrations >100 pM. Increasing the membrane thickness has been shown to shorten $(k_{\bar{G}}^{1})^{-1}$ and to extend the response beyond 1 nM. $\theta = 2$ is satisfied under these conditions at $[A] = 2$ nM, although the response is then dominated by $k^{-1}{}_G$. Approximately three to four decades of dynamic range are accessible, which may be adjusted from the pM to mM range by the appropriate choice of $[MSL_{\alpha}]$, $[G_{\alpha}]$ and $[G_T]$. The choice of $[G_{\alpha}] = 1\times10^8$ molecules cm^{-2} rather than the nominal 3×10^9 molecules cm^{-2} is a result of the use of streptavidin linkers. Because of the multiple biotin sites on streptavidin, many biotinylated channels cross-link to the streptavidin reducing $[G_{\alpha}]$.

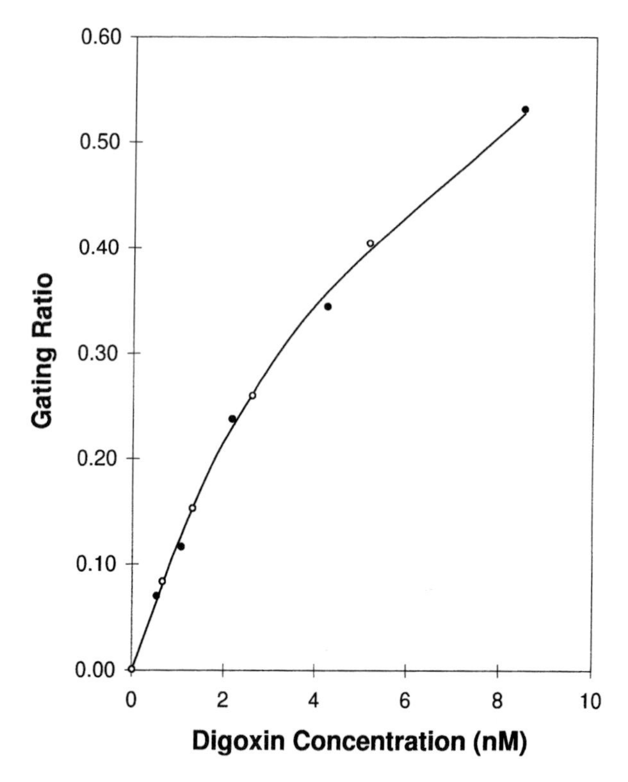

FIG. 7. Digoxin titration. A similar membrane is prepared to that described in the caption to Fig. 6, except that the overall [MSL_α] is increased in concentration tenfold and [G_α] is replaced by [G_h]. The total [MSL] is maintained. Following the self assembly step, the membrane is incubated with 5 μl of 40 nM streptavidin for 10 min, rinsed again and 5 μl 50 nM b-anti-digoxin added. The response is measured as the fractional increase in admittance or gating ratio, consequent upon the addition of digoxin. Superimposed on the data is the predicted response curve based on estimates of the various kinetic and concentration parameters in the membrane. These are: [MSL_α] $= 6\times10^9$ molecules cm^{-2}, [G_h] $= 3\times10^9$ molecules cm^{-2}, $K_{2D} = 1\times10^{-9}$ cm^2 molecule^{-1} and $K_{3D} = 1\times10^9$ M^{-1}.

the two F_{ab}' monomers allowing them to attach randomly to either the tethered lipid or mobile channel species. Six differently targeted antiTSH F_{ab}' fragments have been tested and found to function successfully in the ICS biosensor format. To date, all the matched F_{ab}' pairs we have tested have functioned in the ICS format.

Large analytes. Ferritin is the principal iron-transporting protein in human serum. With a molecular mass of $\sim$450 kDa, ferritin is one of the largest soluble proteins regularly measured clinically. It contains 24 equivalent subunits each accessible to cross linking such that only a single F_{ab}' type is required to elicit a lateral segregation response. Ferritin is a convenient model to characterize the properties of the ICS biosensor challenged by large proteins. The titration curve seen in Fig. 6 shows the response of the sensor to ferritin in the range 1–1000 pM in phosphate-buffered saline (PBS), patient serum and serum calibrator solutions. The limiting response time at high concentrations is thought to arise from the channel lifetime $1/k_G^{-1}$ (Sawyer et al 1990). Modifications to the channel lifetime are possible through altering the membrane thickness or the channel length. The titration curve shown in Fig. 6 has been obtained for a relatively low F_{ab}' capture density of $[MSL_\alpha] = 1.5\times10^{10}$ molecules/cm^2. This is to circumvent a dependence on viscosity. With the introduction of a modest flow, 1 μl/min, to the analyte stream, the mass transfer limitation can be overcome and the capture density $[MSL_\alpha]$ increased by an order of magnitude with a proportionate increase in the response rate.

Single epitope competitive gating

Small analytes. Figure 7 shows a response curve when the system is configured for small analyte detection. The analyte shown here is digoxin, a 781 Da cardiac drug. Anti-digoxin F_{ab}', specifically biotinylated via enzymatically liberated SH groups in the hinge region, is attached through streptavidin to a biotinylated tethering lipid. The mobile channels are now functionalized through flexible tetra aminocaproyl chains to a digoxin hapten linked at the 3-position as shown in Fig. 3b. In the absence of analyte, the mobile channels cross-link to the tether sites, preventing the formation of dimers and turning off the membrane conductance. The introduction of analyte competes with the hapten, increasing the membrane conductance. The response time to analyte challenge is essentially fixed by the off rate, $k_A^{-1\,h}$ of the hapten from the F_{ab}'. In the present example $k_A^{-1\,h}$ is $\sim$50 s. The quantitative measure of analyte concentration may be taken as either the absolute initial rate of admittance increase or the equilibrium fractional admittance change or gating ratio. These are essentially proportional to each other over the sensitive range of the competitive assay. The sensitivity range may be adjusted by varying the tethered lipid/F_{ab}' surface density.

Breadth of applications

The examples given here are far from the total range of uses to which the ion channel lateral segregation switch may be applied. Any sensing application is accessible, including areas such as blood typing, bacterial or viral detection, DNA detection or serology. Receptors other than antibodies may be employed including enzymes, binding proteins and synthetic ligands. The use of ionophores in the tethered membrane extends the range of applications to include electrolytes. A field of particular promise is the use of these functional nanostructures with multielectrode arrays. The data handling and processing power of microelectronics industry coupled with the ICS technology promises a unique level of sensing performance.

Acknowledgements

The authors thank the Australian Industrial Research and Development Board and the Cooperative Research Centres (CRC) Program for support of this work. The partner organizations within the CRC for Molecular Engineering and Technology are the Commonwealth Scientific and Industrial Research Organization, the University of Sydney, and Australian Membrane and Biotechnology Research Institute.

References

Avrone E, Rospars JP 1995 Modeling insect olfactory neuron signaling by a network utilizing disinhibition. Biosystems 36:101–108

Bain CD, Evall J, Whitesides GM 1989 Formation of monolayers by the coadsorption of thiols on gold: variation in the head group, tail group, and solvent. J Am Chem Soc 111:7155–7164

Bufler J, Kahlert S, Tzartos S, Maelicke A, Franke C 1996 Activation and blockade of mouse muscle nicotinic channels by antibodies directed against the binding site of the acetylcholine receptor. J Physiol (Lond) 492:107–114

Dante S, De Rosa M, Maccioni E et al 1995 Thermal stability of bipolar lipid Langmuir Blodgett films by X-ray diffraction. Mol Cryst Liq Cryst 262:191–207

Davies ST, Khamsehpour B 1996 Focused ion beam machining and deposition for nanofabrication. Vacuum 47:455–462

Heysel S, Vogel H, Sanger M, Sigrist H 1995 Covalent attachment of functionalized lipid bilayers to planar wave guides for measuring protein binding to biomimetic membranes. Protein Sci 4:2532–2544

Koeppe RE II, Andersen OS 1996 Engineering the gramicidin channel. Annu Rev Biophys Biomol Struct 25:231–258

Lopatin AN, Makhina EN, Nichols CG 1995 The mechanism of inward rectification of potassium channels: 'long-pore plugging' by cytoplasmic polyamines. J Gen Physiol 106:923–955

Lu XD, Ottova AL, Tien HT 1996 Biophysical aspects of agar gel supported bilayer–lipid membranes: a new method for forming and studying planar bilayer–lipid membranes. Bioelectrochem Bioenerg 39:285–289

Myers VB, Haydon DA 1972 Ion transfer across lipid membranes in presence of gramicidin A. II. Ion selectivity. Biochem Biophys Acta 27:313–322

Philp D, Stoddart JF 1996 Self-assembly in natural and unnatural systems. Angew Chem Int Ed Engl 35:1155–1196

Reiken SR, Vanwie BJ, Sutissna H et al 1996 Bispecific antibody modification of nicotinic acetylcholine receptors for biosensing. Biosens Bioelectron 11:91–102

Sawyer DB, Oiki S, Andersen OS 1990 Single channels formed by gramicidin A in solvent-free bilayers formed from surface monolayers using a tip-dip technique. Biophys J 57:100 (abstr)

Special report 1991 Engineering a small world: from atomic manipulation to microfabrication. Science 254:1300–1342
Steinem C, Janshoff A, Ulrich WP, Sieber M, Galla HJ 1996 Impedance analysis of supported lipid bilayer membranes: a scrutiny of different preparation techniques. Biochim Biophys Acta 1279:169–180
Stetter KO 1996 Hyperthermophilic prokaryotes. FEMS Microbiol Rev 18:149–158

DISCUSSION

Roux: Could you clarify how the electrode is stored?

Cornell: We can store it wet or we can dry it, in which case the membrane-forming material is on the surface and you just add water when you want to use it. The drying process also helps to cope with the problem of antibody stability.

Roux: Once it's wet and you use it, do you then throw it away?

Cornell: You can throw it away, or you can re-use it, especially if you are dealing with a substance that has a rapid 'off rate'. (This is the measure of the survival time of the antibody–antigen complex.)

Roux: Is the process reversible?

Cornell: Yes. It is simply governed by the affinities of the receptor for the analyte. If you're looking at something that comes off quickly, you can use it again.

Woolley: What happens at the edges?

Cornell: That's an interesting question. What happens at the edges was a worry for us, but surprisingly it doesn't seem to be a problem. We imagined that the lipid was going to escape from the upper half of the membrane. It appears that this is not a major problem although it is necessary to have a polarity discontinuity to define the electrode area. We have tinkered with changing the polarity of the surrounding supports, e.g. making it polar or non-polar. If we make it non-polar, the membrane forms everywhere, which isn't good, but if we make it polar things don't seem to escape.

Woolley: How do you obtain an electrical seal at the edges?

Cornell: Because there is such a poor lateral diffusion rate within the reservoir, we only observe ionic diffusion over a fraction of a micrometre, which is small relative to the size of the electrode.

Ring: Would the bilayer still stick to the plate if you made holes in the gold surface?

Cornell: One of the key elements in our success is that the lipids we use intrinsically form hexagonal II phases. This phase becomes lamellar at the electrode surface due to the sulfur–gold chemistry. We achieve a good seal because the hexagonal phase is hydrophobic and coats any exposed inner layer lipid. Lamellar phases will form liposomes and these leak.

Hladky: You said at one point that if you change the percentage of spacers, you can change the volume of the space between the membrane and the electrode. Could you clarify this?

Cornell: If we titrate in ethylene glycol solutions of varying concentrations above the membrane, we find at 10% w/v that the reservoir capacity decreases dramatically, suggesting that the density of spacers within the reservoir space is near 10% w/v.

Hladky: So, are a large fraction of the lipids tethered?

Cornell: Yes, just under 50%. You can only go for about two to three lipids in any one direction before encountering a tethered lipid. It is not possible to support the membrane in the way Avi Ring was referring to, because the phase diagram of the lipid is all important. If you go beyond about three molecules without a tethering lipid, the lipid phase switches back to being that found in the bulk and not the lamellar phase it adopts on the electrode surface.

Ring: How does your detector compare with the glucose meters that measure impedance by enzymatic activity?

Cornell: They operate down to millimolar concentrations, which is 10^9-fold less sensitive than our device. They have been available since about 1987 and have become the big success story in the biosensor world because they have successfully addressed the blood glucose monitoring market for diabetics. They contain glucose oxidase, which is coated on an electrode as a paste in the presence of various electron transporters. What is measured is a glucose concentration dependent redox reaction at the electrode surface.

Smart: How does your system compare with biosensors based on luciferase?

Cornell: Luciferase-based biosensors are sensitive and well suited to many applications.

Smart: Don't you have an advantage because luciferase is inhibited by enormous number of contaminants? What sort of contaminant effects do you see?

Cornell: Yes luciferase-based biosensors do suffer from many inferences that quench fluorescent samples. In our case, we don't have to worry about optical interferences or the optical path length. A common question that we are asked is, what happens when you put detergents or organic solvents in the system? Or what happens if one of the patients is taking an antibiotic that is an ionophore, such as gentamicin? Firstly, we have studied the adverse effects caused by detergents or organic solvents. The damaging effect of these compounds is that they induce changes in the lipid phase. Of course, because the membrane is tethered there are many orders of magnitude more free energy favouring the lamellar phase over all other phase structures. This results in the tethered membrane being much more stable to detergent challenges or the presence of membrane-disruptive compounds such as organic solvents. A further point is that any other ionophore which tries to enter the membrane after all the gold-binding sites have been occupied has to insert into the prepacked membrane. This results in a slow insertion rate for these later arrivers and a low level of interference. Also, since we are not addressing implanting biosensors into the body, and five minutes is the maximum duration over which we need observe the exchange of interferences with the biosensor membrane, the interference effects are minimal.

Hladky: In your water assays, presumably you are assaying what is supposed to be drinking water, but if you went down to the River Thames wouldn't your electrode dissolve pretty quickly?

Cornell: The way in which we would address these longer term data-logging applications is in a batching process. This is like feeding ammunition into a machine gun. Most of these applications do not require a time resolution that is more frequent than once every half an hour.

Huang: Your reaction mechanism depends on the mobility of gramicidin. I wonder if the mobility or diffusion constants of small molecules like gramicidin have been measured in membranes.

Cornell: The diffusion coefficient of gramicidin has been measured by NMR, and is close to the diffusion rate of the lipid. Also, in the biosensor, we get a direct measure of the diffusion coefficient of the outer layer gramicidin species. It is a bit like a fluorescent quenching experiment, except what we measure is the time taken to form or break conducting gramicidin dimers. Also, we can easily modify the molecular species attached to the surface, and determine its effect on the gramicidin diffusion rate. We are putting hex-his tags on proteins and using NTA-labelled gramicidin to look at the aggregation state of the proteins on the membrane surface.

Huang: Is this process sensitive to temperature?

Cornell: The 'on rate' for the gramicidin monomer–dimer interaction between 18 °C and 30 °C appears to closely parallel the 'off rate', resulting in a reasonably constant dimer ratio for the gramicidin. Because the activation coefficient for diffusion and the dimer survival time is about the same, there is $<15\%$ change in conductivity over this range of temperatures. There is, in addition, a twofold change in kinetics of the receptor–ligand 'on rate', but this is simply the temperature dependence experienced by any receptor-based sensor.

Woolley: Is the lateral diffusion dominated by the interaction of gramicidin with the lipid, or is it significantly affected by what is attached to the gramicidin?

Cornell: The lateral diffusion appears to be determined by the intrinsic diffusion coefficient of the outer layer membrane lipid, which is about $10^{-8}\,cm^2\,sec^{-1}$. The limiting response rate for the sensor is not usually the diffusion coefficient of the gramicidin, but depending on the circumstances, either the reaction rate of the ligand binding to the receptor, or the survival time for the gramicidin dimer. Although the ligand 'on rate' is an intrinsic property of the ligand–receptor interaction, you do have control over the thickness of the membrane and thus control over the dimer survival time. It is worth noting that we can devise a tethered membrane that would force gramicidin into configurations that would not normally be accessible in a conventional lipid bilayer. I would be interested to know if there is a major energy barrier that would impede this sort of experiment.

Koeppe: You may need to explore other sequences, rather than the gramicidin A sequence. You would need to reduce the numbers of tryptophans, make some other rearrangements and consider the single- versus double-stranded forms.

Cornell: But we could still play around with the membrane in a manner that might make it possible to create a $\beta^{7.2}$ helix single-stranded form.

Hladky: Would you have to make the membrane thinner to do this?

Cornell: Yes, we could make the membrane thinner, but we could also juggle the area of the molecule in the lipid in a way you can't ordinarily do. We have wondered whether it might be possible to probe different gramicidin structures that were not possible before, especially given the fact that the tryptophans are anchored at the interface, which is the dominant rule for keeping the gramicidin single stranded.

Hladky: Your statement that you can change the surface area of the molecule is interesting because you still have some free lipids in the system, and I'm surprised that they don't redistribute so as to restore the original surface area.

Cornell: We can force the issue. We can go all the way up to total tethering, and can therefore dictate the area per lipid by the spacing at the sulfur–gold interface.

Koeppe: If we shorten the lipids, from C8 to C7 for example, we see a sharp transition from all single-stranded structures to all double-stranded structures (Greathouse et al 1994). These experiments are not in bilayers, but in micelles. Diheptanoyl phosphatidylcholine is able to solubilize gramicidin in aqueous solution, but surprisingly it solubilizes exclusively double-stranded forms that may have multiple conformations.

Cornell: The polar interior for a micelle might result in a different effect to that of forcing a new configuration by altering the chain length of a tethered membrane.

Koeppe: It's not the experiment you described, but in the C7 micelle experiment there's no evidence for a different single-stranded structure with a different pitch. I don't know of any other experiment in which such a conformation has been observed.

Smart: We have a good picture of what we think is the conducting form of gramicidin, but what is the structure of the closed form? We know it to be monomeric, and you show it to be half of the dimeric structure. The closed form is the next model we have to come up with.

Cornell: All the NMR studies are done with such high concentrations of gramicidin that the structure is always dimeric.

Smart: Are there any differences in the circular dichroism (CD)?

Wallace: If you replace the formyl group with an acetyl group, and therefore destabilize the helical dimer, you only see a small change in the CD spectrum (Wallace et al 1981). We assume that this means the backbone structure does not significantly change between the monomeric and dimeric single-stranded helix. However, if you make more drastic changes, such as removal of the formyl group or both the formyl group and the N-terminal valine, there are significant changes in the spectra.

Cornell: We have some evidence that when we change the number density of gramicidin on one side of the membrane, we observe about fivefold more leakage through the monolayer than with no gramicidin, suggesting that there is still some kind of conduction mechanism through the membrane, but we don't know the nature of this. It could be through the monomeric channel.

Heitz: The closed state is probably monomeric. We made an infrared study using transferred monolayers, and we ruled out the possibility that there are double-stranded helices. The spectrum is in agreement with that of a single-stranded species, i.e. with a monomeric form.

Koeppe: Regarding the formyl group, we have made N-BOC-gramicidin that interferes with dimer formation. Huey Huang has shown that the CD of this suggests that it is still folded and it behaves like a monomer (He et al 1994).

Huang: We did X-ray in-plane inference scattering based on the following reason. The scattering intensity is proportional to the form factor squared. The form factor of a monomer is one-half of a dimer, so the square of the form factor is one-quarter. On the other hand, the number of monomers is twice that of the dimers, so the net result is that if BOC-gramicidin is a monomer floating in the monolayer, the scattering intensity should be one-half of the scattering intensity of gramicidin dimers. This result would have proven that the BOC-gramicidin monomer is β-helical and that it floats in a monolayer. When I did the experiment, I was expecting to see a reduction of 50%, but instead we saw a reduction of about 70%. Our explanation for this is that there is a mixture of dimers and monomers. We also think that the BOC-gramicidin monomers may stick together, but not necessarily in a head-to-head dimer form (He et al 1994).

Cross: Are the monomers on the surface or are they still penetrating the bilayer?

Koeppe: They are in the bilayer. One way to interpret these results is that there are still some correlated motions between the two sides, partly because they can't get very far apart and so they're probably still talking to a certain extent to their partner.

Cornell: The K_D is still about 1:10^4, so you have to go a long way down in concentration before you get monomers.

Koeppe: But the X-ray data are consistent with it moving towards monomerization.

Ring: A number of groups made asymmetrical channels, i.e. they used one analogue on one side of the bilayer and another on the other side of the bilayer, and the monomers didn't cross over to the other side, even after a long time. It would be interesting for Bruce Cornell to try this in his membranes.

Cornell: This is a source of the instabilities in our system. We found that gramicidin can flip from one layer to another over hours. If you anchor the gramicidin with, say, streptavidin they don't flip.

Ring: Could you put the gramicidin forms that Roger Koeppe made into the bilayer asymmetrically and keep them there? If you say you have dimers, but they're not channels, couldn't you address whether they are double stranded or not by tethering the monomers?

Koeppe: That may not work because the high concentrations you would need for spectroscopy would cause deformations of the membrane. That experiment can only be done at low concentrations.

Roux: If an uncomplexed monomer is in the bilayer, there are six unsatisfied hydrogen bonds in the middle of the hydrocarbon region. Therefore, it is possible that the β-helix is roughly intact, and there is some degree of reorganization close to the monomer–monomer junction.

References

Greathouse DV, Hinton JF, Kim KS, Koeppe RE II 1994 Gramicidin A/short-chain phospholipid dispersions: chain length dependence of gramicidin conformation and lipid organization. Biochemistry 33:4291–4299

He K, Ludtke SJ, Wu Y et al 1994 Closed state of gramicidin channel detected by X-ray in-plane scattering. Biophys Chem 49:83–89

Wallace BA, Veatch WR, Blout ER 1981 Conformation of gramicidin A in phospholipid vesicles: circular dichroism studies of effects of ion binding, chemical modification, and lipid structure. Biochemistry 20:5754–5760

Final general discussion

Protons

Roux: I would like to know more about various aspects of protons including proton conductance, current–voltage (I–V) curves of proton conductance, NMR in the presence of more protons and proton-binding sites.

Ring: The I–V curves of proton conductance are super linear, and the channel lifetimes are different to those of potassium, sodium or caesium channels.

Roux: Would a proton not stabilize the channel to the same degree as other cations?

Ring: It does. If you started with a relative scale of 1.0 and you increased the voltage, with potassium you would find that the lifetime increases with increased electric field, and with sodium and caesium it would be relatively constant with perhaps a slight increase. With protons, however, the lifetime decreases with increased electric field. The protons also interact differently with hydrogen bonds and with head-to-head dimerization regions.

Roux: Do the conductance curves fit the one proton model or are more complex models necessary?

Ring: If I recall, two is the maximum occupancy.

Busath: The proton conductance is higher in diphytanoyl phosphatidylcholine (DPhPC) bilayers than in monoolein bilayers, which is opposite to sodium and potassium behaviour, and it's much higher if phenylalanines are replaced with tryptophans, which is also opposite to sodium and potassium behaviour. In particular, gramicidin M has exquisitely high proton conductance. Also, in fluorinated tryptophan compounds it is reduced, contrary to sodium and potassium behaviour (R. Phillips & D. Busath, unpublished work 1998). We first thought that this was because negative charge is transporting the current instead of positive charge, but we then dismissed that, and came to a different conclusion after reading Pomès & Roux (1998). For instance, we considered the possibility that OH^- is being transported, and that this is caused by the splitting of a water molecule at the exit of the channel or that the H^+ current is mediated by a free aqueous electron.

Cornell: Does it have to be a discrete OH^-? It has been suggested that hydrogen-bonded chains of water molecules could span the lipid bilayer acting as a wire for the transport of protons. This could also explain why gramicidin is such a good proton transporter.

Roux: The conductance of gramicidin at low pH can be viewed in two ways, but the valence selectivity of gramicidin (cations are preferred over anions) suggests that OH^- is not the charge carrier through the channel.

Jakobsson: The Grotthuss mechanism (Agmon 1995) gives a higher conductance in ice than it does in water. So if the same factors make the conductance of protons go in the opposite direction from the conductance of other cations, this would be consistent with the idea that there are some interactions that make the contents of the channel more or less mobile, and those factors that make the channel contents more mobile are going to increase cation conductance and reduce proton conductance. If this is the mechanism of the conductance modification, then it's a good explanation for why proton conductance and cation conductance can go in opposite directions.

Busath: We invoked that mechanism to explain why mini-channels have increased proton conductance (Busath & Szabo 1988), and it's a plausible explanation. However, the changes we're talking about, rather than being factors that affect water positional stability in the channel, are long-range electrostatic factors. The main difference between DPhPC and monoolein bilayers is the long-range electrostatic interfacial potential. The difference between tryptophan and phenylalanine, or between fluorinated and unfluorinated tryptophan, is the long-range electrostatic dipole potential. These factors could certainly affect the water molecules, because the water molecules have dipole moments, but the idea that they change the mobility of the proton in the same sense as water converted to ice seems strange to me, as does the possibility that OH^- transport is caused by the splitting of water. Instead, if you think about the impact of the dipole potential of an interface or a side chain on the dipole potential on the water column, things fall into place.

Jakobsson: We published in a book chapter (Chiu et al 1992) showing that if you restrict the mobility of the gramicidin side chains, you dramatically reduce the mobility of the channel contents.

Busath: But you are not going to stop the tryptophans from moving around.

Jakobsson: The other issue is that if you change the dipoles outside the channel lumen, you do not observe an effect on the electric fields that are in the lumen.

Busath: In the centre of the channel it is dissipated by 50% at the most, according to Jordan (1984).

Sansom: I can see that the changes in the lipids may affect the mobilities of the tryptophans. It's not just the head groups that are different, but tails are also different in DPhPC bilayers compared to glycerol monoolein (GMO) bilayers. One could imagine that the tryptophans would have less freedom if heavily branched acyl chains are present (as in DPhPC bilayers), as opposed to the unbranched chains of GMO.

Cross: But there is high mobility anyway, and there are no changes in rotameric states.

Roux: All the effects are there, but the zero order effect is that the interfacial dipolar potential is different in these lipids.

Smart: Can't we resolve these issues by looking at exchange with deuterated water?

Cross: I would be surprised if we could sort out all the different effects using this method because the hydrogen bonds will change.

Roux: How rapidly do the NH groups of gramicidin exchange with deuterated water?

Hladky: When you do tritium loading in a clinical measurement of total body water, you can assume that virtually all the tritium stays on the water for an hour or two, because the number of freely exchangeable groups that can take up the tritium is small. You couldn't use tritium as a measure if all the hydrogen in the body could exchange with water. Most of it can't because it is, for example, incorporated into hydrocarbons, so the amount that can is a small proportion of total hydrogen.

Cornell: The area of the diphytanyl molecules is much greater than a diacyl methyl chain phosphatidylcholine, and far more hydrocarbons are exposed. Another intriguing point is the manner in which phosphatidylcholine organizes water on the surface. It is fairly well established that one of the reasons why it is good to have phosphatidylcholines on the outside of cells is that the water is fleetingly not hydrogen bonding to the outside. I was wondering whether natural channels, including gramicidin, have some degree of sympathy with the organization of water on the membrane surface. If you add gramicidin, you find that you interfere with the state of the bound water on the membrane as seen by NMR. If there is a hydrogen-bonded chain of waters that influences proton transport down the gramicidin channel, changing the headgroup of the lipid could significantly influence the manner in which the ordered water on the surface of the membrane is linked down the gramicidin pore.

Separovic: Didn't Dani & Levitt (1981) measure water movement through the gramicidin channel?

Cornell: I don't think so. I don't know how they would have been able to measure this because it would be affected by lateral diffusion.

Roux: Many people have computed energetics for cations in gramicidin using a range of methods. In all cases, the binding site appears to be close to the mouth of the channel. This is in qualitative agreement with rate models, because some of the steps are only weakly voltage dependent. We did a free energy calculation in which we put a proton in the hydrogen-bonded water chain (Pomès & Roux 1998). In this model the proton can move long the chain, so the potential function can account for the Grotthuss mechanism (Agmon 1995). We found that the proton does not want to be close to the mouth, but rather goes to the centre of the single file.

Therefore, I would be curious to see the effect of decreased pH (high concentration of H^+ ions) on the solid-state NMR data. Would you see that the chemical shift anisotropy (CSA) change in the centre of the dimer, or would you see a CSA roughly at the same position as the other cations?

Cross: Do you anticipate that protons spend much of the time at the centre?

Roux: The well is relatively shallow, and I suspect that it would be impossible to see where the protons go in the channel.

Busath: But you neglected the environment in these computations.

Roux: We did not have a membrane, only a channel with about 50 water molecules.

Hladky: Is the image force absent?

Roux: Yes, but the magnitude of that force once you account for the interaction with the channel is not clear. In any case, I would be curious to see if the protons go to the centre of the channel in solid-state NMR experiments.

Cross: Part of the problem is that we can surround the channel with high concentrations of cation, but it is impossible to do this for protons.

Woolley: We did an experiment that might be relevant to this discussion. We made modified gramicidin analogues that had positive charges on one or both ends, and we found that there were differences in the behaviours of those channels towards cations and protons (see Woolley et al 1997). For cations, we saw that a positive charge at the entrance (rather than the exit) dominated the effects on cation flux. For protons it didn't matter whether the charge was present at the entrance or the exit—they both had the same effect on proton flux. So there's a sort of symmetry present for protons but not for cations.

Ring: We measured the conductance of gramicidin in the presence of protons and found that it saturates at about 1 M HCl. The G–C (conductance–concentration) curve also has a clear nick (concave to the *y*-axis). We also did the I–V curves, and did a simultaneous fit with the G–C and I–V curves (Ring & Sandblom 1988a,b). I was surprised to find that the middle peak was so low, giving rise to a shallow well. I thought I could probably fit this with the proton going back and forth in the channel. Does this fit with your idea that the proton sits at the centre of the channel?

Roux: There is about 1 kcal/mol from the centre to the edge of the channel, so it is a shallow free energy well. With Mark Schumaker we have been constructing a model based on our molecular dynamics observations. It is a diffusion model that includes proton shuffling, but it is not like a standard rate model because the proton sits in the middle and does not go to binding site on the left or on the right of the channel. However, once you get to the end of the cycle you have to flip back the orientation of the water chain. Most of the voltage dependence of proton transport in our model arises from flipping back the water chain in the correct direction.

Flipping back the water chain is much slower process than actually transporting the proton.

Ring: It would be interesting to fit this region not with Eyring rate theory but to fit it with the theory that Bob Eisenberg has proposed, i.e. a $D(x)$ (diffusion coefficient as a function of x) with a proton going back and forth in small steps.

Cross: If the protonated lipid surface is positively charged, how does this influence proton conductance?

Busath: Cukierman et al (1997) addressed this. They compensated for protonation in their bilayer when they analysed the I–V relationships. The pK is $\leqslant 1.0$, according to Marsh (1990), so they used that estimate explicitly and assumed there was some protonation of phosphatidylcholine bilayers.

Cross: Would it be better to do the experiments with different lipids that are uncharged at low pH?

Roux: The phosphatidylcholine bilayers seem to be resistant even at pH 1.0.

Cross: I'm sure there will be some protonation at pH 1.0.

Busath: This would affect the conductance, but it would not destroy the bilayer.

Cross: I disagree because in all the experiments performed at low pH, we have completely destroyed the orientation.

Roux: Proton transport in gramicidin seems to be peculiar because the I–V curve differs from that of metal cations. It would be nice to know more about this.

Busath: It will also teach us something about the effect of the water column on alkaline metal cation transport.

References

Agmon N 1995 The Grotthuss mechanism. Chem Phys Lett 244:456–462

Busath D, Szabo G 1988 Permeation characteristics of gramicidin conformers. Biophys J 53:697–707

Chiu SW, Gulukota K, Jakobsson E 1992 Computational approaches to understanding the ion channel-lipid system. In: Pullman A, Pullman B, Jortner J (eds) Membrane proteins: structures, interactions, and models. Kluwer, Dordrecht, p 315–338

Cukierman S, Quigley EP, Crumrine DS 1997 Proton conductance in gramicidin A and its dioxolane-linked dimer in different bilayers. Biophys J 73:2489–2502

Dani JA, Levitt DG 1981 Binding constants of Li^+, K^+, and Tl^+ in the gramicidin channel determined from water permeability measurements. Biophys J 35:485–499

Jordan PC 1984 The total electrostatic potential in a gramicidin channel. J Membr Biol 78:91–102

Marsh D 1990 CRC handbook of lipid bilayers. CRC Press. Boca Raton, FL

Pomès R, Roux B 1998 Free energy profiles for H^+ conduction along hydrogen-bonded chains of water molecules. Biophys J 75:33–40

Ring A, Sandblom J 1988a Evaluation of surface tension and ion occupancy effects on gramicidin A channel lifetime. Biophys J 53:541–548

Ring A, Sandblom J 1988b Modulation of gramicidin A open channel lifetime by ion occupancy. Biophys J 53:549–559

Woolley GA, Zunic V, Karanicolas J, Jaikaran AS, Starostin AV 1997 Voltage-dependent behavior of a ball-and-chain gramicidin channel. Biophys J 73:2465–2475

Summary: what we have learned about gramicidin and other ion channels

B. A. Wallace

Department of Crystallography, Birkbeck College, University of London, Malet Street, London WC1E 7HX, UK

I believe this has been such a successful meeting because the participants, who represent many varying disciplines, have had the opportunity to interact both formally and informally in a constructive and collaborative atmosphere. People using synthetic chemistry and conductance approaches to study functional properties have had the chance to talk with people doing theory and structural studies. I hope this won't be the last time we all get together to talk. This meeting was expanded beyond the original concept, which was to focus just on gramicidin, to include related ion channel-forming polypeptides such as alamethicin and larger ion channels such as acetylcholine receptors and porins, thus giving us a broader perspective. Consequently, we have had a chance to discuss applications of some of the methods and information learnt from this simple model system to those larger channels, and to think of future general developments in the field of ion conduction across biological membranes.

We have compared and contrasted many techniques used to study structures in different environments, and have discussed the differences and similarities amongst the gramicidin structures found in micelles, oriented bilayers, bilayer membranes and phospholipid vesicles, as well as features associated with the environments of crystals and various types of solutions. We have talked about the similarities and differences between the types of data obtained from crystallography, diffraction, circular dichroism, and solution and solid-state NMR spectroscopies, and we have examined the consequences of using different peptide/lipid ratios. One important outcome has been the conclusion that all of these methods and samples seem to be converge to similar answers regarding the double helical and helical dimer forms of gramicidin.

We have also heard of some beautiful studies of what can be done with chemical syntheses and modifications of gramicidin and other polypeptides to enable us to understand their functional properties.

We have covered many aspects of theory and simulation, i.e. rate constants, ion binding, transport and water interactions. People with different views on how to simulate these processes have had a chance to interact with each other and with experimentalists obtaining the information they use as the basis for their theoretical studies.

We have talked a bit about the biological function of gramicidin, and we have spent some time discussing lipid interactions, not only for gramicidin, but also for other natural and synthetic polypeptides. These have helped us to understand lipid influences on peptide conformation and stabilization, as well as peptide influences on lipid phase states.

I would also like to highlight here our discussions of some of the controversies in this field as a significant accomplishment of this discussion meeting. We have had the chance to bring many of these controversies out into the open and to successfully address and reconcile a number of the seemingly different results obtained by different groups using different approaches. For example, we talked about double helical and helical dimer gramicidin structures and the circumstances under which each type of structure predominates, concluding that under most conditions, the helical dimer is the active conducting-channel form. We have compared results obtained with different techniques and have concluded that many of the differences seen are a consequence of differing time scales, lipid ratios, etc. and are not in contradiction with each other. We have discussed the different parameters that go into the various conductance and binding calculations, and as a result have a better feel of what effects they have on theoretical simulations. Thus, this meeting has served an important purpose of helping us to understand the bases for many of the alternative interpretations that exist in the literature.

We have also discussed some new practical applications for gramicidin. We have probably all justified our studies of this molecule as a model system for understanding ion conduction. However, we now have seen that the 'outside world' may also be interested in gramicidin not only as an antibiotic, but also as an important component of a new sensor system that may have commercial value in analytical chemistry.

Finally, with respect to the 'relevance' of doing studies on gramicidin as a model system, we have made comparisons and contrasts between gramicidin and the potassium channel, both functionally and structurally. I suppose the gramicidin community was always a bit worried that once a 'real' (i.e. protein) channel structure was determined, we would be out of work because gramicidin would no longer be considered useful or relevant (especially because its 'unnatural' D-amino acids are integrally involved in its ion binding mechanism and thus were suggested to be unlikely to result in structures found in typical biological channels). We have now learnt, however, that the potassium channel uses glycine residues as 'honorary' D-amino acids to create a binding site for its ions in a similar manner. Thus, with relief, we can conclude that gramicidin is an excellent model system for ion channels and that this simple molecule should continue to be of significant importance for understanding peptide–lipid and peptide–ion interactions and the process of ion conduction in membranes.

Index of contributors

Non-participating co-authors are indicated by asterisks. Entries in bold type indicate papers; other entries refer to discussion contributions.

Subject Index

H

I

Other Novartis Foundation Symposia:

No. 219 **Gap junction-mediated intercellular signalling in health and disease**
Chair: N. B. Gilula
1999 ISBN 0-471-98259-8

No. 221 **Bacterial responses to pH**
Chair: R. K. Poole
1999 ISBN 0-471-98599-6

No. 226 **Transport and trafficking in the malaria-infected erythrocyte**
Chair: H. Ginsburg
1999 ISBN 0-471-99893-1